DEACTIVATION OF HEAVY OIL HYDROPROCESSING CATALYSTS

DEACTIVATION OF HEAVY OIL HYDROPROCESSING CATALYSTS

Fundamentals and Modeling

JORGE ANCHEYTA

WILEY

Published by John Wiley & Sons, Inc., Hoboken, New Jersey
Published simultaneously in Canada

Library of Congress Cataloging-in-Publication Data:

Names:Ancheyta Juárez, Jorge, author.
Title: Deactivation of heavy oil hydroprocessing catalysts : fundamentals and modeling / Jorge Ancheyta.
Description: Hoboken, New Jersey : John Wiley & Sons, 2016. | Includes bibliographical references and index.
Identifiers: LCCN 2016019443| ISBN 9781118769843 (cloth) | ISBN 9781118769812 (epub)
Subjects: LCSH: Petroleum–Refining. | Catalyst poisoning.
Classification: LCC TP690.8 .A53 2016 | DDC 622/.33827–dc23 LC record available at https://lccn.loc.gov/2016019443

Cover image courtesy of GettyImages/avdeev007

Typeset in 10/12pt, TimesLTStd by SPi Global, Chennai, India.

10 9 8 7 6 5 4 3 2 1

TABLE OF CONTENTS

ABOUT THE AUTHOR

Jorge Ancheyta, PhD, graduated with a Bachelor's degree in Petrochemical Engineering (1989), Master's degree in Chemical Engineering (1993), and Master's degree in Administration, Planning, and Economics of Hydrocarbons (1997) from the National Polytechnic Institute (IPN) of Mexico. He splits his PhD between the Metropolitan Autonomous University (UAM) of Mexico and the Imperial College London, UK (1998), and was awarded a postdoctoral fellowship in the Laboratory of Catalytic Process Engineering of the CPE-CNRS in Lyon, France (1999). He has also been a visiting professor at the Laboratoire de Catalyse et Spectrochimie (LCS), Université de Caen, France (2008, 2009, 2010), Imperial College London, UK (2009), and Mining University at Saint Petersburg, Russia (2016).

He has worked for the Mexican Institute of Petroleum (IMP) since 1989 and his present position is Manager of Products for the Transformation of Crude Oil. He has also worked as professor at the undergraduate and postgrade levels for the School of Chemical Engineering and Extractive Industries at the National Polytechnic Institute of Mexico (ESIQIE-IPN) since 1992 and for the IMP postgraduate since 2003. He has been supervisor of more than one hundred BSc, MSc, and PhD theses. He has also been supervisor of a number of postdoctoral and sabbatical year professors.

He has been working in the development and application of petroleum refining catalysts, kinetic and reactor models, and process technologies mainly in catalytic cracking, catalytic reforming, middle distillate hydrotreating, and heavy oils upgrading. He is author and coauthor of a number of patents, books, and about 200 scientific papers, and has been awarded the highest distinction (Level III) as National Researcher by the Mexican government and is a member of the Mexican Academy of Science. He has also been guest editor of various international journals, for example, *Catalysis Today*, *Petroleum Science and Technology*, *Industrial Engineering Chemistry Research*, *Chemical Engineering Communications*, and *Fuel*. He has also chaired numerous international conferences.

PREFACE

Apart from the reactor, catalyst is another important component of a chemical process, and a thorough understanding of the catalytic phenomena occurring during the transformation of reactants into desired products is of vital importance for the development and optimization of the process.

In the case of petroleum refining industry, it is currently immersed in the dilemma of higher production of heavy petroleum compared with that of light petroleum. Producing and refining heavy crude oils is more complicated since they exhibit higher content of impurities (sulfur, nitrogen, metals, and asphaltenes), as well as higher yield of residue with consequent low production of valuable distillates (gasoline and diesel) than conventional crude oils, which in turn are responsible for the low price of heavy petroleum. And not only that, the existing refineries have been designed to process light crude oil and heavy crude oil can only be blended at a certain reduced proportion. To increase the diet of heavy crude oils as feed to a refinery, units require major changes or installation of new plants.

A solution to this problem is the use of heavy crude oil upgrading processes prior to a refinery or conversion processes for the upgrading of bottom-of-barrel (heavy residue) in the refinery. These upgrading processes are able to convert heavy oil into medium/light oil with reduced amounts of impurities and high content of valuable distillates, by two main principles: carbon rejection and hydrogen addition. In the latter case, catalytic hydrotreating is the most used technology in commercial application.

For a proper design of heavy petroleum hydroprocessing reactors, it is required to have simulation tools based on information collected in laboratory experiments, which consist of mathematical models to represent the phenomena occurring during heavy oil conversion. The hydroprocessing reactors are complex and complicated to model and design. The composition and properties of heavy petroleum that is converted in reactors are such that the reaction system can involve various phases, different types of catalysts, reactor configuration, reaction conditions, catalyst deactivation, and so on, making the development of a model a challenging task. Moreover, hundreds of components are present in heavy petroleum that undergo different reaction pathways and compete for the active sites of catalysts, which contribute to increasing the complexity for the formulation of the kinetics, catalyst deactivation, and reactor models.

Deactivation of hydroprocessing catalyst is mainly due to the formation of carbonaceous (coke) and metal depositions, which block the pore mouth leaving unutilized catalytic active sites. This is the most important concern during hydroprocessing of heavy oils, since the life of the catalyst and the entire economy of the process strongly depend on it. Hence, from the industrial point of view, modeling the catalyst deactivation is highly valuable. The deactivation phenomenon is commonly divided into three stages: early deactivation due to coke deposition, middle stage deactivation due to loss of sites by poisoning and pore plugging by metal-sulfide deposits, and total loss of activity by severe diffusional resistances due to almost total pore plugging.

Modeling the kinetics and catalyst deactivation is of great importance for proper reactor and process design as well as to establish suitable operating policies to compensate for the loss of catalyst activity during time-on-stream.

Deactivation of Heavy Oil Hydroprocessing Catalysts: Fundamentals and Modeling deals with this topic of current and future relevance: the loss of activity of catalyst during hydroprocessing of heavy oils. The book is organized in five chapters, each one having individual references. More than 400 references are cited and discussed within the entire book, which cover practically all the previous published literature regarding the fundamentals and modeling of catalyst deactivation during hydroprocessing.

Chapter 1 is dedicated to introduce those readers requiring an in-depth knowledge on topics related to the properties of heavy oils, such as petroleum refining processes, asphaltenes, tendency to coke formation, viscosity of crude oils and blends, stability, compatibility, and assay of petroleum. A brief description of all the petroleum refining processes is given. Detailed experimental data of light, medium, and heavy crude oil are also provided.

Chapter 2 deals with the properties of catalysts for heavy oil hydroprocessing. Particular mention is done to the description of the preparation, characterization, and evaluation of hydroprocessing catalysts using heavy oils. Some of the most important features that must be taken into account when processing heavy feeds are also discussed.

Chapter 3 is devoted to the description and analysis of the deactivation of hydroprocessing catalysts. Detailed descriptions are provided on the reactors used for hydroprocessing of heavy oils, process variables, effect of reaction conditions on catalyst deactivation, mechanisms of catalyst deactivation, and the effect of asphaltenes on it.

Chapter 4 aims at describing the characterization of spent hydroprocessing catalysts. The main characterization techniques are commented. A series of several studies on heavy oil hydroprocessing are reported in detail, which include synthesis of supports and catalysts, and evaluation of hydroprocessing catalyst and its deactivation at different reaction experimental scales.

Chapter 5 is focused on the modeling of catalyst deactivation. Various approaches are described and discussed in detail using different study cases, such as the effect of reactor configuration on the cycle length of heavy oil fixed-bed hydroprocessing, effect of different heavy feedstocks on catalyst deactivation, modeling deactivation by metal deposition, modeling short-term catalyst deactivation, modeling parallel thermal and catalytic hydrotreating including deactivation, modeling catalyst deactivation by the continuous kinetic lumping model, and application of a three-stage approach for modeling the complete period of catalyst deactivation during hydroprocessing of heavy oils. The kinetic modeling approaches, estimation of model parameters, and reactor model and catalyst deactivation models are described.

The development of correlations and models is thoroughly described with the aid of detailed experimental data collected from different reaction scales. Experimental data, explanations of how to determine model parameters, and rigorous treatment of the different topics as well as the step-by-step description of the models formulation and application will make this book an indispensable reference not only for professionals working in the area of modeling reactor and catalyst deactivation but also a textbook for full courses in chemical reaction engineering.

It is anticipated that *Deactivation of Heavy Oil Hydroprocessing Catalysts: Fundamentals and Modeling* becomes promptly an outstanding and distinctive book because it emphasizes a detailed description of fundamentals and modeling of catalyst deactivation, uses laboratory and commercial data for model validations, gives details of results of simulations at different conditions, and, in general, focuses on more practical issues regarding modeling of catalyst deactivation than textbooks published related to the topic in the past.

JORGE ANCHEYTA
Mexico city, Mexico.
2016

INSTITUTO MEXICANO DEL PETRÓLEO

This book has been published as part of the commemoration of the 50th anniversary of the Mexican Petroleum Institute

NOMENCLATURE

a_0, a_1, S_0	parameters of yield distribution function Equation 5.58
A_1, B_1	Riazi's correlation parameters
A_2	Goto and Smith's correlation parameter
A	fitting parameter of Equation 5.13
A_{C_i}	Arrhenius preexponential factor for catalytic kinetic constants (h^{-1})
A_{d_i}	Arrhenius preexponential factor for deactivation constants (h^{-1})
A_{H_2S}	van't Hoff preexponential factor for hydrogen sulfide adsorption constant (cm^3/mol)
a_L	gas–liquid interfacial area (cm^{-1})
a_{MOC}	unit conversion factor of Equation 5.17
A_S	sectional area of the reactor (cm^{-2})
a_S	liquid–solid interfacial area (cm^{-1})
A_{T_i}	Arrhenius preexponential factor for thermal kinetic constants (h^{-1})
B	matrix for orthogonal collocation, fitting parameter of Equation 5.13
C	reactant or species concentration
C_A	reactant concentration of A species
C_{Asph}	asphaltene content (wt%)
C_{A_0}	reactant concentration at entrance of reactor system
C_{BN}	basic nitrogen content (ppm)
C_{H_2}	hydrogen concentration (mol/cm^3)
C_i	molar concentration of compound i (cm^3/mol), content of the compound i at the reactor outlet
$C_{i,0}$	content of the compound i at the reactor inlet

$c(k,\tau)$	concentration of the species with reactivity k at residence time τ
$c(k,0)$	concentration of the species with reactivity k in the feed
C_{H_2S}	hydrogen sulfide concentration (mol/cm^3)
C_{NBN}	nonbasic nitrogen content (ppm)
C_{Ni}	nickel content (ppm)
C_p	molar concentration of compound i (cm^3/mol)
C_S	sulfur content (wt%)
C_V	vanadium content (ppm)
cwt	cumulative weight fraction
D	bulk diffusion
D_e	effective diffusivity of A in the pores of catalyst
$D(k)$	species-type distribution function for hydrocracking reaction
D_r	restrictive diffusion coefficient
d_p	catalyst particle diameter
d_t	reactor diameter
D_0	initial diffusion coefficient within pore catalyst
D_t	diffusion coefficient within pore catalyst
E_{aC_i}	activation energy for the catalytic reactions (kcal/mol)
E_{aT_i}	activation energy for the thermal reactions (kcal/mol)
E_{d_i}	deactivation energy for the reaction i (kcal/mol)
EOR	end-of-run
e	exponential function basis
g	gas mass rate (g/s)
G_L	superficial liquid mass velocity (kg/m^2 s)
H_i	Henry's law constant for compound i (MPa cm^3/mol)
H_2/oil	hydrogen-to-oil ratio (std m^3/m^3)
I	degree of polynomial
I_0	Bessel function of the first kind, zero order
I_1	Bessel function of the first kind, first order
K	reaction rate constant per unit of surface area
k	hydrocracking reactivity of any species (h^{-1})
k_0	global rate constant for hydrocracking of residue
k_1	second-order rate constant for hydrocracking of residue to VGO ($g_T/(g_i\, g_{cat}\, h)$)
k_2	second-order rate constant for hydrocracking of residue to distillates ($g_T/(g_i\, g_{cat}\, h)$)
k_3	second-order rate constant for hydrocracking of residue to naphtha ($g_T/(g_i\, g_{cat}\, h)$)

k_4	second-order rate constant for hydrocracking of residue to gas ($g_T/(g_i\, g_{cat}\, h)$)
k_5	first-order rate constant for hydrocracking of VGO to distillates ($g_{cat}{}^{-1}\, h^{-1}$)
k_6	first-order rate constant for hydrocracking of VGO to naphtha ($g_{cat}{}^{-1}\, h^{-1}$)
k_7	first-order rate constant for hydrocracking of VGO to gases ($g_{cat}{}^{-1}\, h^{-1}$)
k_8	first-order rate constant for hydrocracking of distillates to naphtha ($g_{cat}{}^{-1}\, h^{-1}$)
k_9	first-order rate constant for hydrocracking of distillates to gases ($g_{cat}{}^{-1}\, h^{-1}$)
k_{10}	first-order rate constant for hydrocracking of naphtha to gases ($g_{cat}{}^{-1}\, h^{-1}$)
k^{app}	apparent rate coefficient
$k_{A,B,\ldots,P}$	kinetic constants for the three and four kinetic models ($g_i{}^{1-n}/g_T{}^{1-n}\, g_{cat}\, h$)
$k_{C_{Asph}}$	catalytic kinetic constant for the HDAsph reaction ($wt\%^{-0.503}/h$)
$k_{C_{BN}}$	catalytic kinetic constant for the HDNBN reaction ($ppm^{-0.792}/h$)
k_{C_i}	catalytic kinetic constant for the reaction i
$k_{C_{NBN}}$	catalytic kinetic constant for the HDNNBN reaction ($wt\%^{-1.154}/h$)
$k_{C_{Ni}}$	catalytic kinetic constant for the HDNi reaction ($ppm^{-1.406}/h$)
k_{C_S}	catalytic kinetic constant for the HDS reaction ($wt\%^{-0.503}/h$)
k_{C_V}	catalytic kinetic constant for the HDV reaction ($ppm^{0.290}/h$)
k_d	deactivation rate constant (t^{-1})
k_{d_i}	deactivation constant for the reaction i (h^{-1})
k_{H_2S}	adsorption-equilibrium constant for the hydrogen sulfide (cm^3/mol)
k^{int}	reaction rate constant
k_i^L	gas–liquid mass transfer coefficient for compound i (cm/s)
k_i^S	liquid–solid mass transfer coefficient for compound i (cm/s)
k_i	intrinsic rate coefficient
k_{max}	hydrocracking reactivity of the species with the highest TBP in the mixture (h^{-1})
k^p	particle rate coefficient
$k_{T_{Asph}}$	thermal kinetic constant for the HDAsph reaction ($wt\%^{0.795}/h$)
$k_{T_{BN}}$	thermal kinetic constant for the HDNBN reaction ($ppm^{0.137}/h$)
k_{T_i}	thermal kinetic constant for the reaction i
$k_{T_{NBN}}$	thermal kinetic constant for the HDNNBN reaction ($ppm^{0.137}/h$)
$k_{T_{Ni}}$	thermal kinetic constant for the HDNi reaction ($ppm^{0.350}/h$)

k_{T_V}	thermal kinetic constant for the HDV reaction ($ppm^{0.487}$/h)
k_{T_S}	thermal kinetic constant for the HDS reaction ($wt\%^{0.062}$/h)
L	characteristic catalyst particle size
LHSV, l	liquid hourly space velocity (h^{-1})
M_{ms}	molecular weight of metal sulfide compounds
MOC	concentration of metals-on-catalyst (wt%)
MOR	middle-of-run
m	deactivation rate order
m_i	deactivation order for the reaction i
m_{T_0}	total mass flow
N	reaction order, total number of species in the mixture, total concentration of sites available for any particular reaction
n	reaction order of the hydrocracking of residue
n_{C_i}	reaction order of the catalytic reaction rate i
n_j	order of reaction j
n_{T_i}	reaction order of the thermal reaction rate i
N_A	molar flow across circle area
N_i	concentration of any active sites at any time-on-stream, molar flow of compound i (mol/s)
N_i^0	concentration of any active sites at initial time-on-stream
N_1	concentration of sites type I
N_2	concentration of sites type II
P	total pressure (MPa)
p_i	partial pressure of compound i (MPa)
q	quench fluid mass flow rate (g/s)
R_1	first reactor
R_2	second reactor
R	universal gas constant (kcal/mol K)
r_D	reaction rate of distillates ($g_i/(g_T\, g_{cat}\, h)$)
r_G	reaction rate of gases ($g_i/(g_T\, g_{cat}\, h)$)
$(-r_{HDCCR})$	HDCCR reaction rate (wt%/h)
$(-r_{HDAsph})$	HDAsph reaction rate (wt%/h)
$(-r_{HDBN})$	HDNB reaction rate (ppm/h)
$(-r_{HDNBN})$	HDNNBN reaction rate (ppm%/h)
$(-r_{HDNi})$	HDNi reaction rate (ppm/h)
$(-r_{HDS})$	HDS reaction rate (wt%/h)
$(-r_{HDV})$	HDV reaction rate (ppm/h)

r_j	rate of jth reaction (mol/cm^3 s)
r_{mol}	molecular radius of metal-bearing compounds
r_N	reaction rate of naphtha ($g_i/(g_T g_{cat} h)$)
r_p	instantaneous pore radius
r_R	reaction rate of residue ($g_i/(g_T g_{cat} h)$)
r_{VGO}	reaction rate of VGO ($g_i/(g_T g_{cat} h)$)
S	cross-flow area
SOR	start-of-run
S_p	total geometric external area of particle
t	time
t	time-on-stream (h)
T	absolute temperature
T_0	boiling temperature of the lightest compound in the feed mixture
TBP	true boiling point of any pseudocomponent (K)
TBP(h)	highest boiling point of any pseudocomponent in the mixture (K)
TBP(l)	lowest boiling point of any pseudocomponent in the mixture (K)
T_{mean}	mean absolute temperature
TOS	time-on-stream
t_∞	catalyst life
u_G	gas superficial velocity (cm/s)
u_L	liquid superficial velocity (cm/s)
V_p	total geometric volume of catalyst
v	molar liquid volume
x	hydrocracking reactivity of any species (h^{-1}); variable of integration
x_{MOC}	fraction concentration of metals-on-catalyst (MOC)
y_D	distillates composition
y_G	gas composition
y_N	naphtha composition
y_R	residue composition
y_{VGO}	VGO composition
W_{cat}	weight of catalyst (g)
WHSV	weight hourly space velocity (h^{-1})
wt	weight fraction of species
$wt_{1,2}(\tau)$	concentration in weight fraction of any pesudocomponent with arbitrary boiling point range as function of residence time
z	position within pore, Axial coordinate along the reactor

Subscripts

0	initial
f	feed
Asph	asphaltenes
BN	basic nitrogen
CCR	Conradson carbon residue
HDS	hydrodesulfurization
HDNi	hydrodenickelation
HDV	hydrodevanadization
HDAsph	hydrodeasphaltenization
HDCCR	hydro-Conradson carbon residue conversion
HDM	hydrodemetalization
HDNNBN	hydrodenitrogenation of nonbasic nitrogen
HDNBN	hydrodenitrogenation of basic nitrogen
i	S, Ni, V, Asph, CCR, NBN, BN
in	intrinsic
in	inlet to the following catalytic bed
NBN	nonbasic nitrogen
Ni	nickel
out	outlet of the previous catalytic bed
p	product
q	quench stream
r	restrictive
S	sulfur
t	instantaneous
V	vanadium

Greek Symbols

α	geometry parameter, model parameter in Equation 5.62, fitting parameter of Equations 5.16 and 5.58
α_1	rate constant of diminution on sites concentration type I
α_2	rate constant of diminution on sites concentration type II
β	proportional constant in Equation 5.101, fitting parameter of Equations 5.16 and 5.58
Γ	gamma function
γ	ratio of instantaneous pore radius to initial pore radius
γ	fitting parameter of Equations 5.16 and 5.58
ΔH_{ads}	enthalpy of adsorption of hydrogen sulfide (kcal/mol)

ΔH_R	overall heat of reaction (kJ/kg sulfur)
δ	metal sulfide thickness within catalyst pore, model parameter of hydrocracking yield distribution function ($p(k, K)$)
ϵ	metal sulfide molecules per molecule of reactant
ε_0	bed void fraction
ε_L	dynamic liquid holdup
ζ	dimensionless intraparticle position
η	effectiveness factor
η_0	initial effectiveness factor
η_t	effectiveness factor affected by diffusional resistances
η_{EF}	effectiveness factor
η_{CE}	solid–liquid contact efficiency factor, external catalyst wetting efficiency
φ, ϕ	catalyst activity
φ_A	deactivation function of active sites
φ_D	deactivation function due to deposits
ϕ_j	deactivation function of jth reaction
ϕ^{Coke}	deactivation function for coking reactions
ϕ^{Metals}	deactivation function for metals deposition
θ	deactivation function for metals deposition
λ	ratio of molecule radius to pore radius
μ	dynamic liquid viscosity
ρ_G	gas density at process conditions (g/cm^3)
ρ_{ms}	metal sulfide compound density
ρ_L	liquid density at process conditions (g/cm^3)
ρ_{oil}	heavy oil density
τ	dimensionless time
τ	inverse of space velocity or residence time (h)
ϕ	thiele modulus
χ	dimensionless reactant concentration

Superscripts

app	apparent
G	gas phase
L	liquid phase
Q	quench fluid
S	solid phase

1

PROPERTIES OF HEAVY OILS

1.1 INTRODUCTION

The physical properties and chemical composition of petroleum vary from one source to another. The petroleum fractions, that is, distillates, are separated from each other by fractionated distillation according to boiling points. The lighter fractions, straight-run naphtha and gas oil, are used to produce commercial fuels, gasoline and diesel, respectively. However, bottoms-of-barrel (heavy residua) obtained by distillation of crude oils need further processing. An example of chemical composition of crude oils is given in Table 1.1 in terms of SARA (Saturates, Aromatics, Resins, and Asphaltenes) analysis. It is clearly seen that the composition varies remarkably, for instance, asphaltenes content in crude oils ranges from 5.9 to 23.9 wt%, concentrating more in the heavy petroleum. Another important observation is that heteroatoms (N and S) and metals concentrate in asphaltene fraction. In other words, most of these crude oil impurities are of asphaltenic nature. These aspects and the complex nature of asphaltenes are crucial facts when studying catalyst deactivation.

In the particular case of heavy crude oil, the typical and widely accepted definition is that heavy petroleum is any type of crude oil that does not flow easily. API gravity is the most common parameter to define how heavy or light a crude oil is. API gravity is correlated with the specific gravity (sg) or density by means of the following equations:

$$\text{API gravity} = \frac{141.5}{sg_{60^\circ F}^{60^\circ F}} - 131.5 \qquad (1.1)$$

Deactivation of Heavy Oil Hydroprocessing Catalysts: Fundamentals and Modeling, First Edition. Jorge Ancheyta.

TABLE 1.1 Example of Chemical Composition of Various Crude Oils

	wt%	Crude Oil (wt% of the total)				AR (345 °C+)	
		S	N	V	Ni	Ni	V
			Maya (21°API)				
Saturates	21.2	0.9	3.3	–	–	–	–
Aromatics	27.1	24.4	8.2	0.4	3.3	2.7	2.7
Resins	30.6	38.7	39.6	17.9	17.7	13.0	12.9
Asphaltenes	21.1	36.0	48.9	81.7	79.0	84.3	84.4
			Kern River (13°API)				
Saturates	23.3	0.2	2.7	–	–	–	–
Aromatics	30.7	30.7	4.2	7.5	4.5	1.8	2.7
Resins	40.1	60.3	77.3	52.7	63.0	22.8	16.7
Asphaltenes	5.9	8.8	15.8	39.8	32.5	75.4	80.6
			Arabian Heavy (27°API)				
Saturates	0.2	6.7	–	–	–	–	0.2
Aromatics	29.6	8.4	3.4	10.4	5.2	1.6	29.6
Resins	46.3	43.8	25.2	28.0	14.2	11.8	46.3
Asphaltenes	23.9	41.1	71.4	61.6	80.6	86.6	23.9

AR, Atmospheric residue.

$$sg^{60°F}_{60°F} = \frac{141.5}{\text{API gravity} + 131.5} \tag{1.2}$$

Heavy crude oils possess low API gravity; thus, the API gravity is an inverse measure of the density of petroleum. Heavy crude oils are generally considered as those samples having an API gravity of less than 20°, while extra-heavy crude oils have less than 10° API, both with gas-free viscosity between 100 and 10,000 cP at original reservoir temperature. Heavy oils are then characterized by high viscosities (i.e., resistance to flow) and high densities compared with light crude oil.

In general, heavy oils exhibit a wide range of physical properties. While properties such as viscosity, density, and boiling point may vary widely, the ultimate or elemental analysis varies over a narrow range for a large number of samples. The carbon content is relatively constant, while the hydrogen and heteroatom contents are responsible for the major differences in various heavy oils.

Heavy oils are constituted by heavy hydrocarbons and several metals, predominantly in the form of porphyrins. Heavy feeds also contain aggregates of resins and asphaltenes dissolved in the oil fraction held together by weak physical interactions.

The main problems that heavy crude oil presents during different steps of storage, transportation, and processing are low processing capacity in the refineries, low mobility through the reservoir because of its high viscosity, and difficult and costly transportation from the platform to the ground and to the refineries.

For transportation purposes, viscosity and density (or API gravity) are the most important parameters. Frequently, heavy crude oils with an API gravity of less than 16° cannot be transported without a prior reduction in their viscosity, as this type of crudes come along with viscosities ranging from a few thousands to millions of centipoises (cP) at reservoir temperature, while 250 cSt at 100 °F is a normal maximum desired pipeline viscosity.

Due to the different properties that the various crude oils exhibit around the world, several classifications have been proposed. The classifications of petroleum take into consideration physical properties, distillates properties, chemical structure, origin, and so on. For instance, if the sulfur content in a crude oil is high, the petroleum is classified as "sour," while if this content is low the petroleum is termed as "sweet." On the other hand, if the API gravity is low, the petroleum is termed as "heavy" and if the API gravity is high the petroleum classification is termed "light." Generally speaking, petroleum is classified as follows (Ancheyta and Speight, 2007):

- *Light Crude Oil.* It is also called conventional oil and has an API gravity of at least 20° and a viscosity less than 100 cP.
- *Heavy Crude Oil.* It is a dense and viscous oil that is chemically characterized by its high content of asphaltenes (very complex and large molecules). Its upper limit of API gravity is 20° and a viscosity of 100 cP.
- *Extra-Heavy Crude Oil.* It has an API gravity of less than 10°.
- *Bitumen.* It is also called "tar sands" or "oil sands." It has similar properties to that of heavy oil but is yet more dense and viscous. The main difference between bitumen and heavy oil is that the former does not flow at all. Natural bitumen is oil having a viscosity of greater than 10,000 cP.

Heavy crude oils have low API gravity and high amount of impurities. In general, it is known that the lower the API gravity, the higher the impurities content. Such properties make the processing of heavy feeds different from that used for light distillates, causing several problems such as permanent catalyst deactivation in catalytic cracking and hydrocracking processes caused by metal deposition, temporary deactivation of acid catalysts due to the presence of basic nitrogen, higher coke formation, and lower liquid product yield as a result of high Conradson carbon and asphaltene contents, products with high content of sulfur.

The complex nature of heavy crude oils is the reason why their refining becomes so difficult. Therefore, an evaluation of the overall chemical and physical properties of petroleum feeds is mandatory to determine the processing strategy. Apart from having low API gravity (high density), high viscosity, and high initial boiling point, heavy oils exhibit higher content of sulfur, nitrogen, metals (Ni and V), and high-molecular-weight material (asphaltenes).

1.2 REFINING OF PETROLEUM

A petroleum refinery is mainly designed to produce fuels, for example, gasoline, jet fuel, and diesel. To achieve this goal, the crude oil is subjected to a series of operations and processes. Due to their high amount of impurities (particularly sulfur), straight-run distillates cannot be used directly as fuels; in addition, they possess octane and cetane numbers (naphtha and gas oil, respectively) that are not appropriate for engines. To convert them into suitable materials for fuel production, they need certain treatment, which is carried out in different refining processes. The following sections present a brief description of the main processes used in a petroleum refinery (Ancheyta, 2011, 2013):

1.2.1 Desalting

The first separation process that takes place at the front end of a petroleum refinery is desalting. Its main objective is to prevent corrosion and fouling of downstream lines and equipment by significantly reducing the salt content of oil. Desalting is normally considered a part of the crude distillation unit since heat from some of the streams in the atmospheric distillation is used to heat the crude in the desalting process. Sodium, calcium, and magnesium chlorides, in the form of crystals or ionized in the water present in the crude, are the most frequently found salts in crude oil. The high temperatures found during crude oil refining could cause water hydrolysis if salt is not removed, forming hydrochloric acid (HCl), which will provoke serious corrosion problems in the equipment. Nonremoved salt can also cause fouling problems in pipes, heat transfer equipment, and furnace. Deactivation of catalysts, for example, zeolite-type catalysts used in fluid catalytic cracking (FCC), may be enhanced by metals from salts, particularly sodium. The maximum allowed salt content in the feed to crude distillation units is typically 50 PTB (pounds of salt per thousand barrels of crude oil).

Desalting can be carried out in single stage (dehydration efficiency of ~95%) or in two (dehydration efficiency of ~99%) stages. Dehydration efficiency can be compared with desalting efficiency as most of the salt passed from the organic phase into the water phase if mixing is good. The decision of single or double stage depends on the requirements of the refinery. Typical desalters have an arrangement of two electrodes that generate an electric field among the emulsion causing the droplets vibrate, migrate, and collide with each other and coalesce.

Apart from removing salt, electrostatic desalting also eliminates water and suspended solids in crude oil. Water removal is important to reduce pumping costs and to avoid its vaporization when passing through the preheat train; otherwise, due to high pressure it causes disturbances and vibrations and eventually plant shutdown.

1.2.2 Atmospheric or Primary Distillation

The objective of the atmospheric distillation unit is to fractionate the crude oil into various distillates based on their boiling point ranges. It operates at slightly above atmospheric pressure. The separation is done in a large tower with a number of trays where hydrocarbon gases and liquids interact. The heated desalted crude enters the fractionation tower in a lower section called the flash zone. The unvaporized portion of the crude leaves the bottom of the tower via a steam stripper section, while the distillate vapors move up the tower countercurrent to a cooler liquid reflux stream.

The cooling and condensing of the distillation tower overhead are provided partially by exchanging heat with the incoming crude oil and partially by either an air-cooled or water-cooled condenser. Additional heat is removed from the distillation column by a pump around system, which is simply an internal condenser that ensures a continued reflux stream flow. The overhead distillate fraction from the distillation column is naphtha, which is allowed for leaving the top of the tower to be condensed and collected in the overhead drum. A portion of this stream is returned as reflux, while the rest is delivered to the light end processes for stabilizing and further distillation. The other fractions removed from the side of the distillation column at various points between the column top and bottom are straight-run jet fuel, kerosene, light gas oil, and heavy gas oil, which are steam stripped, cooled by exchanging heat with the incoming crude oil, and sent to other treatments and/or storage. The heavier material, that is, atmospheric residue, is withdrawn from the bottom of the tower. It is important not to subject the crude oil to temperatures above 370–380 °C because high-molecular-weight components will undergo thermal cracking and form coke.

1.2.3 Vacuum or Secondary Distillation

The main objective of a vacuum distillation unit is to recover additional distillates from atmospheric residue. The atmospheric residue is distilled to provide heavy distillate streams used for producing lube oil or as feed to conversion units. The primary advantage of vacuum distillation is that it allows for distilling heavier materials at lower temperatures than those that would be required at atmospheric pressure, thus avoiding thermal cracking. Vacuum distillation is often integrated with atmospheric distillation as far as heat transfer is concerned, which is called combined distillation. Generally, the atmospheric residue is received hot from the atmospheric distillation unit and is sent to the fired heater of the vacuum unit. The vacuum distillation unit is operated at a slight vacuum, which is most often achieved by using multiple stages of steam jet ejectors (absolute pressures as low as 10–40 mmHg). Atmospheric residue is separated into light vacuum gas oil, heavy vacuum gas oil, and vacuum residue.

1.2.4 Solvent Extraction and Dewaxing

Solvent extraction and solvent dewaxing usually remove impurities such as sulfur and nitrogen by dissolving or precipitating at intermediate refining stages or just before sending the product to storage. An important application is the removal of heavy aromatic compounds from lubricating oils. The usual solvents for extraction of lubricating oil are phenol, furfural, and cresylic acid. Solvent dewaxing is used to remove wax from either distillate or residua at any stage in the refining process. Usually two solvents are used: toluene to dissolve the oil and maintain fluidity at low temperatures and methyl ethyl ketone (MEK) to dissolve little wax at low temperatures and act as a wax-precipitating agent. In addition, there is a catalytic process used as an alternative to solvent dewaxing.

1.2.5 Deasphalting

Solvent deasphalting (SDA) is a nondestructive liquid–liquid extraction process (molecular-weight-based separation process), which separates the residue into several fractions on the basis of relative solubility in a solvent (light hydrocarbons such as propane, butane, pentane, or hexane), whereby the last of the molecules that can be refined to valuable products are extracted from the vacuum residue. The application of SDA process has been reported for the production of lube oil feedstocks from vacuum residue using propane as solvent, for the preparation of feedstocks for catalytic cracking, hydrocracking, and hydrodesulfurization (HDS) units, as well as for the production of specialty asphalts. In most of these conversion units, the performance of the catalyst is greatly affected by the presence of heavy metals and the high Conradson carbon content of the residue feed, which are concentrated in asphaltene molecules, so that removing asphaltenes also eliminate these impurities.

SDA process produces a low-contaminant deasphalted oil (DAO) rich in paraffinic-type molecules, and a pitch product rich in aromatic compounds and asphaltenes containing the majority of the feed impurities. The separation of the DAO phase and the pitch phase occurs in an extractor. The extractor is designed to efficiently separate the two phases and minimize contaminant entrainment in the DAO phase. At a constant solvent composition and pressure, a lower extractor temperature increases the DAO yield and decreases the quality. While an increase in solvent ratio the DAO yield remains constant and improves the degree of separation of individual components and results in the recovery of a better quality DAO. The solvent recovered under low pressure from the pitch and DAO strippers is condensed and combined with the solvent recovered under high pressure from the DAO separator, which is then recycled to the initial stage. The produced DAO has lower carbon residue and metal content than the untreated oil; however, SDA is not so effective in lowering sulfur or nitrogen content in DAO.

SDA is also used in refineries to upgrade heavy bottoms streams to DAO that may be processed to produce transportation fuels. The process may also be used in the oil field to enhance the value of heavy crude oil before it gets to the refinery. Thus, SDA is an economically attractive and environmentally friendly process to upgrade heavy petroleum.

1.2.6 Gas and Liquid Sweetening

Gas sweetening is a process used to remove hydrogen sulfide and carbon dioxide from the refinery gas streams. Gas sweetening is generally done with an amine gas treating process, which uses aqueous solutions of various alkanolamines, being MEA (monoethanolamine), DEA (diethanolamine), and MDEA (methyldiethanolamine) the most commonly used amines. MEA has become commercially the most preferred amine due to its high acid gas absorbency.

Apart from amine gas treating, hot potassium carbonate (Benfield) is another process that can be used for acid gas sweetening. There are also other alternatives based on physical solvent processes (Sulfinol, Selexol, Propylene Carbonate, and Rectisol) and dry adsorbents processes (molecular sieve, activated charcoal, iron sponge, and zinc oxide).

Liquid sweetening aims at the elimination of unwanted sulfur compounds (hydrogen sulfide, thiophene, and mercaptans) from crude oil liquid fractions (gasoline, jet fuel, and kerosene) at an intermediate stage in the refining process or just before sending them to storage to improve color, odor, and oxidation stability. The most common process to achieve this target is Merox (mercaptan oxidation). Acids, solvents, alkalis, oxidizing and adsorption agents are the most common materials used for liquid sweetening.

1.2.7 Sour Water Treatment

Sour water (water that contains hydrogen sulfide, ammonia, phenol, and cyanide) is typically treated by a stripping unit with steam whereby H_2S and NH_3 are released at the top of the stripping tower. The H_2S-free water is treated in a biological wastewater treatment plant, where the remaining ammonia is nitrified and then denitrified. In a typical stripping unit, the sour water is fed on the top tray of the tower while steam is introduced below the bottom tray, lending themselves to tray-by-tray mass and heat transfer. The sour water stripping unit is almost always located in the process area of the refinery and can be a single tower with no reflux or a single trayed tower with an overhead reflux stream. Other processes for treatment of sour water are caustic/acid neutralization, caustic oxidization, and oil removal by settling.

1.2.8 Catalytic Reforming

Catalytic reforming is used to convert low-octane straight run naphtha into high-octane gasoline, called reformate, and to provide aromatics (BTX: benzene, toluene, and xylene) for petrochemical plants. The reformate has higher content of aromatics and cyclic hydrocarbons. By means of catalytic reforming, the feed is restructured and cracked to produce a product with hydrocarbons of more complex molecular shapes, whose overall effect is the production of a reformate with higher octane numbers than the feed.

Catalytic reforming also produces very significant amounts of hydrogen gas as by-product, which is released during the catalyst reaction and is used in other processes within the refinery, for example, catalytic hydrotreating (HDT) and hydrocracking.

1.2.9 Isomerization

Isomerization produces gasoline blending component from light paraffins. The objective of isomerization is to convert low-octane *n*-paraffins to high-octane isoparaffins by using a chloride-promoted fixed-bed reactor. As a result of the isomerization reactions, highly branched, high-octane paraffinic blending components are obtained, which by themselves can satisfy the strictest gasoline environmental requirements.

However, the production of this isomerate is low and other streams for gasoline blending are still necessary. Isomerization of *n*-butane is also one of the sources for the isobutane required in alkylation.

1.2.10 Alkylation

The objective of the alkylation process is to combine light olefins (primarily a mixture of propylene and butylene) with isobutane to form a high-octane alkylate gasoline (highly branched C_5–C_{12} isoparaffins), whose major constituents are isopentane and isooctane (2,2,4-trimethyl pentane) with an octane number of 100. Alkylation is an important process that enhances the yield of high-octane gasoline. The reaction occurs in the presence of a highly acidic liquid catalyst (HF: hydrofluoric acid or H_2SO_4: sulfuric acid). As a consequence of the environmental problems associated with the use of these liquid catalysts, solid acid catalysts have also been proposed, with the major problem of rapid deactivation due to coke formation. The function of the acid catalyst is to protonate the olefins feed to produce reactive carbocations, which alkylate isobutane. Alkylation reaction occurs with 100% olefin conversion. It is important to keep a high isobutene-to-olefin ratio to prevent side reactions, which can produce a lower octane product. This is the reason why alkylation units have a high recycle of isobutane.

1.2.11 Polymerization

The objective of a polymerization unit is to combine or polymerize propylene and butylene into molecules of two or three times their original molecular weight. The feeds to this process are propane and butane produced by catalytic cracking. The produced polymer gasoline has an octane number greater than 90 and is an important part of a refinery since polymerization process increases the yield of gasoline possible from gas oil. The polymerization reaction consists of passing the C_3–C_4 hydrocarbon stream with a high proportion of olefins through a reactor containing a phosphoric-acid-supported catalyst, where the carbon–carbon bond formation occurs.

1.2.12 Catalytic Hydrotreating

Catalytic HDT is one of the most important processes in the petroleum refining industry. It is applied to treat almost all of the refinery streams, from straight-run distillates to vacuum residua. The main differences of the HDT processes of each feed are the operating conditions, type of catalyst, reactor configuration, and reaction system. Depending on the feed and the main objective of the treatment, the process can be called HDS, as in the case of straight-run naphtha, which is used as reforming feed where sulfur is the main undesirable heteroatom; HDT for straight-run gas oil because in addition to sulfur removal, aromatic saturation and nitrogen removal are desired for diesel fuel production; hydrodemetalization for the removal of vanadium and nickel from heavy oils; and hydrocracking to change the molecular weight of heavy feeds.

During HDT a number of reactions are carried out, that is, hydrogenolysis, by which C–S, C–N, or C–C bonds are cleaved, and hydrogenation of unsaturated compounds. The reaction conditions of the HDT process vary with the type of feedstock; while light oils are easy to desulfurize, desulfurization of heavy oils is much more difficult. The HDT reactions take place in catalytic reactors at elevated temperatures and pressures typically in the presence of a catalyst consisting of an alumina base impregnated with cobalt or nickel and molybdenum.

1.2.13 Fluid Catalytic Cracking

The FCC increases the H/C ratio by carbon rejection in a continuous process and is used to convert high-boiling, high-molecular weight hydrocarbon fractions (blend of heavy straight-run gas oil, light vacuum gas oil, and heavy vacuum gas oil) to more valuable gasoline, olefinic gases, and other products. The process consists of two main vessels: a reactor and a regenerator, which are interconnected to allow for transferring the spent catalyst from the reactor to the regenerator and the regenerated catalysts from the regenerator to the reactor. During catalytic cracking, the feed is vaporized and long-chain molecules are cracked into much shorter molecules by contacting the feed with a fluidized powdered catalyst at high temperature and moderate pressure.

According to this mechanism, a catalyst promotes the removal of a negatively charged hydride ion from a paraffin compound or the addition of a positively charged proton (H+) to an olefin compound, which results in the formation of a carbonium ion. Carbonium ion is a positively charged molecule that has only a very short life as an intermediate compound and transfers the positive charge through hydrocarbons. This carbonium transfer continues as hydrocarbon compounds come into contact with active sites on the surface of the catalyst that promote the continued addition of protons or removal of hydride ions. This results in weakening of carbon–carbon bonds in many of the hydrocarbon molecules and consequent cracking into smaller compounds. These formed ions also react with other molecules, isomerize, and react with the catalyst to terminate a chain. Coke formation is unavoidable in the catalytic cracking process, which is likely formed by dehydrogenation and condensation of polyaromatics and olefins. Fast deactivation by blocking the active pores of the catalyst is a consequence of coke deposition. During these reactions, the produced catalytic cracked gasoline contains a large amount of aromatics and branched compounds, which is beneficial for the gasoline octane numbers.

1.2.14 Gasification

This process involves complete cracking of residue, including asphaltenes, into gaseous products. The gasification of residue is carried out at high temperature (>1000 °C), yielding synthesis gas or syngas (consisting primarily of hydrogen, carbon monoxide, carbon dioxide, and water), carbon black, and ash as major products. The syngas can be converted into hydrogen or used by cogeneration facilities to provide low-cost power and steam to refineries.

An integrated SDA-gasification facility is an attractive alternative for upgrading of heavy petroleum. The following are some of the benefits obtained by integrating deasphalting and gasification: heavy oils can be upgraded economically, reduction of capital and operating costs of both processes, higher yields of DAO, lower emissions, and increased profit margins of a refinery.

1.2.15 Coking

There are three main coking processes in use:

1.2.15.1 Delayed Coking Delayed or retarded coking can produce shot coke (a type of fuel coke), sponge coke (used to produce anode coke or as a fuel coke), or needle coke. This process accounts for the majority of the coke currently produced in the world. Delayed coking is a semicontinuous thermal cracking process used in petroleum refineries to upgrade and convert bottoms from atmospheric and vacuum distillation of crude oil into liquid and gas product streams, leaving behind a solid concentrated carbon material, petroleum coke, whose value will depend on its properties such as sulfur and metals. The products of a delayed coker are wet gas, naphtha, light and heavy gas oils, and coke. The coke produced in the delayed coker is almost pure carbon and is utilized as fuel or, depending upon its quality, in the manufacture of anodes and electrodes.

In a delayed coker, the feed enters the bottom of the fractionator, where it mixes with recycled liquid condensed from the coke drum effluent. It is then pumped through the coking heater, then to one of two coke drums through a switch valve. The total number of coke drums required for a particular application depends on the quality and quantity of the feed and the desired coking cycle. A minimum of two drums is required for operation, with one drum receiving the heater effluent while the other is being decoked.

A delayed coking unit is frequently designed with the objective of maximizing the yield of liquid product and minimizing the yield of wet gas and coke. The conversion is accomplished by heating the feed material to a high temperature and introducing it into a large drum to provide soaking or residence time for the three major reactions to take place: partial vaporization and mild cracking (visbreaking) of the feed, thermal cracking, and polymerization.

Delayed coking has been selected by many refiners as their preferred choice for upgrading of bottom of the barrel, because of the process's inherent flexibility to handle any type of residua. The process provides essentially complete rejection of metals and carbon while providing partial conversion to liquid products (naphtha and gas oil).

The product selectivity of the process is based on the operating conditions, mainly pressure and temperature. This process is more expensive than SDA, although still less expensive than other thermal processes. The disadvantages of this process are very high coke formation and low yield of liquid products. Despite these disadvantages, delayed coking is the most preferred process for all refiners for residue processing. Advances in delayed coking (low pressure, low oil recirculation) have increased the production of light products and reduced the production of coke.

1.2.15.2 ***Fluid Coking*** Fluid coking produces fluid coke typically used as fuel coke. It is a continuous process that uses the fluidized solids technique to convert residue feedstock into more valuable products. The heated coker feeds (petroleum residua) are sprayed into a fluidized bed of hot, fine coke particles, which are maintained at 20–40 psi and 500 °C. The use of a fluid bed permits the coking reactions to be conducted at higher temperatures and with shorter contact times than in delayed coking. These conditions result in lower yields of coke and higher yields of liquid products. Fluid coking uses two vessels, a reactor and a burner. Coke particles are circulated between these vessels to transfer heat to the reactor. This heat is generated by burning a portion of the coke. The reactor contains a fluidized bed of coke particles, which is agitated by the introduction of steam at the bottom. The residue feed is injected directly into the reactor and is distributed uniformly over the surface of the coke particles, where it cracks and vaporizes. The feed vapors are cracked while forming a liquid film on the coke particles. The particles grow by layers until they are removed and new seed coke particles are added. Coke is a product and a heat carrier.

1.2.15.3 ***Flexi-Coking*** Flexi-coking produces a type of fluid coke that is gasified to generate a low-BTU synthesis gas. It is an extension of fluid coking, which includes the gasification of the coke produced in the fluid coking operation and produces syngas, but the temperature (1000 °C) used is insufficient to burn all coke.

Both fluid coking and flexi-coking are fluid bed processes developed from FCC technology. In both processes, the circulating coke carries heat from the burner back to the reactor, where the coke serves as reaction sites for the cracking of the residua into lighter products. Fluid coking can have liquid yield credits over delayed coking. The shorter residence time can yield higher quantities of liquids and less coke, but the products have lower quality. Fluid coking is a slightly better process than delayed coking because of the advantage of a slightly improved liquid yield, also because delayed coking has higher utilities cost and fuel consumption.

1.2.16 Visbreaking

Visbreaking (viscosity reduction or breaking) is a mature process that may be applied to both atmospheric (AR) and vacuum (VR) residua and even solvent deasphalted pitch, which by means of a mild thermal decomposition, viscosity is improved. The thermal conversion of the residue is accomplished by heating at high temperatures in a specially designed furnace. A common operation is to visbreak residue in combination with a thermal cracker to minimize fuel oil while producing additional light distillates. Visbreaking is a process in which residue stream is heated in a furnace (450–500 °C) and then cracked during a low specific residence time, to avoid coking reactions within a soaking zone under certain pressure and moderate temperature conditions. The cracked product leaves the soaking zone after the desired conversion is achieved and then quenched with gas oil to stop the reaction and prevent coking. The residence time, temperature, and pressure of the furnace's soaking zone is controlled to optimize the thermal free radical cracking to produce the desired products. In general, visbreaking is used to increase the refinery net distillate yield.

The main objectives of visbreaking are to reduce the viscosity of the feed stream and the amount of residual fuel oil produced by a refinery and to increase the proportion of middle distillates in the refinery output.

1.2.17 Residue Fluid Catalytic Cracking (RFCC)

Residue fluid catalytic cracking (RFCC) is an extension of conventional FCC technology that was developed during the early 1980, which offers better selectivity to high gasoline and lower gas yield than hydroprocessing and thermal processes. The RFCC process uses similar reactor technology to the FCC process, in which the catalyst is in fluidized bed at a temperature of 480–540 °C, and is targeted for residual feeds greater than 4 wt% Conradson carbon. Because RFCC requires better feed quality (e.g., high H/C ratio, low content of metals and asphaltenes), it makes this process less likely than hydroprocessing. The need of good feedstock quality is to avoid unreasonable high coke yield, high catalyst consumption, and unit operability. However, such feeds are high in price and limited in refineries.

To control heat balance and to recover part of the heat for steam production, RFCC process design includes two-stage regeneration, mix temperature control, and catalyst cooler. The catalyst properties also play an important role in resisting metal content and carbon deposition. In this respect, catalyst pore structure limits the diffusion of residue on the catalytic sites. The catalyst used for RFCC is of acidic matrix such as crystalline aluminosilicate zeolite in inorganic matrix, which exhibits the required physical–chemical properties.

1.2.18 Hydrovisbreaking Process

1.2.18.1 HYCAR This is one kind of noncatalytic processes, which is based on visbreaking and involves treatment with hydrogen at mild conditions. This process is completed in three reactors: (1) visbreaking, a moderate thermal cracking process in the presence of hydrogen, which leads to more stable products than those obtained with straight visbreaking; (2) hydrodemetalization, which removes impurities particularly metals, before sending to the next reactor; and (3) hydrocracking, where desulfurization and denitrogenation take place along with hydrocracking.

1.2.18.2 Aquaconversion It is a catalytic process that uses catalyst-activated transfer of hydrogen from water added to the feedstock in slurry mode. The homogeneous catalyst is added in the presence of steam, which allows the hydrogen from the water to be transferred to the heavy oil when contacted in a coil–soaker system normally used for the visbreaking process. Reactions that lead to coke formation are suppressed and there is no separation of asphaltene-type material.

The presence of the oil-soluble catalyst and water prevents the coke formation and deposition of sediment that often occurs during visbreaking. In this process, the catalyst may be used as a support or mixed directly with the feedstock. The metals (metal salts) used for hydrovisbreaking are alkali metals such as potassium or sodium. The role of the catalyst is to enhance the dissociation of H_2O to release hydrogen (H^+) ions, which are subsequently consumed in hydroprocessing.

1.2.19 Fixed-Bed Hydroprocessing

Hydroprocessing of residue in fixed-bed reactor (FBR) is well established and reported in the literature. The general characteristic of hydroprocessing is the simultaneous or sequential hydrogenation of hydrocarbon feed in the presence of sulfided catalyst by reacting with hydrogen. The main problem with fixed-bed catalyst is deactivation over time, which can be minimized by a guard-bed reactor, in order to reduce the metal deposition on the downstream reactors. Several combinations using two or three processing steps can be implemented in the refining. The catalyst in the guard-bed reactor is typically an HDM catalyst or large pore catalyst with high metal-retention capacity. Various improvements have been reported in the past decade to increase efficiency of fixed-bed hydroprocessing such as run length, conversion, and product quality. Some of these improvements have been focused on mechanical design such as the use of bunker, swing reactors, guard bed reactors, feed distribution, coke and metal deactivation resistant catalyst, including pore and particle grading and onstream catalyst replacement. In spite of all disadvantages, mainly short catalyst life, up until now most of residue hydroprocessing units have FBRs. The representative technologies of fixed-bed hydroprocessing are RDS/VRDS, Hyvahl-F and Hyvahl-S, IFP hydrocracking process, Isocracking, Mild hydrocracking, MRH Process, Unicracking Process (Unicracking, APCU: Advanced Partial Conversion Unicracking, HyCycle Unicracking Technology), and HIDRO-IMP.

1.2.20 Moving-Bed Hydroprocessing

There are a few types of hydroprocessing reactors with moving catalyst beds, in which the catalyst goes in downflow through the reactor by gravitational forces. In general, catalyst replacement is a batch operation, which is done typically once or twice a week. The fresh catalyst enters at the top of the reactor and the deactivated catalyst leaves the reactor at the bottom, while the hydrocarbon goes either in counter- or cocurrent flow through the reactor. With this moving-bed reactor (MBR), the catalyst can be replaced either continuously or in batch operation. Catalyst transfer is the most critical section. Countercurrent mode of operation seems to be the best configuration since the spent catalyst contacts the fresh feed at the bottom of the MBR, while the fresh catalyst reacts with an almost already hydrodemetalized feed at top of the MBR, resulting in lower catalyst consumption. The moving-bed hydroprocessing technologies are HYCON process, OCR process (Onstream Catalyst Replacement), Hyvahl-M.

1.2.21 Ebullated-Bed Hydroprocessing

In ebullated-bed hydroprocessing, the catalyst within the reactor is not fixed. In such a process, the hydrocarbon feed stream enters the bottom of the reactor and flows upward through the catalyst, which is kept in suspension by the pressure of the fluid feed. The hydrocarbon feed and hydrogen are fed upflow through the catalyst bed, expanding and backmixing the bed, minimizing bed plugging and ΔP. The oil is separated from the catalyst at the top of the reactor and recirculated to the bottom of the bed to mix with the new feed. On the other way, fresh catalyst is added to the top of the reactor and spent catalyst is withdrawn from the bottom of the reactor.

Ebullated-bed reactors (EBR) are capable of converting the most problematic feeds, such as AR, VR, and all other heavy oil feedstocks, which have high content of asphaltenes, metals, and sulfur. Ebullating bed reactors can perform both HDT and HCR functions; thus, these reactors are referred as dual-purpose reactors. Ebullating bed catalysts are made of pellets or grains that are less than 1 mm in size to facilitate suspension by the liquid phase in the reactor.

There are three main ebullated-bed processes, which are similar in concept but different in mechanical aspects: H-Oil, T-Star, and LC-Fining.

1.2.22 Slurry-Bed Hydroprocessing

Slurry-bed reactor (SBR) can also be used for hydroprocessing of feeds with very high metal content to obtain lower boiling products using single reactor. SBR-based technologies combine the advantages of the carbon rejection technologies in terms of flexibility with the high performances peculiar to the hydrogen addition processes. SBR achieves a similar intimate contacting of oil and catalyst and may operate with lower degree of backmixing than EBR. Differently to FBR and EBR, in SBR small amount of finely divided powder is used, which can be an additive or a catalyst (or catalyst precursors). The catalyst is mixed with the feed (heavy oil) and both are fed upward with hydrogen through an empty reactor vessel. Since the oil and catalyst flow cocurrently, the mixture approaches plug-flow behavior. In SBR, the fresh catalyst is slurried with the heavy oil prior to entering the reactor, and when the reaction finishes, the spent catalyst leaves the SBR together with the heavy fraction and remains in the unconverted residue in a benign form. The slurry-bed hydrocracking technologies are CANMET, Microcat-RC, MRH, Veba Combi Cracking (VCC), HDH (Hydrocracking Distillation Hydrotreating), and Eni Slurry Technology (EST).

1.3 PROPERTIES OF HEAVY PETROLEUM

This section describes the most important properties of heavy crude oils that affect deactivation of catalysts.

1.3.1 Physical and Chemical Properties

Heavy crude oil is a thick, black, gooey fluid, harder to handle, and more expensive to refine to produce the most valuable petroleum products. Heavy oil is a type of crude oil that is very viscous, meaning that it does not flow easily. The common properties of heavy oil are high specific gravity (low API gravity), low hydrogen-to-carbon ratio, high carbon residue, high content of asphaltenes, heavy metals (mainly V and Ni), sulfur, and nitrogen.

Because heavy oil is deficient in hydrogen compared with conventional crude oil, either hydrogen must be added to the molecules or carbon be removed to render it a useful feedstock for a conventional refinery.

The physical and chemical properties, as well as exact chemical composition, of heavy crude oil vary from one source to another. Crude oils, especially heavy crude oils, contain large quantities of asphaltenes (high-molecular-weight polar components). Other crude oil components are resins, fatty acids such as naphthenic acids, porphyrins, and wax crystals, which can associate to asphaltenes and affect the oil stability. Particles such as silica, clay, and iron oxides can be present in crude oils.

Table 1.2 reports the properties of several heavy crude oils. Heavy crude oils exhibit a wide range of physical and chemical properties. While the properties such as viscosity, density, boiling point, and color may vary widely, the ultimate or elemental analysis varies over a narrow range. The carbon content is relatively constant, while the hydrogen and heteroatom contents are responsible for the major differences. Nitrogen, oxygen, and sulfur can be present in only trace amounts in some heavy crude oils, which consist primarily of hydrocarbons.

It is the heteroelements that can have substantial effects on the distribution of refinery products. Coupled with the changes brought about to the feedstock constituents by refinery operations, it is not surprising that refining heavy feedstocks is a monumental task. Thus, initial inspection of the feedstock (conventional examination of the physical properties) is necessary. From this, it is possible to make deductions about the most logical means of refining. In fact, evaluation of crude oils from physical property data as to which refining sequences should be employed for any particular crude oil is a predominant part of the initial examination of any material that is destined for use as a refinery feedstock.

The chemical composition of a heavy crude oil is a much truer indicator of refining behavior. Whether the composition is represented in terms of compound types or in terms of generic compound classes, it can enable the refiner to determine the nature of the reactions. Hence, chemical composition can play a major part in determining the nature of the products that arise from the refining operations. It can also play a role in determining the means by which a particular feedstock should be processed.

Thus, initial inspection of the nature of the feedstock will provide deductions about the most logical means of refining or correlation of various properties to structural types present and hence attempted classification of the petroleum. Indeed, a careful evaluation from physical property data is a major part of the initial study of any refinery feedstock. Proper interpretation of the data obtained from the inspection of crude oil requires an understanding of their significance.

1.3.2 Asphaltenes

A crude oil can be separated into two main components: asphaltenes and maltenes. The heavy fractions of petroleum can be defined as molecules possessing more than

TABLE 1.2 Properties of Different Crude Oils

Country	USA	Venezuela	USA	USA	Kuwait	USA	Canada	USA	Canada
Crude oil	California 11	Boscan	California 15	Beta	Eocene	Hondo	Hebron	Carpinteria	Atkinson
API gravity	10.3	10.9	13.2	13.7	18.6	19.6	20.1	22.9	23.7
Total sulfur (wt%)	3.3	5.5	5.5	3.78	4.55	4.3		1.88	0.86
Viscosity at (cP)									
0 °C	220,000	8,826,000	31,000	90,210		3,507		790	65
15 °C	34,000	485,500	6,400	13,380		735	680	164	
30 °C							185		
SARA analysis (wt%)									
Saturates		25	19			33		44	47
Aromatics		35	35			31		30	36
Resins		22	23			24		17	14
Asphaltenes	16	18	23			12		9	3
Metals (wppm)									
Ni	1,606	117	111	112		75	6	49	2
V	245	1,320	266	146		196	1	112	9
Water content (wt%)	8.6		0.1	1.7		1.5		0.1	
Reid vapor pressure (kPa)					8				6
Pour point (°C)	0	21	−9	3	−29	−15	−4	−21	−46

25 atoms of carbon distributed in polar and heavy compounds, such as asphaltenes and resins, having high boiling points (Merdrignac and Espinat, 2007). Since asphaltenes are the main constituents of heavy crude oils and their structure and composition directly affect the whole composition of the petroleum, they deserve special attention. Figure 1.1 shows the physical aspects of asphaltenes precipitated from various crude oils. Although not very clear, it is observed that asphaltenes from heavy crude oils

(a)

(b)

(c)

Figure 1.1 Asphaltenes precipitated from various crude oils: (a) 13°API, (b) 21°API, and (c) 33°API.

are blacker than those from light crude oils, which indicates more complexity of the former.

The most common and widely accepted definition of asphaltenes is based on their solubility properties (Ancheyta et al., 2010):

"Asphaltenes are insoluble in alkanes such as *n*-pentane, *n*-hexane, *n*-heptane or higher but soluble in aromatic solvents (i.e., benzene, toluene)."

The asphaltene content is mostly reported as *n*-heptane insolubles; however, strictly speaking, they must be calculated as the difference between weight percentage of *n*-heptane insolubles and the weight percentage of toluene insolubles.

Asphaltenes are commonly precipitated from crudes by adding *n*-heptane because their properties do not exhibit significant changes when using *n*-heptane or higher carbon number alkanes (Andersen, 1994).

Asphaltenes are the main precursors of sludge or sediments. They are polyaromatic complex compounds with a boiling point higher than 500 °C, which have the following components: aromatic rings carrying alkyl chains up to C_{30}; sulfur present in benzothiophene rings and nitrogen contained in pyrrol and pyridine; ketones, phenols, carboxylic acids; nickel and vanadium complexed with pyrrole nitrogen atoms in porphyrinic rings.

The other component of a crude oil, maltenes, has the following constituents: resins, which are structures similar to asphaltenes but having lower molecular weight; oxygen, nitrogen, and sulfur are not always present in aromatic structures; naphthenes and other saturate hydrocarbons such as straight or branched chains.

The complexity of heavy oils makes compositional analysis very complex, and separation is carried out for compound types instead of individual compounds. When more insolubles (asphaltenes) are removed, the soluble fraction (maltenes) is rich in hydrogen. Sulfur, nitrogen, and oxygen of soluble fractions decrease as more insolubles are removed. Typically, the insoluble fraction contains 7–21 heteroatoms per molecule, and the soluble fraction has 0.8–1.7 (Sharma et al., 2007).

Asphaltenes can be modeled in two different ways:

- Continental type, pericondensed structures in which asphaltene cores are constituted by more than seven aromatic rings.
- Archipelago type, in which asphaltenes are represented by small aromatic cores linked to other cores by means of bridging alkanes, that is, presence of islands of small aromatic cores linked by alkyl and sulfur bonds.

Hypothetical structures of these two asphaltene types are presented in Figure 1.2 (Zhao et al., 2001). The way in which both types of asphaltenes aggregate is different. Archipelago-type asphaltenes form planar aggregates in asphaltenes solutions, whereas continental-type asphaltenes are able to stack forming columns (Murgich et al., 1996).

1.3.3 Tendency to Coke Formation

The tendency to coke formation of crude oils is commonly measured with two analyses: Ramsbottom carbon and Conradson carbon residue. Matsushita et al. (2004)

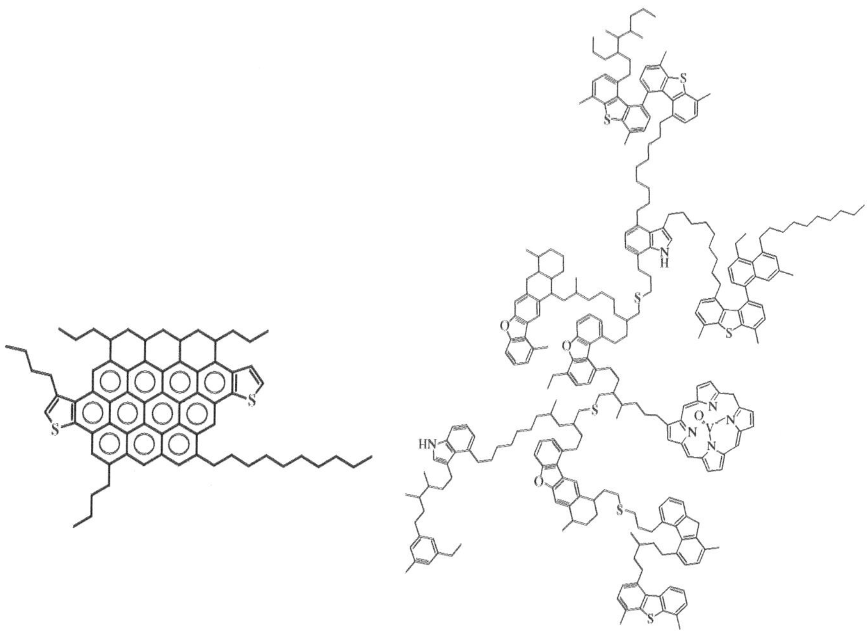

(a) (b)

Figure 1.2 Hypothetical structures of asphaltenes: (a) Continental type and (b) Archipelago type.

also defined the following relationship that takes into account the H/C atomic ratio of asphaltenes and maltenes, which gives certain information about the solubility of asphaltenes and its influence on coke formation during processing of petroleum:

$$\text{Relative solubility index (RSI)} = \frac{\text{H/C of asphaltenes}}{\text{H/C of maltenes}} \tag{1.3}$$

It was reported that the higher the relative solubility index, the lesser the coke formation and that at lower RSI more coke will be deposited, which will have less H/C (more aromatic coke). Based on this index, asphaltenes and maltenes from three crude oils were precipitated and characterized by elemental analysis, and RSI was calculated. The results are presented in Figure 1.3. It can be anticipated that heavy crude oils will have more tendency to form coke during their processing.

1.3.4 Viscosity of Crude Oils and Blends

The problem of determining the viscosity of heavy crude oils is not so critical since most of the times a viscometer is always available in the laboratory. However, sometimes it is required to calculate viscosity of either crude oils or blends of crude oils.

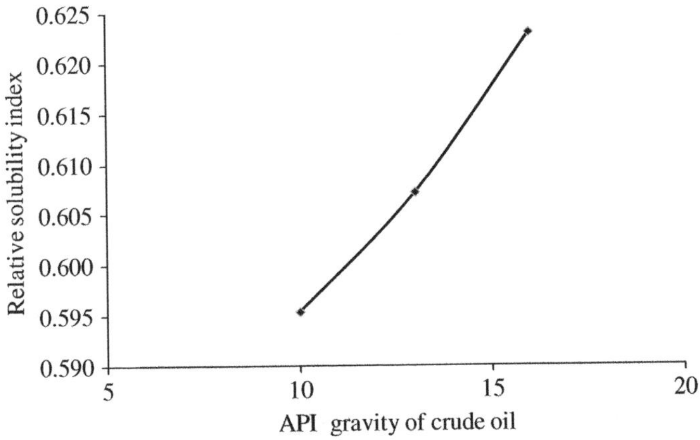

Figure 1.3 Tendency to coke formation of crude oils.

Various standardized methods are available for experimental determination of viscosity for different types of crudes and petroleum products. The most used are ASTM D88, ASTM D445, ASTM D2170, ASTM D7042, ASTM D7483, and ASTM E102. The major differences among them are the type and required amount of sample, experimental setup, time for analysis, operating conditions of equipment, and ranges of viscosity in which the device can be used. In practice, it has been observed that measuring the viscosity of crude oils with low API gravity is complicated due to their own nature and difficulty to handle. This makes the analysis requires more time and greater amount of sample to obtain reliable results.

1.3.4.1 Viscosity of Crude Oils Viscosity is the most important property used for defining if a crude oil is suitable for transportation. In general, heavy crude oils (low API gravity) exhibit high viscosity. As was already stated, the common specification for crude oil transportation is a maximum viscosity of 250 cSt at 100 °F and a minimum API gravity of 16° (Hedrick et al., 2006).

The viscosity of crude oil can be predicted by means of correlations. Several methods have been proposed to predict dead oil viscosity (μ_{od}), saturated or bubble point viscosity (μ_{ob}), and undersaturated oil viscosity (μ_o). Depending on the category, the correlation uses parameters and properties such as API gravity, temperature, pour point, pressure, bubble point pressure, gas–oil ratio, and molecular weight. According to this, two types of correlations for the prediction of crude oil viscosity are reported in the literature: (1) those using conditions and bulk properties such as temperature, pressure, and specific gravity and (2) those using crude oil composition, normal boiling point, and pour point temperature.

All authors reported high accuracy for viscosity prediction with their correlations; however, they were developed and tested with a particular set of experimental data, and extrapolation to other data may change the precision of predictions.

Sánchez et al. (2014) compared the correlations developed to predict the dynamic viscosity of crude oil as a function of API gravity and temperature. The most used correlations have been developed by Beal, Beggs, Glaso, Egbogah, Kartoatmodjo, Elsharkawy, Naseri, Hossain, Alomair, and Petroski (Sánchez et al., 2014). Some of the equations exhibit similarities in the number of parameters included (e.g., Egbogah and Elsharkawy) or in the mathematical structure, such as Beggs and Egbogah (with an additional parameter in the correlation proposed by Beggs). Moreover, Glaso and Kartoadmodjo coincide in using the API gravity value raised to an exponent that depends on the temperature. However, the equations proposed by Kartoadmodjo, Hossain, and Alomair reach higher viscosity values, therefore applicable for heavy crude oils.

Even though the authors of correlations claim that accurate predictions can be obtained with their proposed equations, the correlation usually will not work for a crude oil with different API gravity and temperature range from those that were used to develop it.

The authors used 12 crude oils with a wide range of API gravity (12.4–43°API) for comparison of correlations. It was found that all correlations tend to fail as the viscosity values increase, that is, for heavy oil samples, so that an accurate correlation was needed for calculating viscosity of heavy crude oils. Thus, a new correlation involving two coefficients that are functions of API gravity was proposed, which was validated using viscosity values different from those used to derive it. A comparison of results using literature information and own set of data indicated better predictability over other correlations previously reported. The developed correlation presents a good performance in the range of API gravity evaluated (12.4–43.0°). However, its higher precision was observed for API gravity lower than 21.1°, even higher than other correlations reported in literature. Therefore, the following proposed correlation is recommended to predict the viscosity of heavy and extra-heavy crude oils (Sánchez et al., 2014):

$$\mu = a \exp(b/T^3) \tag{1.4}$$

$$a = 3.9 \times 10^{-5}\,\mathrm{API}^3 - 4.0 \times 10^{-3}\,\mathrm{API}^2 + 0.1226\,\mathrm{API} - 0.7626 \tag{1.5}$$

$$b = 9.1638 \times 10^9\,\mathrm{API}^{-1.3257} \tag{1.6}$$

where μ is the dynamic viscosity in centipoise, T the absolute temperature in Kelvin, and a and b are coefficients, which depend on API gravity.

1.3.4.2 Viscosity of Blends of Crude Oils Crude oil blending is often undertaken to increase the sale price or processability of a lower grade crude oil by blending it with a higher grade, higher price crude, aiming at producing blended crude oil to a target specification at the lowest cost using the minimum higher cost crude oil.

Refiners usually want to know in advance which crude oils should be mixed and in which volumetric ratios, in order to achieve the desired value of viscosity or of other properties. Thus, the prediction of viscosity of crude oil blends becomes a mathematical issue rather than experimental. That is done with the help of mixing rules.

A total of 26 mixing rules are reported in the literature (Centeno et al., 2011), which can be classified according to the number and type of parameters involved for the calculation of viscosity, as well as experimental information required.

- *Pure Mixing Rules.* Pure mixing rules are easy to apply as they only require experimental viscosity of components and composition of mixtures in terms of volume or weight fractions, such as the Arrhenius, logarithmic, Bingham, Kendall, and Monroe, linear, Cragoe, Reid, and Chirinos.
- *Mixing Rules with Viscosity Blending Index.* These rules were developed to predict blend viscosities of all petroleum components from gasoline to vacuum residue. The Refutas index method belongs to this classification, which uses the Refutas index of each component (VBI_i: Viscosity Blending Index or Viscosity Blending Number) and the Refutas index of the blend (VBI_β). The Chevron equation is another example of mixing rule that involves calculation of VBI_β.
- *Mixing Rules with Additional Parameters.* This type of mixing rules includes calculation of extra parameters that are usually obtained by mathematical methods. Among these rules are those proposed by Walther, Latour, Lederer, Shu, Ishikawa, Lobe, power law, Barrufet, and Setiadarma, ASTM D341 method optimized by Twu and Bulls, Panchenkov, Reik, and Lima.
- *Mixing Rules with a Binary Interaction Parameter.* Expressions for estimating viscosity that include a binary interaction parameter have been proposed by several authors such as Van der Wyk, Grunberg and Nissan, and Tamura and Kurata. McAllister proposed a method involving two interaction parameters that are calculated from known values of the mixture viscosity. According to this method, the interaction parameters vary linearly with the inverse of temperature; thus, if the values of these quantities are known at two temperatures, it should be possible to read the values at other temperatures from appropriate graphs.
- *Mixing Rules with an Excess Function.* Ratcliff and Khan calculated the viscosity of a mixture based on the absolute viscosities of components; it is necessary to include an excess function to account for deviation from ideality. Wedlake and Ratcliff reported a model based on an excess quantity that is calculated from a structural constant, the number of groups in molecular species, and the individual group contribution of each group in the mixture.

Mixing rules range from that purely empirical to those that require detailed experimental information and to solve relatively complex mathematical functions. The pure mixing rules are the simplest of the rules. They have been widely used precisely by its simplicity and little experimental information needed (only data of viscosity and proportion of the blend components). This type of mixing rules has a basic mathematical structure (e.g., linear, logarithmic, inverse, power law) that oversimplifies the problem, so their successfulness in predicting viscosity of heavy oils is nearly random. The mixing rules with additional parameters try to compensate the simplicity of the pure mixing rules by adding constants, which in fact are correlation parameters whose values have been calculated from experimental data and are valid only within the range of conditions used for their determination. The authors of these rules claim more accurate results than the pure mixing rules, which is expected, as the number of parameters increases how well a correlation fits the data is also better.

It is well known that viscosity does not blend linearly, but a function of viscosity, the Viscosity Blending Index, does. Blending linearly means that the value of viscosity of a blend is simply the arithmetic mean of the values of viscosity for the components weighted by the quantity of the component in the blend. The mixing rules based on VBI (Refutas and Chevron) should, in principle, behave better for predicting viscosity. This indeed happens but only for low-viscosity samples. The reason behind this underperformance is because viscosity is a nonadditive property that behaves differently from those based on mass, such as impurity content and density.

The mixing rules with a binary interaction parameter or with an excess function are supported by thermodynamic concepts and may achieve better performance than the others. In any case, the parameters have to be estimated from a wide range of experimental data and blends, so that the application of these mixing rules can be extended to other samples different from those used to derive the parameter values.

It is worth to mention that petroleum is generally close to Newtonian fluids; however, it strongly depends on oil composition. It is quite likely that at low temperatures the oil is so viscous that Reynolds number drops drastically (by virtue of the log–log relationship). Viscosity variability is more related to composition than density/temperature. Due to this, viscosity is one of the most unreliable properties to determine by equations. Commercial process simulators offer several correlations, but they are usually not accurate for heavy oils.

Centeno et al. (2011) concluded that the pure mixing rules estimate crude oil viscosity with high standard error. The most-known and used Refutas and Chevron mixing rules show general overestimation of experimental viscosity, while only those rules that include an additional parameter such as the Walther, Einstein, and power law predict viscosity with relative low error. No rule shows capability to accurately estimate the viscosity for all crude oils, as the disagreement of estimated and experimental values increased with the API decrement. The analysis of light petroleum distillate blends confirms that the capability of each mixing rule to predict viscosity decreases as the sample becomes heavier. The results show that predicting viscosity of heavy crude oils and their blends is still a challenging task, and a proper mixing rule for such a purpose is needed.

More recently, Centeno et al. (2015) also examined the method developed by Ratcliff and Khan for calculating viscosity of crude oil blends by using data from the literature. To apply this method, it is necessary to include an excess function to account for deviation from ideality, by means of the following equation:

$$(\ln \nu)_{\text{real}} = \sum x_i \ln \nu_i \pm (\ln \nu)^E \tag{1.7}$$

where the excess function is defined by

$$(\ln \nu)^E = a_{\text{AB}} w_{\text{A}} w_{\text{B}} \tag{1.8}$$

For instance, for ternary mixtures, there are three possible combinations for binary blends, that is, three parameters to be estimated: a_{12}, a_{13}, and a_{23}. Applying

Equation 1.8 to this system results in

$$(\ln \nu)_1^E = a_{12}x_1^1x_2^1 + a_{13}x_1^1x_3^1 + a_{23}x_2^1x_3^1 \tag{1.9}$$

$$(\ln \nu)_2^E = a_{12}x_1^2x_2^2 + a_{13}x_1^2x_3^2 + a_{23}x_2^2x_3^2 \tag{1.10}$$

$$(\ln \nu)_3^E = a_{12}x_1^3x_2^3 + a_{13}x_1^3x_3^3 + a_{23}x_2^3x_3^3 \tag{1.11}$$

To solve this system of equations, it is necessary to know the composition of each crude in the blend (x_i^j) and the excess function $(\ln \nu)^E$. (x_i^j) is defined by experimental blends and $(\ln \nu)^E$ can be calculated with the experimental information and Equation 1.7. The authors concluded that in all cases better prediction is obtained with the method using binary interaction parameters. Apart from this, the values of the parameters resulted to have an inverse dependency with temperature, so that a correlation between each parameter with temperature can be established, and viscosity of blends at different temperature can be easily computed.

1.3.4.3 Other Properties Following the same approach described for viscosity of blends, to calculate other properties the following approaches can be used:

1. *Additive Properties*. Those that depend on mass, such as API gravity and impurity content (sulfur, metals, nitrogen), which use the following linear mixing rule:

$$\beta_b = \sum_{i=1}^{N} y_i \beta_i \tag{1.12}$$

where β_b is the property of the blend, β_i is the property of component "i," and y_i is the volume or mass fraction of component "i" in the blend.

2. *Nonadditive Properties*. Those that do not depend on mass, such as Reid vapor pressure and flash point. To calculate nonadditive properties, it is necessary to have the blending index of the property, which allows for linearization of the behavior of such a property, by means of the following equation:

$$\mathrm{BI_b} = \sum_{i=1}^{N} y_i \mathrm{BI}_i \tag{1.13}$$

where BI_i is the blending index of component "i."

The following blending indexes are found in the literature for various properties:

Reid vapor pressure (RVP): $$\mathrm{BI_{RVP}} = (\mathrm{RVP})^{1.25} \tag{1.14}$$

Flash point (FP): $$\log_{10}\mathrm{BI_{FP}} = -6.1188 + \frac{2414}{\mathrm{FP} - 42.6} \tag{1.15}$$

Pour point (PP): $$\mathrm{BI_{PP}} = \mathrm{PP}^{1/0.08} \tag{1.16}$$

Aniline temperature (AT): $$\mathrm{BI_{AT}} = 1.124[\exp(0.00657 * \mathrm{AT})] \tag{1.17}$$

1.3.5 Stability and Compatibility

The definitions of stability/instability and compatibility/incompatibility are as follows:

- *Stability*. Ability of the oil to remain in an unchanged condition despite circumstances that may tend to cause change, while instability is the tendency of a crude oil to produce a deposit of asphaltenic sludge as a function of time and/or temperature.
- *Compatibility*. Degree to which two or more crude oils of different origins can be mixed without causing flocculation (precipitation) of asphaltene or wax deposition. As a general rule, crude oils of the same type are always compatible. However, even within the same group or type of crude oil, incompatibility can occur. In such cases, the prediction of incompatibility problems requires considerable experimental work.

Once precipitated, it is difficult to redissolve asphaltenes quickly. Meanwhile, precipitated asphaltenes can cause stable oil–water emulsions, plugging of transportation lines, fouling of heat exchangers, and catastrophic coking of distillation furnace tubes.

To be considered as stable and compatible, a blend of various crude oils must have the following characteristics:

- Must be homogeneous immediately after preparation
- Must remain homogeneous in normal storage
- Must not produce or tend to produce sludge on a significant scale.

Asphaltenes play an important role in controlling the viscous behavior of crude oil and the stability and compatibility during storage and transportation. This has been extensively revealed in the literature by rheological studies with crude oils from different sources. Although crude oils are generally considered Newtonian fluids, crude oils with high amount and more complex asphaltene molecules will exhibit pronounced pseudoplastic, non-Newtonian behavior.

Asphaltenes are believed to exist as micelles in petroleum. Resins, which can be considered as low-molecular-weight asphaltenes, are also known as maltenes and act as a solvent for the other constituents. Thus, heavy oil is generally considered to contain a disperse phase of asphaltenes complexed with high-molecular-weight components of the maltenes (resins) and liquid hydrocarbons in the form of a micelle. An equilibrium state exists under these conditions, and the micelles are considered to be peptized (i.e., colloidally dispersed). However, if the C/H ratio of the maltenes is lowered, the resins that are absorbed in the asphaltenes are partially desorbed. The consequence of this change is an alteration of the equilibrium, which results in the asphaltene particles not being completely surrounded by resins and they are mutually attracted. This ends up to precipitation of asphaltenes, which appear as sludge.

The most common methods to evaluate the stability/compatibility of crude oils and blends of crude oils can be classified as direct and indirect as follows (Speight, 2001):

1.3.5.1 Direct Test Methods

- *Spot-Test*. It gives a visual determination of oil stability and hence the results are slightly subjective.
- *Hot-Filtration Test*. It gives the total amount of sediment in the oil but does not give information about the stability of the oil.
- *Xylene Equivalent Test*. It determines the necessary proportion of xylene in a blend with isooctane to obtain a homogeneous spot on a chromatographic paper when blended with an equal quantity of the specified sample. The lower the proportion of xylene, the better the compatibility of the sample.
- *Bottle Test*. This test is gravimetric in nature and easy to perform and is the traditional standard method for crude oil–completion brine compatibility testing.
- *Gum Stability*. It is an approximate measure of the tendency to form gum during storage.
- *Water, Sediment, and Salt Contents*. The higher the bottom sediment and water content, the higher sludge and deposit formation rates that can be expected.

1.3.5.2 Indirect Test Methods

- *Ultimate or Elemental Analysis (C, H, O, N, S, Metals, and Ash Contents)*. The higher the nitrogen and sulfur contents in crude oils, the higher the sludge-formation tendencies.
- *Specific Gravity/API Gravity*. Crude oils having low API gravity (heavy oils) are more susceptible to sludge formation due to their higher content of polar/asphaltic constituents.
- *Viscosity*. The viscosity of the crude oil generally increases due to the presence of products of instability or compatibility processes.
- *Pour Point*. Crude oils with high pour point are more waxy and tend to form wax-like materials that enhance sludge formation.
- *Asphaltenes*. The higher the asphaltenes content, the greater the tendency of crude oils to form sludge, especially when blended with other noncompatible stocks.
- *Total Acidity (Acid Number)*. Crude oils of higher acidities may exhibit a tendency of instability.
- *Characterization Factor*. If the characterization factor is above 12 (paraffinic), the crude oil is expected to form waxy deposits.

1.3.5.3 Methods Based on SARA Analysis

- *Colloidal Instability Index (CII)*. This index expresses the stability of asphaltenes in terms of SARA (Saturates, Aromatics, Resins, Asphaltenes) fractions and is defined as the mass ratio between the sum of asphaltenes and unfavorable components of its stability in the oil, that is, its flocculants (Saturates) and the sum of peptizing agents (Aromatics and Resins), which are components favorable to the stability of asphaltenes present in a specific oil (Asomaning, 2003):

$$\mathrm{CII} = \frac{\text{Asphaltenes } (\%) + \text{Saturates } (\%)}{\text{Resins } (\%) + \text{Aromatics } (\%)} \tag{1.18}$$

CII can also be represented in a plot of (Resins+Aromatics) versus (Asphaltenes+Saturates). The lower the CII value, the greater the stability of asphaltenes in the oil. CII gives a relative way to assess whether the oil is more or less stable compared with others. If CII ≥ 0.9, asphaltenes are unstable in the oil. If CII < 0.7, the asphaltenes are stable in the oil. For values between 0.7 and 0.9, there is uncertainty about the stability.

- *Stankiewicz Plot* (Stankiewicz et al., 2002). It uses a cross-plot of two ratios: (Saturates/Aromatics) versus (Asphaltenes/Resins). This type of plot gives a quick identification of asphaltene risk by separating the stable and unstable area.
- *Stability Cross-Plot* (*SCP*) (Sepulveda et al., 2010). It uses four plots to determine the stability of asphaltenes for any type of petroleum:
  1. (Resins/Asphaltenes)/(Saturates/Aromatics) versus (Aromatics/Asphaltenes)
  2. (Resins/Asphaltenes)/(Saturates/Aromatics) versus (Resins/Asphaltenes)
  3. (Saturates/Aromatics) versus (Resins/Asphaltenes)
  4. (Resins/Asphaltenes)/(Saturates/Aromatics) versus. (Aromatics/Saturates/Asphaltenes).

  Better confidence of the SCP method (92%) has been claimed compared with other methods (CII: 72%, Stankiewicz: 86%).

1.3.5.4 Other Methods

- *De Boer Plot* (de Boer et al., 1995). It uses a plot of *in situ* oil density versus undersaturation pressure. Higher risks are observed in case of light oil under high undersaturation pressure.
- *Heithaus Titration Method* (*P-value*). It gives the total amount of *n*-heptane that can be added before the oil becomes instable. The Heithaus titration is performed only with the toluene-soluble components of the oil. Three toluene solutions at different concentrations are titrated with a weak solvent such as isooctane. The weight of oil (W_a), the volume of toluene (V_s), and the volume of isooctane titrant (V_t) are recorded at the flocculation point, where asphaltenes just begin to precipitate for each solution. The flocculation ratio (FR) and dilution concentration (C) are calculated as follows (Heithaus, 1962):

$$\mathrm{FR} = \frac{V_\mathrm{s}}{V_\mathrm{s} + V_\mathrm{t}} \tag{1.19}$$

$$C = \frac{W_\mathrm{a}}{V_\mathrm{s} + V_\mathrm{t}} \tag{1.20}$$

A plot of FR versus C is made and the intercepts are determined (FR_{max} and C_{min}). The Heithaus parameters are defined as follows:

$$\text{Peptizability of asphaltenes:} \quad p_a = 1 - FR_{max} \tag{1.21}$$

$$\text{Solvent power of maltenes:} \quad p_0 = FR_{max}(1/C_{min} + 1) \tag{1.22}$$

$$\text{Overall compatibility of residuum:} \quad P = p_0/(1 - p_a) = 1/C_{min} + 1 \tag{1.23}$$

Larger values of p_a indicate peptizable asphaltenes and larger values of P indicate an overall compatible system. A larger p_o value is subject to a mixed interpretation. Oil with $P < 1$ is considered as instable oil.

- *Intrinsic Stability* (S-value). It describes a procedure for quantifying the intrinsic stability of the asphaltenes in an oil matrix, by using an optical device. Similarly to P-value method, three parameters are determined:
  - S: It defines the state of peptization of the asphaltenes in oil and characterizes the global stability of the sample. Oil with a low S-value is likely to undergo flocculation of asphaltenes. Oils with high S-values are likely to maintain asphaltenes in a peptized state and do not lead to asphaltene flocculation when blended.
  - Sa: It represents the solvency requirements of the peptized asphaltenes and characterizes the aromaticity of asphaltenes. In other words, it gives the ability of asphaltenes to remain in colloidal dispersion. The lower the Sa value, the higher the aromaticity.
  - So. It identifies the power of peptization of oil and characterizes the aromaticity of resins and their capability to maintain asphaltenes in solution. The higher the value of So, the higher the aromaticity.

1.4 ASSAY OF PETROLEUM

A crude oil assay is a compilation of laboratory (physical and chemical properties) and pilot plant (distillation and product fractionation) data that characterize a specific crude oil. Assay analyses of whole crude oils are carried out by combining atmospheric and vacuum distillation units, which when combined will provide a TBP (True Boiling Point) distillation. The values of the distillation ranges of the distilled fractions are usually defined depending on the refinery product classifications. The most common distillation ranges used in international assays of crude oils are as follows:

- Light straight-run naphtha (LSRN): IBP–71 °C
- Medium straight-run naphtha (MSRN): 71–177 °C
- Heavy straight-run naphtha (HSRN): 177–204°C
- Jet fuel (JF): 204–274 °C
- Kerosene (K): 274–316 °C

- Straight-run gasoil (SRGO): 316–343 °C
- Light vacuum gasoil (LVGO): 343–454 °C
- Heavy vacuum gasoil (HVGO): 454–538 °C
- Vacuum residue (VR): 538 °C+

Sometimes, the atmospheric residue (343 °C+, blend of LVGO, HVGO and vacuum residue) is also obtained and characterized. A detailed hydrocarbon analysis of crude oil off-gas is also run.

The most common applications of petroleum assays in different areas in petroleum refining industry are as follows:

- To provide extensive detailed experimental data for refiners to establish the compatibility of a crude oil for a particular petroleum refinery
- To anticipate if the crude oil will fulfill with the required product yield, quality, and production
- To determine if during refining the crude oil will meet with environmental and other issues
- To help refiners make decisions about changes in plant operation, development of product schedules, optimize the refining process, and examination of future processing ventures
- To supply engineering companies with detailed crude oil analyses for preparing the process design of petroleum refining plants and detailed refinery engineering
- To facilitate the companies' crude oil pricing and to negotiate possible penalties due to impurity content and other undesired properties.

An assay varies in depth and complexity depending on the crude oil type and its end use. The assay can be an inspection assay or comprehensive assay. There are various types of assays that vary considerably in the amount of determined experimental information:

- Assays that include yields and properties of those streams used as feed for catalytic reforming (naphtha) and catalytic cracking (gas oils)
- Assays that give more details for potential production of lubricant oil and/or asphalt
- Assays that include minimal information such as a distillation curve (typically TBP distillation, true boiling point) for the crude and a specific gravity curve
- Assays that are complete and include experimental characterization of the whole crude oil and of various boiling range fractions, curves of TBP, specific gravity, and sulfur content.

REFERENCES

Ancheyta, J. 2011. Modeling and simulation of catalytic reactors for petroleum refining, John Wiley & Sons, Inc., New Jersey

Ancheyta, J. 2013. Modeling of processes and reactors for upgrading of heavy petroleum, CRC Press – Taylor & Francis, New York.

Ancheyta, J., and Speight, J.G. 2007. Hydroprocessing of heavy oil and residua, CRC Press – Taylor & Francis, New York.

Ancheyta, J., Trejo, F., and Rana, M.S. 2010. Asphaltenes: chemical transformations during hydroprocessing of heavy oils, CRC Press – Taylor & Francis, New York.

Andersen, S.I. 1994. Dissolution of solid Boscan asphaltenes in mixed solvents. Fuel Sci. Technol. Int. 12(11–12):1551–1577.

Asomaning, S. 2003. Test method for determining asphaltene stability in crude oils. Pet. Sci. Technol. 21:581–590.

de Boer, R.B., Leeriooyer, K., Eigner, M.R.P., and van Bergen, A.R.D. 1995. Screening of crude oils for asphalt precipitation: theory, practice, and the selection of inhibitors. SPE Prod. Facil. 10(1):55–61.

Centeno, G., Sánchez-Reyna, G., Ancheyta, J., Muñoz, J.A.D., and Cardona, N. 2011. Testing various mixing rules for calculation of viscosity of petroleum blends. Fuel 90:3561–3570.

Centeno, G., Sánchez-Reyna, G., and Ancheyta, J. 2015. Calculating the viscosity of crude oil blend by binary interaction parameters using literature data. Pet. Sci. Technol. 33(8):893–900

Hedrick, B.W., Seibert, K.D., and Crewe, C. 2006. A new approach to heavy oil and bitumen upgrading, UOP LLC, Meta Petroleum.

Heithaus, J.J. 1962. Measurement and significance of asphaltene peptization. J. Inst. Pet. 48:45–53.

Matsushita, K., Marafi, A., Hauser, A., and Stanislaus, A. 2004. Relation between relative solubility of asphaltenes in the product oil and coke deposition in residue hydroprocessing. Fuel 83(11–12):1669–1674.

Merdrignac, I., and Espinat, D. 2007. Physicochemical characterization of petroleum fractions: the state of the art. Oil Gas Sci. Technol. 62(1):7–32.

Murgich, J., Rodríguez, J., and Aray, Y. 1996. Molecular recognition and molecular mechanics of micelles of some model asphaltenes and resins. Energy Fuels 10(1):68–76.

Sánchez, F., Sánchez, G., Ancheyta, J., and Marroquín, G. 2014. Comparison of correlations based on API gravity for predicting viscosity of crude oils. Fuel 138:193–199.

Sepulveda, J.A., Bonilla, J.P., and Medina, Y. 2010. Stability prediction for asphaltenes using SARA analysis for pure petroleum. Rev. Ing. Region 7:103–110.

Sharma, B.K., Sharma, C.D., Bhagat, S.D., and Erhan, S.Z. 2007. Maltenes and asphaltenes of petroleum vacuum residues: physico-chemical characterization. Pet. Sci. Technol. 25(1–2):93–104.

Speight, J.G. 2001. Handbook of petroleum analysis, John Wiley & Sons, New York.

Stankiewicz, A.B., Flannery, M.D., Fuex, N.Q., Broze, G., Couch, J.L., Dubey, S.T., and Iyer, S.D., 2002. Prediction of asphaltene deposition risk in E&P operations, Proceedings of 3rd International Symposium on Mechanisms and Mitigation of Fouling in Petroleum and Natural Gas Production, AIChE 2002 Spring National Meeting, New Orleans, USA, March 10–14, paper 47C, pp. 410–416.

Zhao, S., Kotlyar, L.S., Woods, J.R., Sparks, B.D., Hardacre, K., and Chung, K.H. 2001. Molecular transformation of Athabasca bitumen end-cuts during coking and hydrocracking. Fuel 80(8):1155–1163.

2

PROPERTIES OF CATALYSTS FOR HEAVY OIL HYDROPROCESSING

2.1 INTRODUCTION

Crude oil produced by different petroleum fields differs importantly in properties such as viscosity, API gravity, impurities content, and yield of distillates. More viscous and high-sulfur-content crudes are called "heavier" or "sour," whereas low-sulfur-content crudes are called "sweet." Most of the conventional refineries are not capable of processing heavy or sour crudes due to their complex nature. On the other hand, stricter environment legislation on fuel quality has been implemented or is being planned throughout the world. These increasing stringent standards reduce the yield of gasoline and diesel per barrel of crude even if the quality of the crude inputs is not declining. Since the quality of petroleum feed, configuration, and complexity of refinery processes can vary substantially, each refinery has its own strategy to meet the new goals of fuel specification. Each country develops new product specifications, but the time of their introduction varies depending on the region and origin of fuel oils. Apart from several impurities, the pressure on crude oil and oil product prices is linked to a deficit of upstream (oil production) and downstream (oil refining) capacities to fulfill the demand over an extended period of time.

Hydroprocessing has been preferred by many refiners to be the technology of choice that can achieve the target by producing upgraded oil and high-quality fuels. This technological route has several options for its commercial application according to the type of reactor used.

Fixed-bed reactors are advantageous due to their simplicity and with the use of multiple beds of catalysts it is possible to obtain acceptable catalyst life. For high

Deactivation of Heavy Oil Hydroprocessing Catalysts: Fundamentals and Modeling,
First Edition. Jorge Ancheyta.

metal content and asphaltenes content feeds, in which deactivation of the catalyst is faster, other alternatives are moving-bed, ebullated-bed, or slurry-phase reactors, in the last two reactors the catalyst is maintained in a fluidized state by the upward liquid and gas stream. The use of fixed-bed hydroprocessing has been extended to cover feeds with higher content of impurities and lower API gravity by a combination of catalysts with improved stability working under moderate conversion regime to minimize sediment formation (Ancheyta et al., 2010).

In moving-bed reactors, the catalyst goes in downflow through the reactor by gravitational forces. The fresh catalyst enters at the top of the reactor and the deactivated catalyst leaves the reactor at the bottom, while the hydrocarbon goes either in counter- or cocurrent flow through the reactor. With this moving-bed system, the catalyst can be replaced either continuously or discontinuously.

In ebullated-bed type of reactor, the addition and extraction of catalyst can be performed in normal operation. Therefore, in this reactor, catalyst life is not an issue. However, because of the hydrodynamics inside the reactor, recycling of liquid is substantial and, therefore, the reactor operates close to a continuous stirred tank reactor, which presents a lower rate of reaction than a fixed-bed reactor. Another disadvantage of this type of reactor is that because of the backmixing of solids, fresh and used catalysts are taken out each time the catalyst is withdrawn from the reactor.

Slurry-phase reactor achieves a similar intimate contact of oil and catalyst and may operate with a lower degree of backmixing than ebullated-bed reactor. Small amount of finely divided powder is used (typically from 0.1–3.0 wt%), which can be an additive or a catalyst (or catalyst precursors). The catalyst is mixed with the feed (heavy oil), and both are fed upward with hydrogen through an empty reactor vessel. The slurry-phase reactor is free of internal equipment and operates in a three-phase mode. The solid additive particles are suspended in the primary liquid hydrocarbon phase through which the hydrogen and product gases flow rapidly in bubble form. Since the oil and catalyst flow cocurrently, the mixture approaches plug-flow behavior. The fresh catalyst is slurried with the heavy oil prior to entering the reactor, and when the reaction finishes, the spent catalyst leaves the reactor together with the heavy fraction and remains in the unconverted residue.

Usually, fixed-bed reactor is preferred when the catalyst stability is more than a year as in the case for middle distillate hydrotreating. For heavy oil processing, where the expected stability is less than a year, the use of moving-bed or ebullated-bed reactor technology is preferred. Nevertheless, the process selection depends on the feed composition and required conversion.

Ebullated-bed reactors are capable of converting most problematic feeds, such as atmospheric residue, vacuum residue, and heavy oils (all of which have high contents of asphaltenes, metals, sulfur, and sediments precursors) into lighter more valuable products. In addition, this type of reactors can perform both hydrotreating and hydrocracking functions; thus, they are referred as dual-purpose reactors. One of the problems for predicting the performance of ebullated-bed reactors is that in addition to data on feed composition and catalyst properties, knowledge of the hydrodynamic characteristics of the bed (incipient fluidization velocity, bed expansion, and bubble flow regime) at the severe operating temperature and pressure is necessary. Although these properties have been extensively studied at ambient conditions, little has been reported for systems at high temperature and pressure (Ruiz et al., 2004, 2005).

Ebullated-bed catalysts are made of pellets or grains that are less than 1 mm in size to facilitate suspension by the liquid phase in the reactor (Colyar and Wisdom, 1997; Courty et al., 1999). Moving- and ebullated-bed catalysts are chemically quite similar to those used in fixed-bed reactors, except that the mechanical strength, size, and shape of the catalyst particle must meet more demanding specifications.

Slurry-phase reactors require the separation of the catalyst from the products, which needs additional technology.

In general, the hydrotreating catalyst life in fixed-bed reactors is of several years if the feedstock is light oil (middle distillate). However, for heavy oil hydroprocessing, catalyst stability with time-on-stream (TOS) is low. The relationship between approximate deactivation timescale and type of process reactor is given in Table 2.1 (Moulijn et al., 2001).

The major concern for fixed-bed heavy oil hydroprocessing catalyst is its deactivation, which decreases catalytic activity with TOS, which means loss of money and time. The expected deactivation is caused by the metals (mainly Ni, V) and coke deposition on the surface of the catalyst (Absi-Halabi et al., 1991; Muegge and Massoth, 1991; Furimsky and Massoth, 1993; Kim and Massoth, 1993; Ancheyta and Rana, 2005), making it necessary to raise the severity of the reactor (i.e., increase of reactor temperature) to compensate for catalyst deactivation and maintain the conversion at the desired level, and ultimately terminate the operation and replace the catalyst. Thus, catalyst deactivation or catalytic activity loss over time is a continuing concern in heavy oil hydroprocessing. In general, deactivation is either by coke or by metal deposition over the active sites and/or at the pore mouth. It is likely that the deactivation by carbon deposition is due to the various types of carbon deposits on the catalyst surface, which accumulate and adhere to the catalyst surface during hydrocarbon processing. On the other hand, deactivation by metal sulfides is believed to proceed via blockage of surface active sites or pore plugging. Both coke and metal sulfides contribute to the loss of activity either burying the catalytic sites or causing diffusion limitations. Therefore, to determine the rate and mechanisms of deactivation, it is essential to characterize the spent catalysts for coke and metal deposition.

TABLE 2.1 Approximate Deactivation Timescale and Type of Reactor

Time of Deactivation	Type of Reactor	Type of Regeneration	Feedstock
Years	FBR	No regeneration	Middle distillate
Months	FBR	Off-line	Heavy oil
Weeks	FBR-SM, MBR	Alternative	Heavy oil
Minute-days	FDBR, SPR	Continuous	Heavy oil
Seconds	EFR (riser)	Continuous	Heavy oil

FBR, fixed-bed reactor; SM, swing mode; MBR, moving-bed reactor; FDBR, fluidized-bed reactor; SPR, slurry-phase reactor; EFR, entrained-flow reactor.
Source: Adapted from Moulijn et al. (2001).

The deactivation can also occur due to the competitive adsorption of organic species on the catalytic sites. It is also well known that spent catalysts contain appreciable amounts of nitrogen compounds, which appears to be strongly adsorbed (Zeuthen et al., 1991). Thus, catalyst deactivation occurs by three simultaneous mechanisms: coking, metal deposition, and substrate interactions. The specific rate of deactivation for each mechanism and each process depends, to a large extent, on the properties of the catalyst, hydrogen partial pressure, properties of the feedstock, and process conditions.

2.2 HYDROPROCESSING CATALYST

2.2.1 Catalyst Support

Due to high content of metals and asphaltenes, to upgrade heavy crude oil, catalysts need to have wide pores to allow for diffusion of such large molecules to the catalytic sites. However, increasing catalyst porosity leads to reduction in surface area and consequently decrease in specific activity. For heavy feeds, adequate porosity is essential since it is necessary to achieve a large metal-retaining capacity to extend the catalyst life, which decreases rapidly due to coking and metal deposition. Therefore, the textural properties of the catalyst could be of greater importance than surface area and chemical composition of the surface.

For hydrotreating and hydrocracking of heavy feeds, the catalyst support plays an important role. Because of this, there is a continuous effort to develop improved support formulations through suitable methods of preparation leading to adequate surface composition and textural properties that allow for appropriate diffusion of the feed molecules to take full advantage of the deposited active phase. For instance, Luck (1991) stated the inherent support properties that fulfill the various technical and economical criteria that the ideal hydrotreating catalyst must have as follows:

- Stabilization of group VI and VIII oxides in highly dispersed or microcrystalline phases without interactions forming inactive compounds
- Stabilization of the corresponding sulfides in highly dispersed phases
- High purity or at least absence of any negative interference of adventitious impurities with the active phase
- No parasitic reactions of the support with the feedstock to be treated
- Positive contribution of the active sites of the carrier to the catalyst performance
- Easily tailored pore structure and specific surface area (SSA)
- Thermal stability under reaction and regeneration conditions
- Easy to form into the desired shape (pellets, extrudates, etc.) with good mechanical strength
- Low cost
- No adverse effects from metal recovery from used catalysts.

For hydroprocessing catalysts, alumina is being the most commonly used support because it combines virtually all of the above-listed characteristics. A large variety of aluminas exist; γ-alumina, which is the most versatile and extensively used catalyst support, can be prepared by thermal decomposition of gibbsite, boehmite, or by precipitation of colloidal gels (Luck, 1991). Gamma alumina has high surface area, is stable at the temperatures used in hydroprocessing, easily formed in spheres or extrudates and inexpensive. Microporosity, surface area, pore distribution, and pore volume of alumina can be finely tuned, each aspect almost separately, through the physical parameters of the oxyhydroxide crystallites. The surface area–pore volume region that aluminas can cover is clearly shown in Figure 2.1. The acid–base character of alumina plays an important role in impregnation of the active phase precursor. Acidic properties of alumina are attributed to incompletely coordinated aluminum cations (Lewis acid centers).

The alumina support can be easily prepared in the laboratory or can be obtained in sufficiently large quantities from the commercial companies. Usually, the preparation of alumina in the laboratory is carried out with aluminum nitrate (Rana et al., 2004a). Aluminum sulfate has also been used for several studies but due to the presence of the sulfate ion high acidity is generated on the surface, leading to coke formation and poor catalyst stability. In general, the following steps are followed for preparation of alumina in the laboratory:

1. Precipitation
2. Hydrothermal transformation (aging)

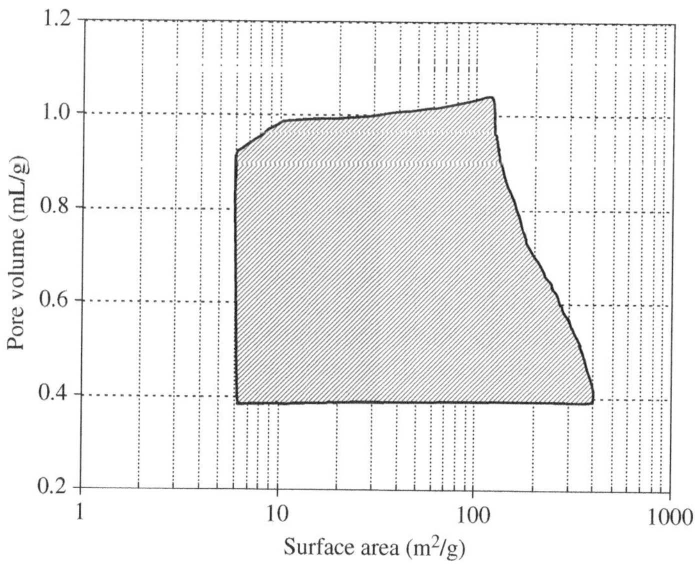

Figure 2.1 Pore volume versus surface area for commercial alumina supports (Adapted from Luck (1991)).

3. Centrifugation and filtration
4. Washing
5. Drying
6. Shaping of the support (crushing, grinding, extrudates, etc.)
7. Drying
8. Calcination.

Each of the above steps has a particular effect on the properties of the final support; for example, the textural properties are affected by variations in the pH of the precipitating medium (Perego and Villa, 1994), and the aging period and hydrothermal conditions also affect the properties of the final solid (Moulijn et al., 1993).

2.2.2 Chemical Composition

The chemical composition and physical properties of the catalyst are crucial when high removal of impurities is desired (Furimsky, 1998). For catalyst design, in hydroprocessing of heavy feeds, the amount of asphaltenes and metals is of high importance. Porosity determines the activity and life of the catalyst. Moreover, the catalyst needs to have enough acidity to perform hydrocracking but not so much as to produce excessive coking. Since heavy feeds vary considerably depending on their origin, there is not a common catalyst for hydroprocessing of such feeds. Chemical composition and porosity are more important in the case of heavy feeds. Hence, support acidity and porosity should be carefully designed to achieve optimum catalyst performance.

The support of hydrotreating catalyst interacts with the active component, leading to beneficial or detrimental interactions to catalytic activity, that is, the rate and selectivity of hydrotreating reactions over sulfide catalysts are affected by the support's nature (Scott et al., 1987). Such interactions between support and precursor active phases depend on several factors, such as the solution chemistry of precursor salts, method of deposition of the active phases, isoelectric point of the support, impregnation pH, and calcination temperature (Prins, 1992; Wachs et al., 1993; Kabe et al., 1999). For instance, Ramírez et al. (2004) demonstrated that increasing the amount of Ti in the Al_2O_3–TiO_2 support increases hydrodesulfurization (HDS) and hydrogenation rates of NiMo and NiW catalysts supported on Al_2O_3–TiO_2 mixed oxides due to an electronic promotion effect of the Ti present in the support over the Mo or W sulfided active phase. Thus, an electronic interaction between the support and the active component can alter the activity and selectivity of catalysts. The support can also influence the active component by favoring the exposure of some crystallographic planes in preference to others. Morphological changes of the active phase induced by the support may also cause a change in the active metal–support interaction.

Although a wide variety of supports have been proposed in the literature for hydrotreating catalysts, CoMo and NiMo catalysts supported on γ-Al_2O_3 continue to play, probably for economic reasons, the workhorse role ever since the hydrotreating processes entered into petroleum refineries. Different amorphous and crystalline materials have been proposed as supports for hydrotreating catalysts. Among them, the use of mixed oxides has been reported to offer an interesting range of textural properties and different interactions with the active phases.

Although most of the work on the support effect has been carried out using model molecules, few different support systems have been tested in hydroprocessing of crude oil and coal-derived liquids. Among them, the following have shown interesting features:

- TiO_2–Al_2O_3 (Nishijima et al., 1986; Rayo et al., 2004; Ramírez et al., 2005; Rana et al., 2005a; Maity et al., 2003a)
- SiO_2–Al_2O_3 (Maity et al., 2003b)
- MgO–Al_2O_3 (Caloch et al., 2004)
- ZrO_2–Al_2O_3 (Rana et al., 2005b).

The use of mixed-oxide-supported catalysts in hydrotreating is not new; initially, mixed oxides such as SiO_2–ZrO_2–TiO_2 (Hansford, 1964) and Al_2O_2–TiO_2 (Jaffe, 1968) were developed for hydrocracking applications due to their acidic nature. These supports were generally prepared by coprecipitation of appropriate Al and Ti salts followed by washing, drying, and calcination. Hydrocracking was later dominated by zeolitic supports due to the uncomplicated way to control or modify the acidity (number and strength of acid sites). FAU, ZSM-5, erionite, and mordenite have been used. Among them, Y-zeolite is the most commonly used one because it has the widest pores (van Veen, 2002) although ZSM-5, erionite, and mordenite have been used for shape-selective reactions controlled by pore geometry (Scherzer, 1990). The use of these microporous zeolites is however limited for heavy oil processing due to their small pore diameter. Nevertheless, incorporation of a small amount of zeolite into alumina can have important effect on the product selectivity and conversion.

Recently, silica-based mesoporous materials such as MCM-41 and SBA-15 have gained some attention in view of their interesting textural properties (surface area $\sim$1000 m^2/g and pore diameter of $\sim$3.0–10 nm); although due to its poor hydrothermal stability, they are not likely to have commercial impact as supports for heavy oil hydroprocessing catalysts.

2.2.3 Shape and Size

Shape and size of supports and catalysts are important parameters in the preparation of commercial hydroprocessing catalysts. To achieve good catalyst performance, it is essential to match the size and shape of the catalyst with the properties of the feed, process technology, and type of reactor. In the case of heavy feeds, special attention has to be paid to the shape and size of the catalyst in view of the diffusion problems encountered during hydroprocessing of large molecules contained in these feeds. It is well known that in diffusion-controlled reactions, the rate of diffusion is smaller than the rate of reaction. Therefore, a steep radial concentration profile of the reactants is established along the pellet radius, and the use of large catalyst particles can result in the central part of the catalyst particle not being used because the reactants are exhausted by the chemical transformation before they reach the center of the catalyst particle. This problem can be avoided in some cases by decreasing the diffusion path of the molecules, increasing in this way the effectiveness factor of the catalyst. In practice, this is achieved by decreasing the particle diameter. However, there is a limit to the decrease of particle size after which particles disintegrate. Apparently, this limit is reached when the diameter of the particles reaches 0.8 mm (1/32 in.). It must be considered nonetheless that such small catalyst particles will cause high pressure drops in fixed-bed reactors.

The problem of a long diffusional path of reactants from the external surface to the center of the catalyst particle can be partially circumvented by the use of different shapes of catalyst (grading).

The definition and evaluation of particle size is useful for comparing the relative activities of catalyst particles with different shapes. This is especially important when heavier feeds are being processed. In this case, more severe process conditions are necessary and many existing hydrotreaters are underdesigned with respect to catalyst inventory. It is then important to obtain more catalytic activity in the same reactor volume. The introduction of more active catalysts appears to be the solution. However, since hydroprocessing reactions in the case of heavy feeds are diffusion limited, improvement of the catalyst shape to obtain higher activity per volume of bed is necessary. For many refiners, the change from cylinder-shaped to multilobed catalysts has been the solution. Different techniques are used for shaping the catalyst into various forms (Stiles, 1983; Richardson, 1989; Le Page et al., 1987; Moulijn et al., 1993).Generally, multilobed catalyst shapes presenting big pores, small particle diameter, and large external area are preferred (Cooper et al., 1986; Bartholdy and Cooper, 1993).

Cooper et al. (1986) compared the relative HDS activities of various catalyst shapes in hydrotreating of a heavy Arabian VGO. They concluded that catalysts with the same volume-to-surface (V_p/S_p) ratio exhibit the same effectiveness factors and therefore the same activity per pound of catalyst, provided that the catalysts have the same physical and chemical properties. Concerning the relative volume activity (RVA), noncylindrical catalyst shapes exhibit higher void fraction and therefore lower bed volume activities than cylindrical catalysts with the same V_p/S_p ratio. However, pressure drop across the catalytic bed is also important for industrial applications. Since 3-lobe catalysts have a higher void fraction and a larger hydrodynamic particle size, they present a lower pressure drop per foot of catalyst bed. In some cases, it is relevant to compare the activities of particles giving the same pressure drop. By examining the plots of RVA versus V_p/S_p with those of pressure drop versus V_p/S_p, it is possible to compare different catalyst shapes on the basis of equal pressure drop.

The results in Table 2.2 show that when low diffusion restrictions exist, a decrease in effective particle size has little effect on increasing activity (Cooper et al., 1986). Thus, in this case, it is better to pack the reactor with as much catalyst as possible.

TABLE 2.2 Effect of Particle Size and Shape on Hydrodesulfurization Activity

Shape	Cylinder	Cylinder	Cylinder	Ring	Ellipse	Three-Lobe	Crushed
OD[a] (mm)	0.83	1.2	1.55	1.62	1.9	1.0	0.25–0.45
ID[b] (mm)	–	–	–	0.64	1.0		
Length (mm)	3.7	5.0	5.0	4.8	5.0	5.0	
V_p/S_p (mm)	0.189	0.268	0.345	0.233	0.262	0.295	~0.04
Activity[c]	9.7	7.9	5.7	8.7	8.4	8.2	14.0

[a]Outside diameter.
[b]Internal diameter.
[c]On catalyst weight basis (heavy gas oil, 622 K, 7 MPa).
Source: Adapted from Cooper et al. (1986).

This means that dense loading is preferable and that cylinder-shaped catalysts with lower bed porosity are preferred. In contrast, when severe diffusional limitations are present, small particles with a high external surface will give the best performance for a given pressure drop. In the other case, highest activities, significantly higher than those of cylinders, were obtained with sock-loaded-shaped catalysts. In general, shaped catalysts will give an improved performance with respect to cylinders when a large increase in effectiveness factor can be achieved with a relatively small decrease in particle size.

2.2.4 Pore Size Distribution

It is important to remember that under severe diffusion limitations, the catalyst performance can also be improved by increasing the catalyst pore size. For HDS reactions involving small molecules, the optimum average pore diameter (APD) is about 10 nm, while for asphaltenes hydrocracking and vanadium removal the optimum APD ranges between 15 and 20 nm. However, it must be considered that this will change with the type of feed. Figure 2.2 shows an example of the pore size distribution required for HDS and hydrodemetalization (HDM) catalysts. It appears that no easy solution is available since small catalyst particles decrease diffusion paths, but they cause high bed pressure drop. On the other hand, catalysts with wide pores are less susceptible to diffusion limitations, but they necessarily have lower surface area. Therefore, to determine what option one must use in a particular hydroprocessing operation, an integrated approach that takes into account size, shape, pore size, and catalyst loading must be considered (Cooper et al., 1986). It must be remembered that for hydroprocessing of heavy feeds, stability of the catalyst is as important as initial catalyst activity.

Increased pressure drop and fouling of the catalytic bed can be partially solved with effective bed grading and the use of guard beds on the top of the reactor to trap particulate matter from impure reactor feeds. In this way, it will be possible to minimize pressure drop during operation and make a more efficient use of downstream catalyst bed. Porous solids, such as activated bauxite or alumina, are used in the guard beds (Wolk and Rovesti, 1974; Wolk et al., 1975; Howell and Wilson, 1987;

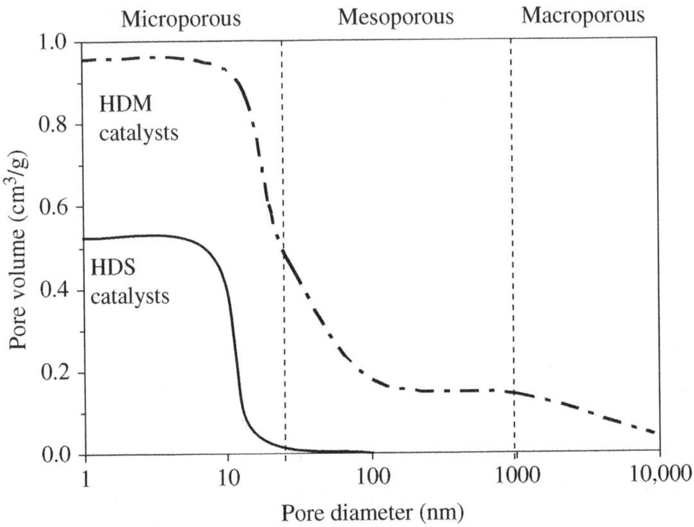

Figure 2.2 Typical pore size distribution for HDS and HDM catalysts (Adapted from Peries et al. (1986)).

Hilbert et al., 1996). Other low-cost solids such as magnesium silicate have been also proposed (Shirota et al., 1979).

2.2.5 Mechanical Properties

Mechanical properties of the catalyst are of industrial importance. Catalyst particles must have adequate mechanical strength to stand the weight of the catalyst bed itself to assure proper operation. In this respect, catalysts for hydrotreating of heavy feeds suffer more than those used for hydrotreating of light feeds since mechanical strength of a macroporous pellet is less than that of a microporous one. The breaking of the catalyst particles in a fixed-bed reactor can cause critical problems such as higher pressure drop, maldistribution of flow, and shutdown of the operation due to collapse of the catalyst bed. Unsatisfactory resistance to attrition can cause similar problems in fixed beds and great loss of catalyst in ebullated beds. Plugging of the filters is also another consequence of low catalyst resistance to attrition. For ebullated beds, spherical particles have improved attrition resistance and offer lower pressure drop.

Mechanical and textural properties of the catalyst, mostly provided by the support, can be controlled during the preparation stage. Factors that can affect the mechanical properties of the support and catalyst are (Ono et al., 1983; Snel, 1984a, 1984b, 1984c; Trimm and Stanislaus, 1986) as follows:

- Precipitate aging
- pH
- Number of pH swing cycles

- Type of binder
- Calcination temperature and time
- Method of incorporating the active components to the support.

Hydrotreating catalysts are usually prepared by filling the support pores with a basic or acid solution containing the active metals and additives salt precursors. However, it is possible that although the chosen pH could be adequate to induce the formation of the desired active metal species in the impregnating solution, partial dissolution of the support pore walls can take place reducing the mechanical strength of the final catalyst.

2.2.6 Active Metals

For heavy feeds containing high concentrations of heavy metals and asphaltenes, physical properties of the catalyst (size, shape, porosity, etc.) can be even more important than chemical composition since in this case the presence of meso- and macropores is critical to achieve full utilization of the catalyst. Nevertheless, some modification of the chemical composition may be needed to improve the HDM function of the catalyst.

The detrimental effect of the presence of large concentration of metals such as Ni and V in the feed is well known, and significant efforts have been made to develop catalysts that can stand high concentration of metals by providing high metal retention capacity without losing too much activity and life (Decroocq, 1997). To remove metals, the HDM catalyst must transform the resins and asphaltenes where most of the metals are found. For a microporous catalyst, the metal deposits will occur close to the external surface of the catalyst pellet, leading to pore mouth poisoning. On the other hand, catalysts with a wide range of porosity or multimodal catalysts with wide pores avoid diffusion and plugging problems and are better suited for HDM. Supports with "chestnut-burr"-like pores are reported to avoid the problem of having internal pore walls thus allowing a homogeneous distribution of the metals on the catalyst surface with a 100 wt% metal retention capacity referred to the weight of the fresh catalyst (Morel et al., 1997; Kressmann et al., 1998; Toulhoat et al., 1990). Catalysts with a large pore volume and appropriate pore size distribution allow diffusion of asphaltenes and resins into the pores and their adsorption on the catalyst sites. However, to obtain high metal capacity in HDM catalysts, it is necessary to tune the activity provided by the NiMo active phase. If the activity is too high, diffusion of resins and asphaltenes will be rate controlling and the metal deposit will be less homogeneous with more metal deposit close to the external surface of the pellet. Nevertheless, sufficient hydrogenation activity is necessary to reduce the carbon deposits on the catalyst surface and improve HDM activity. So that activity and porosity must be properly combined to achieve maximum utilization of the catalyst, acceptable activity and life.

As for the formulation of the active phase, it is found that in general CoMo catalysts are more effective for HDS, whereas NiMo formulations have good activity for hydrogenation (HYD) and hydrodenitrogenation (HDN).

Usually, hydroprocessing of heavy feeds requires the use of mixed or multiple beds of catalyst. The catalyst formulation will depend on its place in the catalytic bed system. Furimsky and Massoth (1999) reviewed the literature on the use of mixed and layer beds, multiple bed and multistage systems. Several works have been published on the use of a mixed bed consisting of two or more catalysts. A mixture of two catalyst particles in the same bed, one having high HDS and HDN and the second with high hydrocracking activity offered good selectivity to liquid products and low deactivation (Habib et al., 1995, 1997). Incorporation of a bed of alumina particles impregnated with magnesium salts on top of the catalyst layer was claimed to give better performance than the bed without the alumina layer (Gardner and Kukes, 1989).

For multiple bed systems, the first bed or catalyst layer is always designed to provide high HDM activity, the second less HDM but significant HDS, and the third is responsible for hydrocracking as well as for HDS and HDN (Furimsky, 1998). The first bed or layer contains a large pore catalyst, while the second and third beds require smaller pores and larger surface areas. Sufficient surface area must be supplied by mesopores greater than 120 Å to allow the heavy fractions to react while most of the surface, comprised by pores between 80 and 120 Å, will be highly accessible to the lighter fractions of the feedstock but much less for residual asphaltenes (Decroocq, 1997). Clearly, the composition of each bed must take into consideration the feed composition and the different levels of catalyst deactivation expected for each bed.

A number of works have been published on the use of graded catalyst systems consisting of multiple beds with different compositions. When the beds are in the same reactor, they operate at the same temperature and pressure (Furimsky and Massoth, 1999). For heavy feeds, sometimes three beds are needed (Hensley and Quick, 1984; Gardner et al., 1987).

The use of multiple beds located in different reactors is more versatile but also more costly. In this case, the operating conditions (pressure and temperature) of each bed can be varied to suit the feed that is being processed. One of the advantages of this scheme is that part of the effluent from one reactor can be removed before introducing the remaining feed to the next reactor. For example, with this scheme it would be possible to remove light fractions from the effluent of the first stage, reducing in this way the paraffinic character of the feed to the second reactor; this would reduce the possibility of precipitation of heavy components such as asphaltenes in the second reactor.

Sometimes, heavy feeds are so problematic that they cannot be processed in fixed-bed reactors. To solve this issue, several patents have been disclosed on the use of ebullated-bed reactors in series (Mounce, 1974), or combinations of ebullated-bed reactors with fixed-bed reactors (Hammer and Clem, 1975; Kunesh, 1982). The use of multiple reactor systems allows for improving hydroprocessing of the feed by setting a processing scheme that allows efficient contact of the feed with hydrogen and catalyst. For example, in the multibed reactor described by Harrison et al. (1994), the make-up hydrogen is fed to the second bed while the H_2-containing gas exiting from the second bed is supplied to the first bed. The gas coming from the first bed is purged and introduced to the third bed.

2.3 CHARACTERIZATION OF CATALYSTS

To understand the performance of a particular catalyst formulation, deep characterization of the fresh and used catalysts is necessary. The catalysts used for heavy oil hydroprocessing are Mo and W sulfides promoted by Co or Ni supported mostly on γ-alumina although some other supports such as zeolites, silica, and silica–alumina are also used. In some cases, modifiers such as Ti, P, B, or other metals are added to tune the surface properties.

2.3.1 Activity

The most important catalyst property is its ability to achieve the desired transformation of reactants at an acceptable industrial rate. However, it is also necessary that this transformation is carried out with the desired selectivity, maintaining stable operation for long time and with the possibility to regenerate the catalyst at the end of the run cycle. The following are important concepts that need to be clearly understood to evaluate the catalyst performance based on characterization techniques:

- *Activity*. It is characterized by the magnitude of the overall rate constant or by the product of the intrinsic rate constant times the adsorption coefficient. Since frequently the rate equation is not known, the activity can be evaluated as the variation in concentration, pressure, number of moles, or molecules per unit time at a given set of operating conditions for a specific type of reactor.
- *Selectivity*. For consecutive and parallel reactions, selectivity can be evaluated from the ratio of activities or can be expressed as the yield at fixed conversion of main reactant.
- *Stability*. The life of a catalyst depends on its stability and regenerability. Catalyst stability is normally expressed as the time between two regenerations for a given catalyst load and feed input.
- *Regenerability*. It is the number of regenerations the catalyst can undergo without losing unacceptable levels of activity, or before reaching too low operating times in the duration of the cycles between successive regenerations.

All the above-listed factors are evaluated in different types of reactors mainly batch, continuous stirred tanks, and continuous tubular reactors. The important aspects concerning the evaluation of activity, selectivity, mass transfer limitations, and so on have been reviewed by Wijngaarden et al. (1998).

For industrial applications, it is of great importance to determine the morphological characteristics of the pellet grain in order to estimate the catalytic bed dimensions. Among the morphological characteristics of the catalyst pellet, shape, grain size distribution, skeletal density, and bed density must be evaluated. In addition, it is also important to establish mechanical properties such as resistance to crushing, abrasion, and attrition. These properties are easy to measure using different ASTM methods in laboratory (Le Page et al., 1987).

2.3.2 Textural Properties

The study of the physicochemical properties of a catalyst requires knowledge about the texture of the support and catalyst (fresh and spent), composition of the chemical elements, nature and structure of the catalytic chemical species, and state and dispersion of the active phases. Table 2.3 lists some of the techniques frequently used for catalyst characterization (Montarnal and Le Page, 1967).

TABLE 2.3 Methods for Analyzing Physicochemical Properties of Catalysts

Properties	Methods
Composition of chemical elements	Standard chemical analysis
	X-ray fluorescence
	Emission spectrometry
	Atomic absorption
	Flame spectrometry
	Neutronic activation
	EDX
Nature and structure of the catalytic chemical species	X-ray diffraction
	Electron diffraction
	Nuclear magnetic resonance (NMR)
	Electron paramagnetic resonance (EPR)
	Infrared and Raman spectroscopy
	Visible and ultraviolet spectroscopy
	Magnetic methods
	Thermogravimetric analysis (TGA)
	Differential thermal analysis (DTA)
	Mössbauer spectroscopy
	X-ray photoelectron spectroscopy (XPS)
Textural properties	BET methods
	Porosimetry
State of the dispersed active agents	Chemisorption
	X-ray diffraction
	Electron microscopy
	Magnetic methods
	Chemical methods
	Electron microprobe analyzer
Quality of the active surface	Chemisorption kinetics
	Flash desorption
	Heats of adsorption
	Color doping
	EPR
	Infrared spectroscopy
Electronic properties	EPR
	Conductivity, semiconductivity
	Electron extraction work functions

Source: Adapted from Montarnal and Le Page (1967).

Heavy feedstocks contain several metals such as Ni, V, Fe, and so on, which are present as porphyrins or chelating compounds. During hydrotreating, these metals are transformed to sulfides (Ni_3S_2, V_3S_2, and V_3S_4) and deposited on the catalyst surface as crystallites of 2–30 nm in size. These deposited transition metal sulfides decreases the number of catalytic sites, hinder the transport of reacting molecules to the internal catalyst surface, and eventually cause the complete plugging of the catalyst pores. For these reasons, in the case of heavy oil hydroprocessing, characterization of the catalyst textural properties is of great importance given the large molecules to be processed. The textural properties of catalysts are evaluated by physical adsorption of inert gases, usually nitrogen at liquid nitrogen temperature. Total pore volume and pore size distribution are also calculated through the physical adsorption–desorption nitrogen isotherm. However, physical adsorption techniques are less likely to be used for the evaluation of textural properties of heavy oil catalysts because they contain a large fraction of macropores, which cannot be deduced effectively from physisorption experiments. For these large-pore-diameter catalysts, the pore size distribution must be estimated by mercury porosimetry (Gregg and Sing, 1982; Leofanti et al., 1994).

Mercury porosimetry is utilized to obtain the pore size distribution of pores in the range of $7.5–1.5 \times 10^4$ nm (2000 atm). When working with this technique, some breakage of the pore walls may occur in some materials such as high porous silica with a pore volume of greater than 3.0 cm^3/g (Ritter and Drake, 1945; Lowell, 1980).

The method of incipient wetness is simple and reliable to estimate the catalyst pore volume; the solid is impregnated with a volume of a nonsolvent liquid, usually water or hydrocarbons, enough to fill the pores. This method can be used for high porous silica for which other methods fail (McDaniel and Hottovy, 1980).

2.3.3 Surface Properties

Although chemical reaction occurs on the catalyst surface, not all the surface is active because only a fraction of the total surface area is occupied by the active phase. In fact, not even the complete surface of the active phases is active since the reaction occurs on specific centers of the active phase surface. In the case of acid catalysts, the acid centers not only occupy a small fraction of the surface, but they can also differ in strength and nature (Lewis or Brönsted).

To understand the way a catalyst works, it is necessary to know the number of active sites and their nature. In general, the methods that can render information on the characteristics of the catalyst surface at the atomic or molecular level can be divided into two groups:

1. Methods using probe molecules
2. Methods of direct study of the catalyst surface.

Volumetric, gravimetric, static and dynamic chemisorptions, adsorption calorimetry, spectroscopies of adsorbed molecules, and temperature programmed adsorption (TPA) and temperature programmed desorption (TPD) methods belong to the first

TABLE 2.4 Main Techniques for Surface Characterization Using Probe Molecules

Technique	Information	Quantitative Analysis
Volumetric adsorption	Amount of adsorbate as a function of pressure	Yes (high precision)
Gravimetric adsorption	Amount of adsorbate as a function of weight	Yes (high precision)
Dynamic adsorption	Amount of irreversible adsorbate	Yes
Calorimetry	Heat of adsorption as a function of coverage	Yes (high precision)
FTIR spectroscopy	Surface and adsorbate functional groups	Possible
Raman spectroscopy	Surface and adsorbate functional groups	Possible
UV–vis spectroscopy	Local environment of surface groups perturbed by adsorption	Possible
TPA, TPD	Amount of adsorbed (TPA) or desorbed (TPD) species as a function of temperature	Yes
TPD-MS	Amount and composition of desorbed species as a function of temperature	Yes
TPD-IR	Composition of surface species as a function of temperature	No

Source: Adapted from Leofanti et al. (1994).

group. Table 2.4 gives a remainder of the information obtained with each technique and the possibility of quantitative analysis.

Part of the above techniques such as FTIR (Fourier transform infrared spectroscopy) can be used for *in situ* studies for direct observation of the surface under reaction conditions (Topsøe, 2006).

In general, the techniques of the second group are more expensive and specialized. These techniques consist of measuring the response of the solid when it is exposed to a radiation. The response may be the scattering or absorption of radiation or absorption followed by emission of another type of radiation. Table 2.5 presents the characteristics of some of these techniques.

Other techniques such as low-energy electron diffraction (LEED), high-resolution electron energy loss spectroscopy (HREELS), and surface extended X-ray absorption fine structure (SEXAFS) are also used. A classification and description of these techniques according to the incident radiation and the response has been given by Delannay (1984).

Among the surface spectroscopic techniques, X-ray photoelectron spectroscopy (XPS) is widely used because it can give valuable information on the number of the different atoms on the surface and their oxidation state.

The development of electron microscopy in the past years has enabled catalytic and other researchers to observe and analyze particle shapes, size distributions, and structural features and chemical information of the support and supported active phase species. The interaction of a high-energy electron beam with a solid surface gives rise to a variety of signals that can be used to study the nature of the solid

TABLE 2.5 Main Spectroscopies and Related Techniques for Surface Characterization

Technique	XPS	AES	SIMS	ISS
Excitation	Photons	Electrons	Ions	Ions
Response	Electrons	Electrons	Ions	Ions
Information	Surface atomic composition Nature and binding of the atoms	Surface atomic composition	Surface atomic composition Short distance order	Surface atomic composition
Monolayers sampled	2–20	2–20	1–3	1
Quantitative analysis	Yes	Possible	Possible	No
Gaseous atmosphere	Difficult	Difficult	No	No

XPS, X-ray photoelectron spectroscopy; AES, Auger electron spectroscopy; SIMS, secondary ions mass spectrometry; ISS, ion scattering spectroscopy.
Source: Adapted from Leofanti et al. (1994).

specimen. Scanning electron microscopy (SEM), used in the analysis of bulk solids, uses an electron beam of low energy (typically 20–30 keV) and is used to obtain a high-magnification imaging of the sample. In transmission electron microscopy (TEM), a thin solid (<200 nm) is irradiated with an electron beam of sufficiently high energy to propagate through the specimen. Diffracted electrons are used to obtain a diffraction pattern of the specimen and thus obtaining information on the structure of the material. Transmitted electrons are used to form images, which are analyzed to obtain information about the atomic structure of the solid and the defects present in the solid sample. Additional information on these and other surface techniques used for the analysis of sulfided and other types of catalysts can be obtained from reference textbooks (Delannay, 1984; Wachs, 1992; Topsøe et al., 1996).

Many of the these techniques have been used for the characterization of sulfided hydroprocessing catalysts; structural characterization of sulfided catalysts has been obtained using TEM (Delannay, 1985; Hensen et al., 2001; Eijsbouts et al., 2005), XPS (Prins et al., 1989), EXAFS (Boudart et al., 1983), STEM (Lauritsen et al., 2004; Carlsson et al., 2004).

The dispersion and number of coordinatively unsaturated sites (CUSs) in sulfided catalysts have been evaluated using temperature programmed reduction of sulfides (TPR-Ss) and adsorption of probe molecules such as oxygen, NO, and CO (Massoth and Muralidhar, 1982; Portela et al., 1995; Travert et al., 2001; Hédoire et al., 2003; Ramírez and Gutiérrez-Alejandre, 1998; Oyama et al., 2004; Choi et al., 2004). A correlation between the number of CUS associated to molybdenum or the promoters with HDS catalytic activity has been obtained in some cases. The chemical and/or spectroscopic characterization of fresh heavy oil hydroprocessing catalysts is quite similar to other hydrotreating catalysts (Topsøe et al., 1996).

Characterization of heavy oil hydrotreating spent catalysts has been performed to gain insights into the deposition of coke and metal sulfides on the catalyst surface.

Deposits of metal sulfides have been confirmed by X-ray diffraction (XRD) and TPR for spent catalysts used in Maya crude hydrotreating (Rana et al., 2004a, 2005c). Deposited metal sulfides have been characterized by Smith and Wei (1991a) using TEM, EDX (energy dispersive X-ray), STEM, and XPS for the HDM of model molecules on a deactivated catalyst, and by Takeuchi et al. (1985) by means of XRD, ESR (electron spin resonance), and SEM, on catalysts used in HDM of heavy oil, where the V_2S_3 phase was observed in acicular or rod-shaped crystallites about 100 nm in length. Toulhoat et al. (1987) reported that nickel was always associated with vanadium [$Ni(V_3S_4)$] and its crystallite grew perpendicular to the support surface. Apart from these metals, Ca, Mg, and Fe sulfides have also been reported to deposit on the exterior surface of the catalyst (Gosselink, 1998) and may play an important role in the plugging of catalyst pores.

The activity of spent catalysts has been characterized by using model molecule reactions over spent hydroprocessing catalysts in order to elucidate the role of deposited metals on the performance of active sites (Mitchell and Valero, 1982; Morales and Galiasso, 1982; Weitkamp et al., 1984; Ware and Wei, 1985; Mitchell et al., 1985; Dejonghe et al., 1990). Dejonghe et al. (1990) confirmed that part of deposited vanadium is able to hydrogenate toluene by using vanadium deposited on pure alumina; XPS was used to estimate the dispersion and chemical states of the deposited elements. Ledoux et al. (1987) reported HDS and HDN activity over a vanadium-deposited NiMo commercial catalyst. In another work, it was confirmed by the ESR and NMR (nuclear magnetic resonance) that a small amount of vanadium and nickel porphyrins have a drastic effect on the thiophene HDS activity of a conventional NiMo catalyst (Ledoux and Hantzer, 1990).

The characterization of a series of commercial alumina-supported nickel–molybdenum and cobalt–molybdenum hydrotreating catalysts exposed to different degrees of reaction length and severities using a variety of naphtha and gas oil feeds was reported by Weissman and Edwards (1996). Fresh and used catalysts were characterized in their HDS and HDN activities using gas oil as feedstock, chemical analysis of deposits, porosimetry, and high-resolution transmission electron microscopy (HRTEM). In addition, the deposits on the catalyst were characterized by ^{13}C NMR. It was found that deactivation occurred primarily by carbon deposits on the support. Changes in the structure of the MoS_2 phase as observed by TEM did not appear to contribute to activity changes. Two types of carbon were found depending on the feed used; using gas oil as feedstock led to deposits of a compact carbon, resulting in significant loss of HDN activities but only a slight loss in HDS activities. Processing with light feed (naphtha) led to the deposition of less dense carbon, which reduced nitrogen and sulfur removal proportional to the carbon loading. Some other recent studies have been conducted on the performance and characterization of catalysts for hydroprocessing of heavy oil and other real feeds (Galarraga and Ramírez de Agudelo, 1992; Caloch et al., 2004; Maity et al., 2004; Rana et al., 2004a, 2005b, c, d; Ramírez et al., 2005).

2.4 GENERAL ASPECTS FOR DEVELOPING CATALYSTS FOR HYDROPROCESSING OF HEAVY CRUDE

The composition of catalyst is defined based on the properties of heavy oil feedstock, for example, low metal loading and large pore diameter. However, the selection of the support nature (acidic or basic) is based on the selectivity and stability of a catalyst with TOS. In addition to the fundamental catalyst properties, external morphology (shape and size) and mechanical properties are essential. Apart from these properties, the technical and economic aspects behind the industrial scale are of great importance. Thus, the development of industrial catalysts is an evolutionary process that requires significant pieces of information, which cannot be obtained just with few formulations prepared in the laboratory. There are many works dealing with fundamental research regularly published in different journals, but most of them do not explain commercial applicability of these catalysts, that is, most likely because either these catalysts are not at all applicable to the industrial process or the cost of the catalysts is too high. Therefore, the preparation of industrial catalysts remains as an art. A large amount of basic information of a catalyst formulation is hidden in support preparation, active metal impregnation, and its drying and calcination. Most of the catalyst properties are not independent; thus, if one of them is varied with a view to improvement, other properties may also be modified, which do not necessarily have the same effect. In hydroprocessing, the stability of a catalyst is less important if the feedstock is in the middle distillates range, but when using heavy crude oil, the catalyst stability becomes crucial due to deactivation by carbon and metal deposition during the operation. Therefore, the support plays an important role since it provides the textural properties of catalyst and surface for dispersion of active metals.

2.4.1 Preparation of Supports

The catalytic performance of a catalyst depends on its different components such as support and active metal. Usually, the behavior of the active phase is influenced by the type of support material; therefore, it becomes necessary to select an optimum support material according to the criteria of interaction with active phases, acid–base properties, textural properties, stability, extrusion properties, and cost at commercial scale. These properties become critical when the catalyst is being used for heavy oil hydroprocessing. Therefore, taking all properties into consideration, the most used support for hydroprocessing of heavy crude has been alumina. Alumina combined with small amounts (5–10 wt%) of other oxides such as TiO_2, ZrO_2, SiO_2, and MgO has also been used as catalyst support. The major component of these supports is γ-alumina due to its outstanding mechanical properties, nature of interaction with active phases (Mo, W, Co, Ni), and cost. In contrast to middle distillate or model molecules hydrotreating catalysts, heavy oil hydroprocessing catalytic activity depends not only on the active phases but also on the textural properties of the solid.

Catalyst particle size, pore diameter, and pore size distribution are the main issues in the development of a suitable catalyst for hydroprocessing of heavy crudes. These characteristics of a catalyst are endowed by support materials. The desired textural properties, pore volume, and surface area of a catalyst can be controlled during the preparation of the alumina support during the mixing of aluminum source (commonly aluminum nitrate) and precipitating agents. The control of textural properties is also possible by using additives, aging, washing, drying, and calcination. The effect of seeding, pH swing, and different precipitating agents are also reported for such a purpose. In general, alumina supports are precipitated at around $pH \approx 8$ to obtain pseudo-boehmite with moderate pore volume and unimodal pore diameter, while precipitation at $pH > 10$ forms barite, having bimodal pore size distribution. The bimodal type of support can also be prepared by using additives or combustible fibers during the extrusion of boehmite, while the effect of steaming agents on the texture of the alumina extrudate has been studied (Absi-Halabi et al., 1993; Walendziewski and Trawczynski, 1993, 1994). These supports preparation methods were observed at laboratory scale, whereas very few have been reported at commercial scale using an integrated mini pilot plant (Kaloidas et al., 2000). The impurities trapped in the precursor may influence the support pore structure as well as chemical properties such as isoelectric point (IEP), which are responsible for active metal dispersion and so the catalytic activity. The source of the alumina precipitating agent also requires attention such as traces of Na and SO_4 ions must be removed during washing.

Bimodal alumina is prepared by variation of pH from acid to basic or vis-à-vis. In this method, aluminum sulfate and sodium aluminum oxide salts are used to prepare acid and basic solutions, respectively. A basic salt solution is added to acid salt solution to change the pH of the mixture from the acid to the basic side and then the mixture is aged several minutes. In this way, the pH of the mixture is changed several times and finally the precipitate is aged in basic media. The common pore size distribution of the calcined support obtained by this method is shown in Figure 2.3. However, when urea is added into this mixture at the beginning of the precipitation, a wide pore range of alumina is observed instead of bimodal alumina (Maity et al., 2005).

The effect of hydrolyzing agents such as ammonia, ammonium carbonate, and urea on alumina preparation has been studied (Rana et al., 2004a). Changing the hydrolyzing agent in the alumina preparation caused changes in the pore structure of alumina-supported CoMo catalysts (Figure 2.4), indicating that alumina prepared by ammonium carbonate has ample pore structure compared with other hydrolyzing agents. The resulting wide range of pore diameters was explained by the presence of CO_3^{2-} ions, which are coming from ammonium carbonate or urea, trapped in between aluminum oxyhydroxide (Rana et al., 2004a). These trapped carbonate ions lead to wide pore alumina during calcination.

Among the mixed oxides, alumina–titania support is widely used for the preparation of hydrotreating catalysts due to its favorable acid–base properties. The different procedures employed to prepare this binary oxide have been discussed in the literature (Rana et al., 2005b; Maity et al., 2006). In these methods, Al_2O_3–TiO_2 mixed oxide supports are prepared by coprecipitation method using different precipitating agents, urea, and ammonia. It is reported that Al_2O_3–TiO_2 supports prepared by sodium aluminate and titanium chloride salts have higher pore volume and larger pore diameter

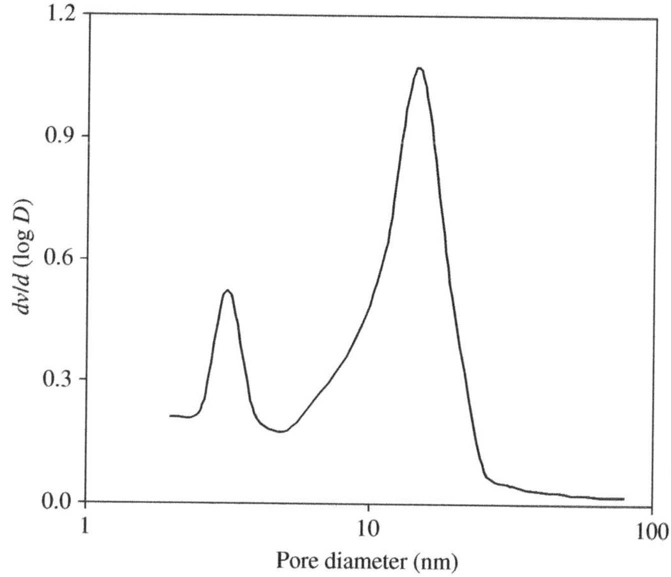

Figure 2.3 Bimodal pore size distribution of CoMo/Al_2O_3–SiO_2 using pH swing method for support preparation.

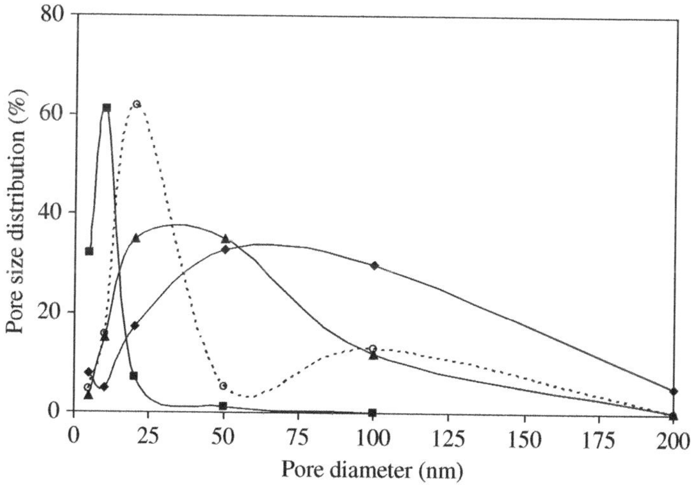

Figure 2.4 Effect of different hydrolyzing agents on pore size distribution of CoMo supported catalysts. (♦) Al_2O_3-u, (▲) Al_2O_3-acs, (■) Al_2O_3-am, (O) reference catalyst.

solids compared with the same support prepared by other methods (Maity et al., 2006). A simplified flow diagram of different supports and their preparation methods is presented in Figure 2.5.

The second step of support preparation is shaping and sizing, which also affect the mechanical and pore texture of the carrier. Thus, at the kneading stage, a balanced use

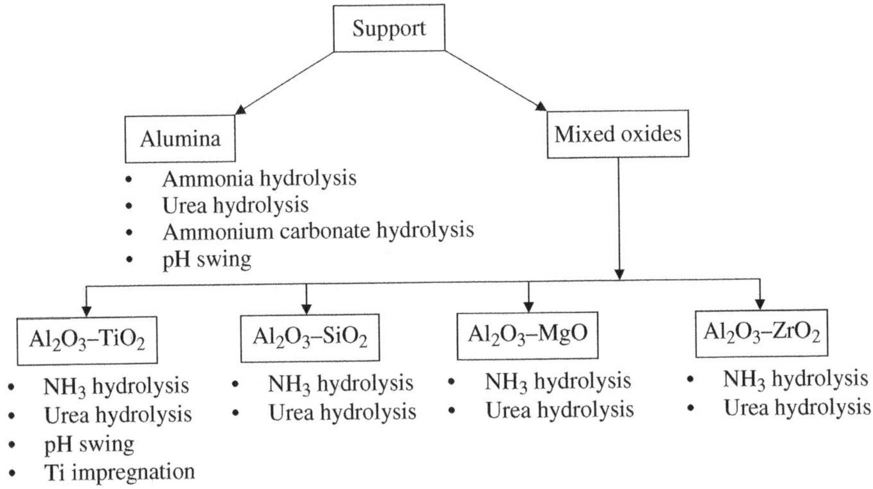

Figure 2.5 Methods used for preparation of mixed oxide supports.

of precursor, additive, peptizing agent (HNO_3), and its concentration is required to produce a homogeneous paste to form suitable extrudates. Moreover, extrudates drying and calcination play an important role to generate suitable textural and mechanical properties (e.g., crushing strength and attrition), which can be affected during the removal of water, additive, and peptizing agent.

2.4.2 Preparation of Catalysts

A hydroprocessing catalyst is composed by support and active metals such as cobalt (nickel) and molybdenum (tungsten) in sulfided state. Various parameters such as impregnation procedure, sequence of impregnation, and presence of additives during the preparation strongly affect the morphology or surface structure of the catalyst. Thus, the catalytic performance of the working sulfided catalyst (CoMoS or NiMoS) is strongly dependent on the catalyst preparation and its activation conditions. Molybdenum is introduced as ammonium heptamolybdate, while cobalt and nickel nitrates are the usual precursor salts for the promoters. Usually, CoMo catalyst is more active for HDS while NiMo is for HDN. The more expensive NiW catalyst is used where high saturation and moderate cracking of low sulfur feedstock are desired. A number of other transition and noble metals have been tested in laboratory scale for hydrotreating (Ru, V, Fe, Mn, Cr, Pt, Rh, and Pd) but do not appear to be used commercially. The effects of the amounts of these or other active precursors with their optimum concentration (monolayer formation) over different supports are given elsewhere (Pecoraro and Chianelli, 1981; Topsøe et al., 1996). The choice of active metal (CoMo or NiMo) and support selection may depend on the feed composition. The method of metal loading is one of the most important parameters with respect to the metal dispersion to obtain high catalytic activities. Usually, impregnation of the active metal is carried out by wetting the support either by an excess or incipient wetness method. Typically, catalyst is prepared by introducing Mo(W) first and then Co or Ni promoter atoms. However, coimpregnation method can be applied when an additive, such as P, urea, chelating agents, and so on, is used. A small quantity of additives frequently used in hydroprocessing catalyst to stabilize the alumina support and avoid nickel and cobalt diffusion into the support. To prepare hydroprocessing catalysts for heavy crude oil, both subsequent and coimpregnation methods are used.

2.4.3 Characterization of Catalysts

Hydrotreating catalysts are characterized using different techniques; most of the characterization is physicochemical as well as surface characterization of active sites. Apart from this, heavy oil hydroprocessing catalysts must have superior mechanical as well as morphological properties due to the complex nature of the feedstock or pressure drop along the reactor. Abrasion and attrition problems are complicated and can be eliminated by using the required characteristics of catalysts, which are part of the know-how of commercial catalyst formulation. Usually, these properties are related, for example, the larger the pore diameter, the greater the attrition loss and low crushing strength.

Different γ-Al_2O_3 support preparation methods have been employed to vary the porosity and pore size distribution, which play an important role in stability as well as in the metal retention capacity of the catalyst. The French Institute of Petroleum has reported a microcrystalline form "chestnut-burr"-like pores and nonacidic material. This catalyst has a large pore volume and an appropriate pore size distribution that allow the diffusion of resins and asphaltenes into the catalyst pellet and their adsorption on the active sites. It is reported that this catalyst has better metal retention capacity and asphaltenes elimination than traditional HDS catalysts. The catalyst has high dispersion of metal over acicular alumina platelets, large pore volume, and an appropriate pore size distribution, which prevents plugging of the pore network and a high metal retention up to 100% in relation to its weight as fresh catalyst (Kressmann et al., 1998).

Characterization of catalysts before and after reaction provides important information about the role of textural properties in metal deposition. Moreover, the HDM catalyst behavior can be observed in two ways, which depends on the textural properties of the catalyst:

- Microporous and low-range mesoporous catalysts decrease activity rapidly due to pore mouth blockage and diffusion limitations of complex molecules into the pores.
- The catalysts containing macropores in which deposited metal sulfides (V and Ni) behave as catalytic sites showed comparative more stability with TOS.

Apart from the textural properties, the nature and quantity of the supported species, their structure and size, the chemical state of the elements, their dispersion on the support, and the interaction between the molybdenum entities with the support or promoter can be characterized by using several spectroscopic techniques.

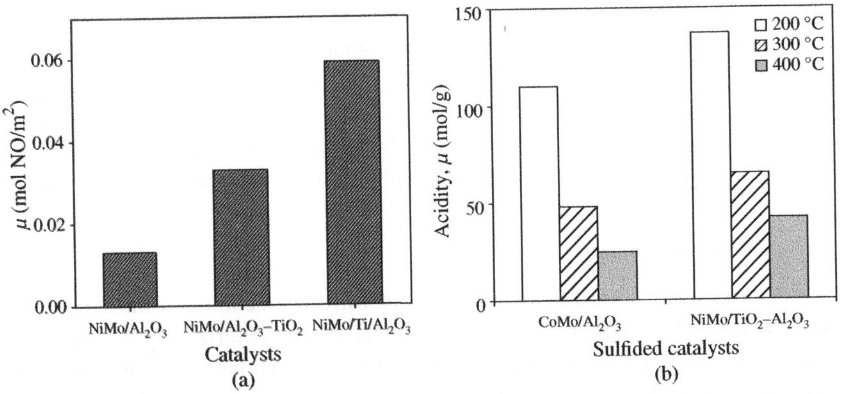

Figure 2.6 (a) NO adsorption over various NiMo-supported catalysts and (b) Lewis acidity of sulfided catalysts with different support composition (Adapted from Rana et al. (2005a) and Ramírez et al. (2005)).

The use of mixed-oxide-supported catalyst in heavy oil hydroprocessing modifies the metal support interaction and facilitates sulfidation of the active phases and consequently the number of catalytic sites (i.e., CUSs or anion vacancies), which are proportional to the increased catalytic activity. To analyze the state of dispersion of the sulfided phase in different catalysts used for hydroprocessing of heavy crude oil, NO adsorption is on the sulfided samples. It is well known that NO titrates sulfur vacancy sites or coordinated unsaturated sites, which are related to the hydrotreating activity. Figure 2.6 shows the results of NO adsorption over various NiMo-supported catalysts normalized per square meter of catalyst surface (Ramírez et al., 2005). Not surprisingly, this trend is the same as the one observed for catalytic activity. It appears that incorporation of Ti into the catalyst support increases the number of sulfur vacancies associated with the active sites. This can also be an indication of increased dispersion of the sulfided phase. It was reported that NO chemisorption correlates better with catalytic activity than O_2 adsorption (Agudo et al., 1984; Zmierczak et al., 1982; Redey et al., 1988) because NO chemisorption is more selective than O_2. Moreover, the mixed-oxide-supported catalysts have higher acidity than alumina. In Figure 2.6, the differences between two supported catalysts (Al_2O_3 and Al_2O_3–TiO_2) are relatively small, but they appear to be sufficient to explain the behavior of catalysts. Nevertheless, the sulfided catalysts showed a negligible amount of Brönsted acidity at 1541 cm^{-1} after 200 °C pyridine desorption, while an adequate number of Lewis acid sites remains up to 400 °C. The IR bands at 1598 and 1445 cm^{-1} are specific to pyridine interaction with the Lewis acid sites, and quantification of these bands is also shown in Figure 2.6 (Rana et al., 2005a).

2.5 CATALYST FOR MAYA CRUDE OIL HYDROPROCESSING

This section exemplifies the different steps for the development of a catalyst for hydroprocessing of heavy oil, particularly for Maya crude oil. First of all, details of feed properties, catalyst loading to the reactor and pretreatment, that is, activation, are given. Then, the characterization techniques are provided, followed by a discussion on the active sites and catalytic activity required for hydroprocessing of heavy petroleum. Next, results of experiments carried out with different feeds, for example, Maya crude oil diluted with naphtha and with diesel, as well as pure Maya crude, are discussed.

2.5.1 Composition of Maya Crude Oil

The composition of crude oil classically depends on the origin, for example, Boscan (10.1°API), Kern River (13.6°API), and Maya (21°API). These crudes are paraffinic, naphthenic, or aliphatic in nature. Boscan is aliphatic and Kern River is naphthenic, whereas Maya is intermediate. Contents of metals and heteroatoms of Maya crude, which are the principal concern when upgrading, are reported in Table 2.6 together with properties of its AR and VR.

The foremost problem of upgrading of crude oil is asphaltene due to its complex nature. In general, asphaltene molecules are difficult to process because they are very large, polar, noncrystalline, and have multiple stacked structures, and also contain high amount of heteroatoms, which can easily deactivate the hydroprocessing catalysts during the operation. Furthermore, the chemistry of asphaltene is not completely understood, and its molecules are precursors to the most problematic organic deposits and sediment formation. The organic deposits or solid precipitation occurs due to asphaltene flocculation and formation of random aggregates (Mansoori,

TABLE 2.6 Physical Properties of Maya Crude Oil and Its Residua

Properties	Maya Crude	Maya AR, °F. 651+	Maya VR, °F. 1,000+
API gravity	21.31	7.14	1.43
Density 20/4 °C	0.9232	1.0177	1.0615
Sulfur (wt%)	3.52	4.60	5.7
Nitrogen (ppm)	3200	5086	6591
Nickel (ppm)	49.5	79.3	125.2
Vanadium (ppm)	273.0	407.7	649.4
n-C_7 insolubles (wt%)	12.7	17.74	26.3
Ramscarbon (wt%)	10.87	17.66	26.21
Distillation (°C)	TBP	ASTM D-1160	
IBP/5 vol%	19	340	
10 vol%	131	410	
20 vol%	201	454	
30 vol%	273	497	
40 vol%	352		
50 vol%	430		
60 vol%	509		
70 vol%	586		

1997), which is likely to form steric colloids in the presence of resins and paraffins (Branco et al., 2001). These colloids and floc aggregates rarely can convert into some other products at hydrotreating conditions. Moreover, these aggregates deposit over the catalytic sites as well as near the pore mouths, leading to diffusion limitations. Asphaltene molecules carry a core of stacked flat sheets of condensed aromatic rings linked at their edges by aliphatic and/or naphthenic chains. The condensed aromatic rings exist in the form of a nonhomogeneous flat sheet. Due to the biogenic origin of petroleum, most of the metals are associated with metal–porphyrin structure of chlorophyll type and the stability of these complexes vary in this order V > Ni > Mg (Falk, 1964).

During hydroprocessing, asphaltene molecules are allowed to diffuse into the catalyst pore and their metal atoms (V and Ni) are removed from the ring structures. The asphaltene molecules may be partially hydrogenated or cracked over the catalyst surface depending on the reaction conditions and produce relatively small-sized molecules, called maltenes, but it is also likely that these molecules may thermally decompose and polycondense to coke precursors. These lead to instability and finally to dry sludge, which is the dead point for any kind of catalyst stability. The structure of asphaltenes has been the subject of several investigations and is now believed to consist of polycyclic aromatic clusters, substituted with varying alkyl side chains. The composition becomes more complex in the bottom-of-barrel such as atmospheric and vacuum residues that contain large percentages of heteroatoms (O, S, N) and organometallic constituents (Ni, V, Fe). The residue oils generally have a relatively lower percentage of paraffin and higher percentage of metals, sulfur, nitrogen, and asphaltene than virgin Maya crude oil have.

2.5.2 Catalyst Loading and Pretreatment

Typically, hydroprocessing catalysts, for example, CoMo and NiMo, are loaded to the reactor in oxide state, which are sulfided *in situ* or *ex situ* before the activity test. In the experiments conducted here, oxide catalyst is loaded with an equal volume of diluent (SiC), dried at 120 °C, and atmospheric pressure and then allowed for soaking with straight-run gas oil. The sulfiding agent, a mixture of dimethyldisulfide (DMDS) with gas oil having 2.7 wt% total sulfur, is introduced after soaking. The increase in temperature during sulfidation was 1.5 °C/min, the first sulfiding step was performed at 260 °C for 3 h and subsequently at 320 °C for 5 h. Both low- and high-temperature sulfidations are performed at 28 kg/cm^2 pressure. A detailed catalyst activation procedure for testing catalyst activity during Maya heavy oil hydroprocessing is reported elsewhere (Marroquín et al., 2004). Usually, the better or higher the degree of sulfidation corresponds to higher catalytic activity. After sulfidation, the flow was switched to the corresponding feedstock (Table 2.7) and the operating conditions were adjusted as shown in Table 2.8.

2.5.3 Feedstocks and Characterization Techniques

Different catalyst formulations were tested to find out optimum catalyst by using diluted as well as pure Maya crude feeds, whose properties are shown in Table 2.7.

TABLE 2.7 Properties of the Feed for Different Experimental Setup

Properties	Bench-Scale Reactor Feed		Microflow Reactor Feed		
	Maya	HDT Maya[a]	HDM[b]	HDS[c]	HDS[d]
Density 20/4 °C	0.925	0.877	0.88	0.865	0.8401
API gravity	21.31	31.14	29.29	32.10	
Pour point (°C)	−30		−15	−24	
Ramscarbon (wt%)	10.87	8.0	5.45	5.54	
Viscosity (g/(cm s))					
at 50 °C			3.08	2.63	
at 100 °C			9.45	8.29	
Elemental analysis (wt%)					
C	86.9	85.5	84.2	83.2	
H	5.3	7.2	8.8	9.5	
N	0.32	0.1852	0.184	0.118	0.0873
S	3.52	1.217	2.21	0.648	0.437
Metals (wppm)					
Ni	49.5	36.76	26.21	18.9	14.93
V	273.0	107.98	124.78	81.66	61.59
C_7 insolubles (wt%)	12.7	6.87	8.43	–	4.35

[a]Partially hydrotreated Maya crude.
[b]HDM feed (Maya+diesel, 50/50, wt/wt).
[c]HDS feed (HDT Maya+diesel, 50/50, wt/wt).
[d]HDS feed (HDT Maya+naphtha, 50/50, wt/wt).

TABLE 2.8 Reaction Conditions for Fixed-Bed Integral Reactors

Conditions	Microflow Reactor	Bench-Scale Reactor
Temperature (°C)	380	400
Pressure (MPa)	5.4	7.0
Hydrogen flow (L/h)	4.6	90
Flow of Maya crude (mL/h)	10	100
LHSV (h^{-1})	1.0	1.0
Hydrogen/oil ratio (m^3/m^3)	356	891.0
Mode of operation	Upflow	Downflow
Time-on-stream (h)	120	200
Catalyst volume, mL (g)	10 (≈8.5)	100 (≈85.0)
Catalyst shape	Cylindrical extrudate	Cylindrical extrudate
Catalyst size (in.)	1/16	1/16
Feed composition tested		
Pure Maya crude	–	✓
HDT Maya	–	✓
HDM feed	✓	–
HDS feed	✓	–

The dilution of Maya heavy oil was used to minimize the experimental problems occurring in microreactor processing due to high viscosity of Maya crude and difficulty for handling.

The analyses of feed and products were carried out by different ASTM techniques: metals were analyzed by flame atomic absorption spectrometry (ASTM D 5863-00a), total sulfur content was determined with an HORIBA model SLFA-2100/2800 equipment using energy dispersive XRF (X-ray fluorescence) X-ray beam, total nitrogen content was measured by oxidative combustion and chemiluminescence (ASTM D 4629-02), while asphaltene was defined as the insoluble fraction in *n*-heptane.

2.5.4 Active Sites and Catalytic Activity

Comprehensive catalytic studies of sulfided catalysts have been carried out with Maya heavy crude oil using different diluents as well as pure crude oil to determine the HDS, HDN, HDM, and HDA activities. Generally, these reactions take place on the sulfided catalyst, sulfur ion vacancy that has electron deficiency, which adsorbs sulfur, nitrogen organic molecules through unpaired electrons, for example, porphyrin of metal chelates in the case of HDM. On the other hand, adsorption of organic molecule also competed by the presence of H_2S or H_2 dissociation species (HS^-; H^+; H^-) for the same site. The adsorption of dissociated species convert CUS into the saturated sites (i.e., sulfhydryl group), which is a Brönsted acid sites center (Topsøe and Topsøe, 1993; Rana et al., 2000, 2004b, c; Breysse et al., 2002). Thus, these saturated sites depend on the partial pressure of H_2S (P_{H_2S}), so the optimum concentration of H_2S is a critical parameter for maintaining the number of CUS. The role of sulfhydryl (–SH) groups for heavy oil hydroprocessing is confirmed and it is a source of hydrogen for hydroprocessing reactions (Breysse et al., 2002), which are based on the transfer of hydrogen to reactant. Since heavy oil feedstock is also rich in nitrogen content, poisoning of the active sites also takes place by competitive adsorption. On the other hand, it is also reported that H_2S enhances HDM (Rankel, 1981; Ware and Wei, 1985; Bonné et al., 2001). According to Bonné et al. (2001), H_2S coordinates to the central transition metal atom (Ni or V) and as a result weakens the metal–nitrogen bond, while Rana et al. (2007a) proposed that sulfhydryl groups are responsible to the destabilization of the metal nitrogen bond. It is also expected that the increased presence of –SH groups enhances the cracking of asphaltene molecule or HDAs. Hence, the formation of coke and metal deposits nearby CUS is expected. The loss of catalyst activity during hydroprocessing of heavy feeds will be caused mainly by coke and metals (Furimsky and Massoth, 1999).

Apart from the hetero-metal atoms, a major problem in Maya heavy crude oil hydroprocessing is the high content of asphaltenes, which affects the conversion and makes the characterization process more difficult due to its complex nature. Recently, small-angle neutron scattering characterization of Maya crude asphaltene indicated that a wide range of molecular changes occurs with temperature and its fractal network as shown in Figure 2.7 (Tanaka et al., 2003). Thus, it is a theme of further studies that asphaltene conversion follows catalytic conversion or thermal cracking mechanisms. Comparison of Maya with other crude asphaltenes such as Iranian light

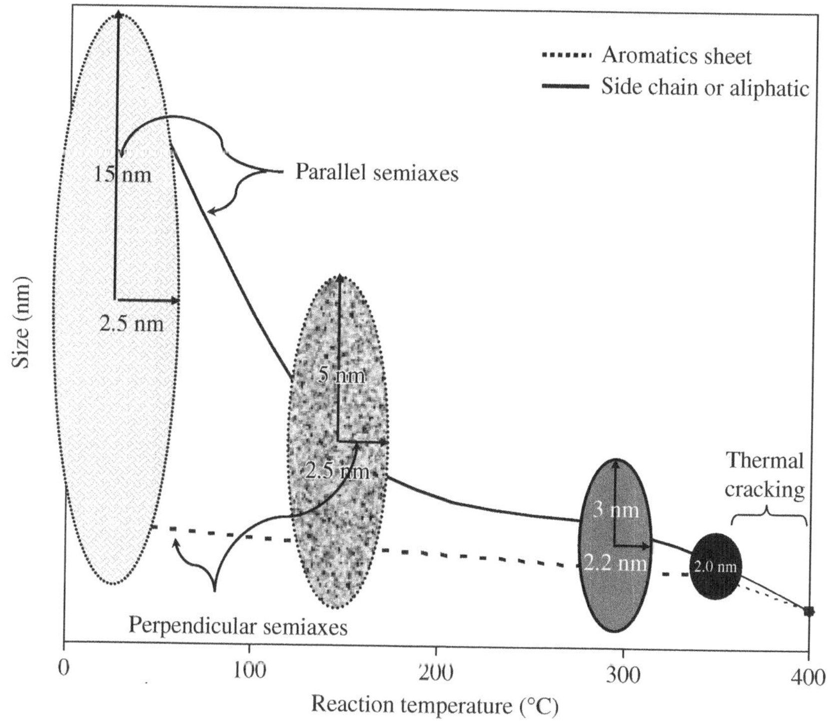

Figure 2.7 Effect of temperature on the shape and the size of Maya asphaltene (Adapted from Tanaka et al. (2003)).

oil and Arabian heavy oil (Khafji) showed that Maya asphaltene is more refractory in nature. The decalin-precipitated Maya asphaltene showed a fractal network even at 350 °C. Hence, thermal cracking starts at around these temperatures, which further indicates high coke-making tendency of Maya asphaltene. The size of asphaltene in Maya crude is about 2 nm in radius. Moreover, metals in the asphaltene fraction are believed to be present as organometallic compounds associated in the form of micelles. The metal chelate molecules in crude oil mostly exist in the form of etioporphyrin, which have been extensively characterized, and their diameter is reported to be approximately 1.6 nm (Fleischer, 1963). Therefore, the required catalyst pore diameter is necessarily larger than 2 nm to allow for diffusion of these molecules to the catalytic sites.

2.5.5 Experiments with Naphtha Diluted Feedstock

Maya crude diluted with desulfurized naphtha was used as feedstock to evaluate the catalyst activity by Maity et al. (2003a; b; c) and Rayo et al. (2004). HDS, HDN, HDM, and HDA activities of four different catalysts are compared in Figure 2.8. The properties of the catalysts compared in Figure 2.8 are presented in Table 2.9.

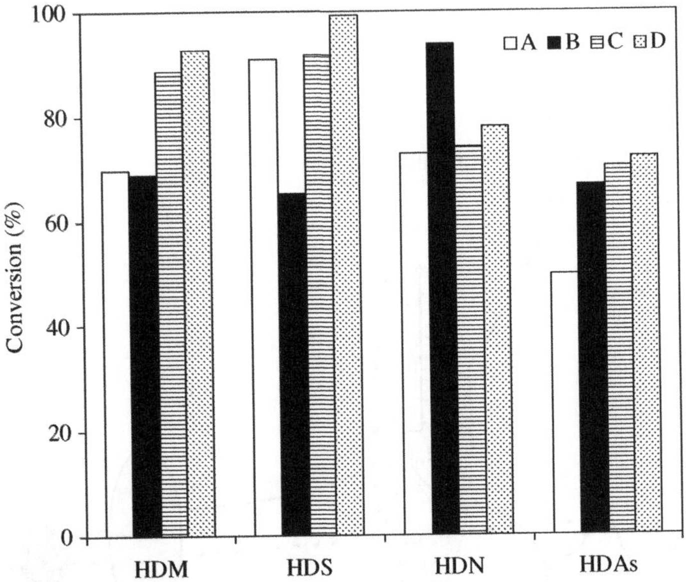

Figure 2.8 Effect of catalyst composition on Maya crude+hydrodesulfurized naphtha feedstock. (A) CoMo/AT-1, (B) CoMo/AT-2, (C) CoMo/Al_2O_3–TiO_2, (D) PCoMo/Al_2O_3–TiO_2 (Adapted from Maity et al. (2003b)).

It is observed that the HDM and HDAs activities are higher for the catalyst having higher APD. It is then stated that the internal active surface of the catalyst having bigger pore diameter is accessible to the metal and asphaltene molecules. On the other hand, HDS and HDN activities do not directly correlate with pore diameter. It is suggested that the lack of correlation of HDS and HDN activities with pore diameter was due to the presence of sulfur and nitrogen compounds of significantly different size. One type of sulfur compounds is bigger in size and these are generally attached to asphaltene structure. The other sulfur compounds are smaller such as benzothiophene and dibenzothiophene (Gorbaty et al., 1990; Kelemen et al., 1990; Mullins, 1995). Therefore, the conversion of the sulfur compounds may not only depend on pore diameter but also on the active site of the catalyst.

The effect of phosphorous on Maya crude hydrotreating catalysts was studied by Maity et al. (2003a). Catalysts C and D (see the last column of Table 2.9) are both supported on the same Al_2O_3–TiO_2 support, but the latter has 0.8 wt% P. Although the P containing catalyst exhibited high initial activity, rapid decrease was observed with TOS. Similar results were also found for catalysts supported on alumina (Maity et al., 2003c). It was assumed that P increases acidity of the catalyst and hence its initial activity. The increase in catalyst acidity also increases coke formation and therefore catalyst activity decreases rapidly with TOS. Li alkali metal was introduced to the catalyst to reduce acidity. However, the addition of Li did not improve catalyst stability. The synergetic effect of phosphorous, however, is not always the same and may depend on the support and preparation method. Though P increases significantly the

TABLE 2.9 Effect of Support and Catalyst Preparation on Textural Properties

Support	Preparation Method	Catalysts Textural Properties SSA (m^2/g)	TPV (mL/g)	APD (nm)	Catalysts Composition (wt%) Mo (W)	Co (Ni)	P	Catalyst Name
TiO_2–Al_2O_3 (5/95, w/w)	AlSul–TiCl–NH_3	236	0.24	3.6	6.7	2.36		A
	AlSul–Ti–iso-NH_3	216	0.25	4.4	6.7	2.36		B
	AlSul–Ti–iso-urea	258	0.62	9.6	6.7	2.36		C
	AlSul–Ti-iso-urea	250	0.67	10.4	6.7	2.36	0.8	D
SiO_2–Al_2O_3 (5/95, w/w)	AlSul–NaSi–urea	216	0.30	5.4	6.7	2.36		E
	AlSul–NaSi–urea	154	0.31	7.9	6.7	2.36	0.8	F
	AlSul–NaSi–urea	167	0.31	7.4	6.7	(3.36)	0.8	G
Al_2O_3	$AlNO_3$–NaAl-pH	244	0.78	12.2	10	3.44		
		240	0.78	12.3	10	(3.44)		
		213	0.70	12.8	(15.8)	3.14		
		225	0.73	12.5	(15.8)	(3.14)		
		243	0.82	12.9	6.7	2.36	0.8	
Al_2O_3	Commercial Al_2O_3	223	0.64	11.4	6.7	2.36		
	$AlNO_3$–NaAl-pH	262	0.49	7.3	6.7	2.36		
	$AlNO_3$–NaAl-pH	257	0.49	7.4	6.7	(2.36)		
Al_2O_3		241	0.75	12.0	6.7	2.36		
Al_2O_3–SiO_2		218	0.30	5.6	6.7	2.36		
Al_2O_3–TiO_2		257	0.62	9.6	6.7	2.36		

AlSul, aluminum sulfate; TiCl, titanium chloride; Ti-iso, titanium isobutoxide; NaSi, sodium silicate; $AlNO_3$, aluminum nitrate; NaAl, sodium aluminate; SSA, specific surface area; TPV, total pore volume; APD, average pore diameter.
Source: Adapted from Maity et al. (2003a, b, c).

activities of Al_2O_3- and Al_2O_3–TiO_2-supported catalysts (Maity et al., 2003a, b), it did not show any synergetic effect on the Al_2O_3–SiO_2-supported catalyst. For this catalyst, the effects of promoters (Co or Ni) on hydrotreating activities of Maya crude are compared in Figure 2.9. Catalyst F, promoted by Co, shows higher HDM and HDS activities compared with catalyst G promoted by Ni.

The variation of catalyst activity with TOS during hydroprocessing of heavy crude indicated that severe catalyst deactivation takes place and most probably near the pore mouth. These results were observed by the analysis of metal and coke deposition in the spent catalyst. It is reported in the literature that phosphorus decreases the number of strong acid sites, which impede coke formation on the catalyst (Stanislaus et al., 1988). Thus, this may be the reason for inhibition of coke formation in P catalysts. Kushiyama et al. (1990) stated that phosphorus may interact with vanadium in heavy feed preventing deactivation of the catalyst.

The above activity results for Maya crude hydroprocessing over mixed-oxide-supported catalysts indicated high initial activity but showed rapid deactivation with

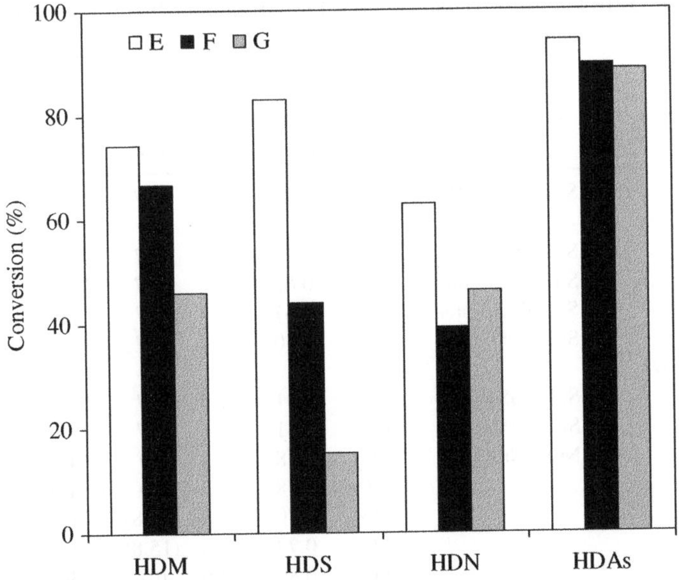

Figure 2.9 Effect of Al_2O_3–SiO_2-support preparation and catalyst composition on Maya crude activities (Adapted from Maity et al. (2003a)).

time. The reason behind the fast deactivation could be explained by the following two ways:

- The preparation method of the support, which contains high acidity, particularly supports prepared with aluminum sulfate. When sulfate ions remain on the solid they could produce super active acid sites during or after the sulfidation, which can have very high conversion especially of asphaltenes at the initial stage but with time acid sites deactivate, consequently a drop in conversion is observed.
- The diluent used (i.e., naphtha), which contains a significant amount of paraffins that induce the precipitation of asphaltenes over the catalyst surface, leading therefore to a continuous drop in catalyst activity. In this regard, a detailed study of the diluent effect (HDS naphtha and HDS diesel) on Maya crude hydrotreating indicated that diluent has important role in HDS and HDAs conversions over NiMo/TiO_2–Al_2O_3-supported catalyst as shown in Figure 2.10. The authors concluded from their results that naphtha diluent leads to an increase in the poisoning of catalyst surface by carbon deposition caused by the insolubility of asphaltene in naphtha (Rayo et al., 2004).

These results indicate that careful attention must be paid to the type of diluent, and, in general, the type of feed used when studying hydroprocessing of heavy crude oils. In this case, dilution of Maya crude oil with hydrodesulfurized diesel avoided the insolubility problem. Thus, this dilemma leads to think switching the dilution with a heavy hydrocarbon, that is, Maya crude dilution with hydrodesulfurized diesel.

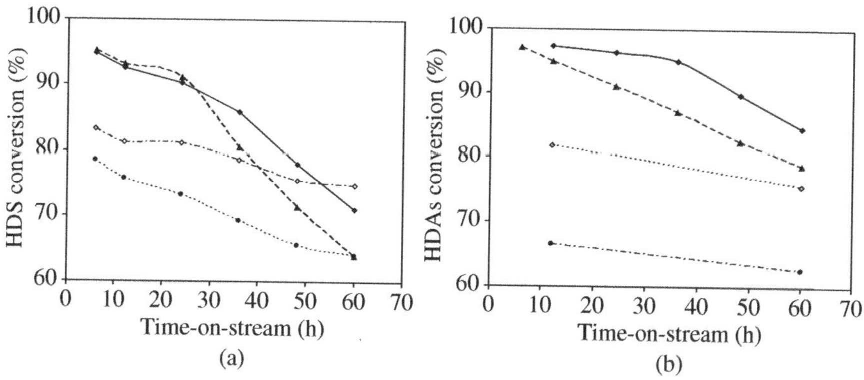

Figure 2.10 Effect of diluent on (a) HDS and (b) HDAs of Maya crude oil. (▲) $NiMo/Al_2O_3$ (HDT Maya+naphtha), (♦) $NiMo/Al_2O_3$–TiO_2 (HDT Maya+naphtha), (◊) $NiMo/Al_2O_3$–TiO_2 (HDT Maya+diesel), (●) $NiMo/Al_2O_3$–TiO_2 (Maya+diesel) (Adapted from Rayo et al. (2004)).

2.5.6 Experiments with Diesel Diluted Feedstock

It has been shown earlier that the support properties are crucial for defining the desirable porosity of catalysts. The advantage of using mixed oxides supports for hydrotreating is their acid–base properties. This in turn improves hydrocracking activity, which is required for achieving a desirable conversion of asphaltenes and resins to distillate fractions. At the same time, more acidic supports have an adverse effect on other catalyst functionalities. Therefore, the activity of catalysts for hydroprocessing of heavy feeds must be optimized to achieve a desirable level of hydrocracking with considerable activities of HDM, HDS, and HDN. In this regard, literature reports show that the catalyst functionalities can be modified by incorporating small amount of TiO_2 into Al_2O_3, using different incorporation methods (Rayo et al., 2004; Rana et al., 2005b; Maity et al., 2005, 2006; Ramírez et al., 2005). In the case of neutral supports such as carbon, a CoMo catalyst showed to be less sensitive to poisoning by N-bases and better ability to restrict coke or sediment formation compared with alumina (Fukuyama et al., 2004); however, their hydrocracking activity is low, which is a subject to the unique modification of carbon.

Catalyst activity with Maya crude oil using different supports nature (acid or base) is reported in Figure 2.11. Initially, the TiO_2–Al_2O_3-supported catalyst was more active due to its higher acidity. However, CoMo catalysts on basic supports (e.g., MgO–Al_2O_3, ZrO_2–Al_2O_3) were slightly more stable and more selective for HDM than TiO_2–Al_2O_3-supported catalyst (Caloch et al., 2004; Rana et al., 2005c). The variation in activity was attributed to the small amounts of MgO, ZrO_2, and TiO_2 (7.2, 9.2, and 9.0 wt%, respectively) incorporated to γ-Al_2O_3. On the other hand, the use of catalysts supported on basic supports in hydroprocessing of heavy feeds may be limited because of their low hydrogenation and hydrocracking activities. A comparison of HDS and HDM at 120 h TOS over different TiO_2–Al_2O_3- and alumina-supported CoMo catalysts is shown in Figure 2.12 (Rana et al., 2005b). In

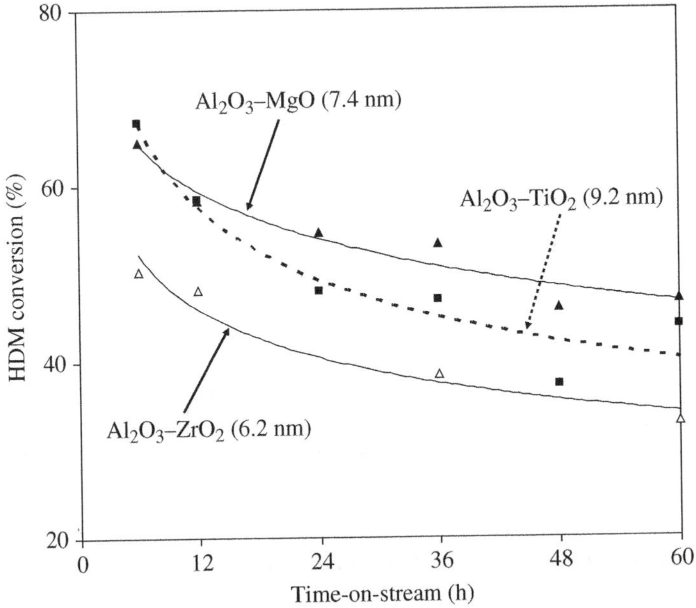

Figure 2.11 Effect of catalyst support on HDM (Adapted from Caloch et al. (2004) and Rana et al. (2005b)).

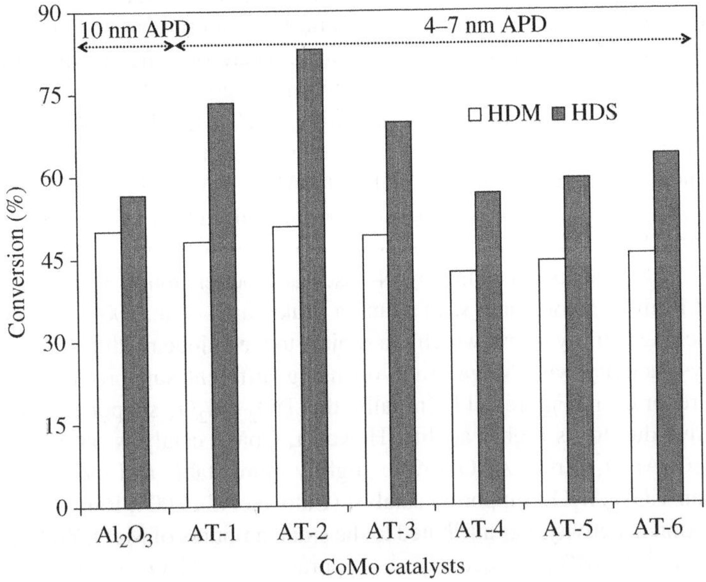

Figure 2.12 Comparison between HDS and HDM for CoMo catalyst over Al_2O_3 and TiO_2–Al_2O_3 (AT) as function of TiO_2 (10 wt%) precursor ($TiCl_4$ = AT-1, AT-2, and Ti isopropoxide = AT-3, AT-4, AT-5, and AT-6) at 120 h TOS and its incorporation method in alumina (Adapted from Rana et al. (2005b)).

all cases, the initial conversion was higher for the TiO_2–Al_2O_3-supported catalysts than for the alumina-supported catalyst, but the results of initial conversion are not justifiable for heavy oil processing. The TiO_2 containing catalyst showed enhanced HDS activity due to the structural promotional effect of TiO_2, while HDM conversion either remains the same as in alumina or slightly less, which seems to be an effect of APD or stability of the catalysts. The TiO_2–Al_2O_3-supported catalysts have an APD in the range of 4–7 nm, while the alumina-supported catalyst has 10 nm, which might have a corresponding effect on the stability of the catalyst.

To study the effect of pore diameter on different catalysts for Maya crude hydroprocessing, various aluminas were prepared using different preparation methods to vary the pore diameter. The composition and textural properties of these catalysts along with support preparation method are reported in Table 2.10 (Rana et al., 2004a). A steady-state conversion comparison for these catalysts is made in Figure 2.13. The results with a CoMo/γ-Al_2O_3-22.2 catalyst having a pore diameter of 22.2 nm show that pore diameter is not the only factor because activities of this catalyst are comparatively lower (except HDAs) than the other catalysts having smaller pores probably due to the lack of catalytic sites on the low surface area catalyst ($\approx$60 m^2/g). Since the composition of all catalysts is similar, these results reveal that asphaltene conversion is less affected by the metallic function (Co–Mo active sites) while largely affected by the pore diameter of the catalyst. The comparison between HDM and HDS shows opposite trends, which reveals

TABLE 2.10 Textural Properties and Composition of Catalysts

Support	Preparation Method	Textural Properties			Composition (wt%)	
		SSA (m^2/g)	TPV (mL/g)	APD (nm)	Mo	Co (Ni)
Al_2O_3-am (6.4 nm)	Ammonia	164	0.27	6.4	7.8	(2.4)
Al_2O_3-am (6.5 nm)	Ammonia	169	0.27	6.5	7.5	2.5
Al_2O_3-u (12.9 nm)	Urea	136	0.39	12.9	7.5	2.5
Al_2O_3-ac (6.0 nm)	Ammonium carbonate	183	0.37	6.0	7.3	2.5
Al_2O_3-acs (17.3 nm)	Ammonium carbonate-pH swing	160	0.47	17.3	6.8	2.5
Al_2O_3-22.2	Catapal-200	60.0	0.48	22.2	7.2	2.3
Al_2O_3-u-ac	Urea–ammonium carbamide	184	0.48	10.0	5.93	2.1
Ti–Al_2O_3	Ti/Catapal-1	176	0.40		6.56	(4.1)
Al_2O_3–MgO	Ammonia	169	0.31	7.4	7.9	2.3
Al_2O_3–ZrO_2	Ammonia	193	0.33	6.2	7.2	3.6
Al_2O_3–TiO_2	Ammonia	163	0.320	6.3	7.5	2.5
Al_2O_3–B_2O_3	Ammonia	48.0	0.34	24.7	7.1	2.2

APD, average pore diameter; SSA, specific surface area; TPV, Total pore volume.
Source: Adapted from Rana et al. (2004a, 2005a, 2005c).

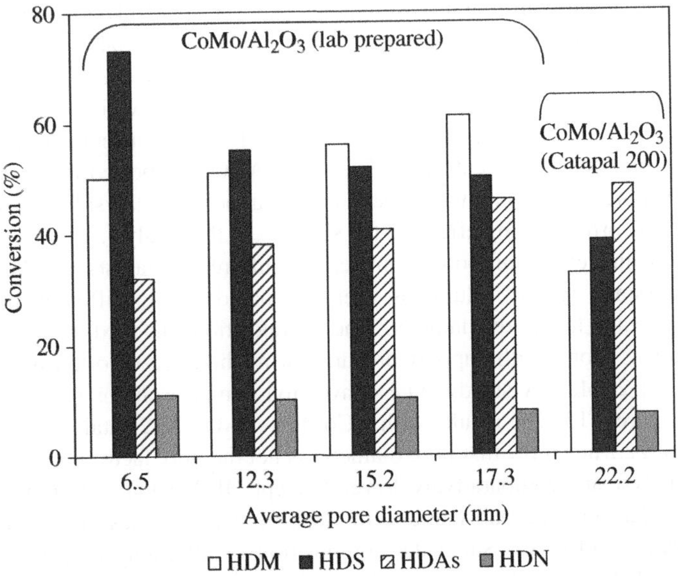

Figure 2.13 Effect of average pore diameter of catalyst on Maya crude hydroprocessing at 60 h TOS (feedstock: Maya+HDS diesel, microreactor) (Adapted from Rana et al. (2004a, 2007b)).

that HDM catalyst should be essentially macroporous in nature. It is observed that HDM and HDAs significantly depend on the catalyst pore structure, while the HDS activity may depend on the dispersion of the active metal. The result in Figure 2.13 indicates that the HDS activity distinctively differs from the HDM and HDAs. Most probably, the HDS activity appreciably depends on the surface area and the number of active sites. $CoMo/Al_2O_3$-6.5 (6.5 nm) catalyst shows the lowest activity for HDM and HDAs while higher activities can be seen for HDS. Thus, HDM and HDA conversions are limited due to the diffusional limitations of complex metalloid and asphaltene molecules for the catalyst having a pore diameter less than 10 nm. Thus, the performance of heavy oil hydroprocessing with respect to different functionalities, such as HDM, HDS, and HDAs, is clearly linked to the catalyst porosity and nature of the heavy crude oil. Hence, the effect of support preparation on the pore size distribution and APD apparently controls the catalytic activities along with the metal dispersion of active phases.

2.5.7 Experiments with Pure Maya Crude Oil

The previous results indicated that mixed-oxide-supported catalysts gain great interest for heavy oil hydroprocessing. It might be due to the textural properties and modified acid–base sites. The next step for developing a suitable catalyst for hydroprocessing of Maya crude oil is to study the catalyst stability with TOS for a prolonged period of time using pure Maya crude, to this end a bench-scale reactor

(100 mL catalyst) was used. It is worth to mention that in the previous sections, catalyst activities were studied with diluted (naphtha or diesel) Maya crude oil and the primary focus was oriented on the screening of catalysts.

In addition, to avoid dry sludge formation with prolonged temperature and TOS or compatibility of asphaltene and maltenes (Takatsuka et al., 1988; Inoue et al., 1998), controlled hydrogenation of asphaltene is mandatory. It is then a good idea to use a two-stage fixed-bed reactor system, where the combination of the two reactors can be effectively used to handle the real feedstock (Ancheyta et al., 2010).

In this respect, an alumina support was prepared by using urea hydrolysis for the first reactor (CoMo, HDM, catalyst), while Catapal C1 and TiO_2 (5 wt%, impregnated) was prepared and considered the second reactor catalyst (NiMo, HDS catalyst) (Rayo et al., 2011). The composition and textural properties of these catalysts are given in Table 2.10. Thus, the aim is to compare the behavior of catalysts for Maya crude hydrotreating with adequate stability with TOS at bench-scale reactor. The comparison is made considering a CoMo catalyst with high porosity (for HDM) and a NiMo catalyst that is more prone to the deep HDS and hydrogenation using Al_2O_3 and TiO_2/Al_2O_3 as supports, respectively, as shown in Figure 2.14. Thus, the HDM catalyst (first reactor) is designed with high porosity and low surface area, while the HDS catalyst (second reactor) is characterized by higher surface area and moderate pore size distribution.

The results for different activities against TOS are shown in Figure 2.15 for CoMo and NiMo catalysts. It is shown that all activities marginally decrease with TOS. As expected, the CoMo catalyst shows much lower activity for the hydrogenation of asphaltenes than for HDM and HDS. However, the decay in activity with TOS for the three reactions is very similar, which indicates that the catalytic sites remain constant or vary similarly with time for all catalytic activities. Surprisingly, a more or less similar deactivation trend is observed for the microflow and the bench scale reactors, while the compositions of feedstocks were different (Rana et al., 2005a). This consistency of activity with TOS could be related to the porosity of the catalyst, in which hydrotreating of heavy oil is directly proportional to the stability of the catalysts (Rana et al., 2004a). On the other hand, stability of the NiMo catalyst is not similar for all the

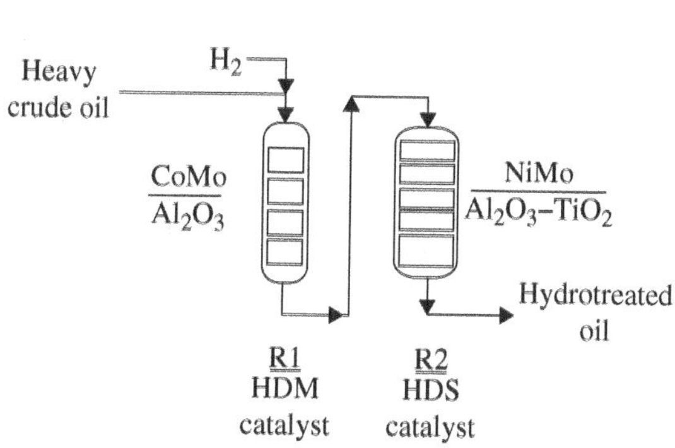

Figure 2.14 Simplified diagram of the two-stage reactor for hydrotreating of heavy oil crude (Adapted from Ancheyta et al. (2010)).

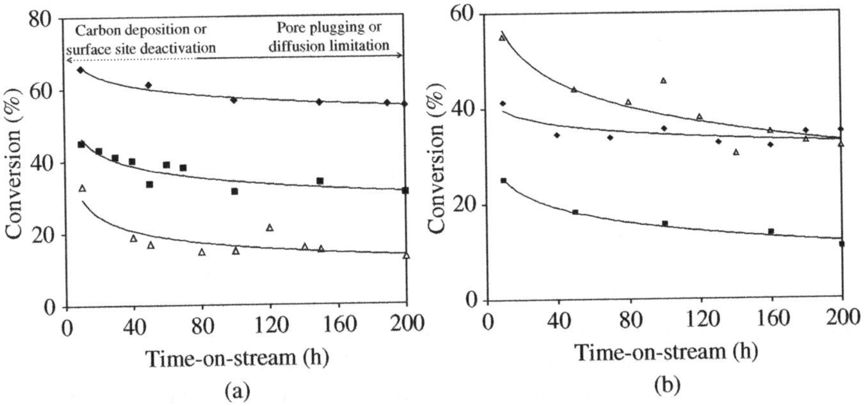

Figure 2.15 Catalytic activities of (a) $CoMo/Al_2O_3$ and (b) a $NiMo/TiO_2–Al_2O_3$ catalysts with time-on-stream during hydroprocessing of Maya crude at bench-scale reactor ((♦) HDS; (■) HDM; (Δ) HDAs) (Adapted from Rana et al. (2005a)).

functionalities; for example, HDAs decrease faster than HDS and HDM. Obviously, in this case HDAs activity is higher than HDS and HDM, which further reveals that the NiMo catalyst enhances the hydrogenation of asphaltenes. The unusual deactivation for HDAs in the case of $NiMo/TiO_2–Al_2O_3$ indicates a role of catalyst acidity, which is higher in the case of this catalyst. Thus, the faster deactivation can be understood by the presence of acidic sites generated due to the mixing of TiO_2 and Al_2O_3, which favors the cracking of asphaltene molecules and the subsequent deactivation by coke formation. One more plausible explanation in this respect is the stability for HDS and HDM activities with TOS, which indicates that the metallic sites in the hydrotreating catalyst are more stable than the acidic sites of the support or active phases (–SH groups). However, catalyst deactivation during heavy oil processing is not only due to the carbon deposition, but it is also equally possible that metallic sites (Co–Mo–S) have been poisoned by Ni and V deposition during the HDM, although the acidic sites will be deactivated by the coke deposition. The deposition of metals or carbon is most likely at the entrance of the pores, which plugs the pore mouth or restricts diffusion of big, complex asphaltene molecules into the pores (Toulhoat et al., 1990).

2.5.8 Characterization of Spent Catalysts

2.5.8.1 Deposition of Metals Characterization of spent catalyst shows that carbon is deposited at around 7–16 wt% as coke on the catalyst. Metal deposition is found to depend on the feed composition, that is, Ni of around 0.3 wt%, while V of around 1.4 wt% for 200 h TOS, as shown in Table 2.11. These results are determined over the spent catalyst, which is washed with toluene and dried at 200 °C. The effect of washing was negligible, which slightly decreases the carbon and S/Mo ratio that could be due to the loss of soft coke and the exchange of S atoms with oxygen

TABLE 2.11 Characterization of Spent Catalysts

Catalysts	wt%						S/Mo (mol/mol)
	Mo	Ni	V	C	S	Fe (wppm)	
			HDS feed				
CoMo/γ-Al_2O_3-ac	4.9	0.05	0.19	11.3	5.19		3.1
NiMo/γ-Al_2O_3-am	5.4	2.01	0.13	11.0	4.64		2.5
NiMo/TiO_2–Al_2O_3	4.8	3.86	0.06	7.10	4.42		2.72
			HDM feed				
CoMo/γ-Al_2O_3-u	5.3	0.11	0.69	13.9	5.03		2.8
CoMo/γ-Al_2O_3-acs	4.7	0.11	0.65	16.1	6.23		3.9
CoMo/γ-Al_2O_3-am	5.4	0.08	0.43	10.5	4.45		2.4
CoMo/Al_2O_3	4.7	0.140	0.558	8.66	5.10	201.5	3.25
			HDT Maya feed				
NiMo/TiO_2–Al_2O_3	3.8	3.8	0.171	7.53	4.30	320.1	3.34
			Maya feed				
CoMo/Al_2O_3	4.301	0.302	1.393	13.3	3.81	804.5	2.66

Source: Adapted from Rana et al. (2004a, 2005a)

(Rana et al., 2005b). In the fixed-bed reactor, the metal and carbon deposition over the catalyst is generally affected by the bed length of the reactor; broadly, the maximum deposition of Ni and V occurs at around 15–20% bed depth, while carbon deposition profile is relatively uniform or slightly increases along the reactor depth (Fleisch et al., 1984). The coke and metal deposition on the spent catalyst also depends on the catalyst as well as on the scale and type of the reactor (Thakur and Thomas, 1985).

It has been established that HDM and HDAs activities depend on the pore diameter, while HDS activity is distinctively different, which appears to be more likely dependent on metal dispersion. Not only catalytic activities but also metal retention capacity increases with large pore diameter of the catalyst. The deposited foulant significantly alters the textural properties of the catalyst, for example, surface area and pore volume decrease by 30–50% and 40–70%, respectively; this decrease depends on the composition of feedstock and total time on run. The loss of surface area and pore volume is due to the decrease in micro- and lower mesopores (<10 nm), while pores with a diameter greater than 25 nm remain unaffected after 200 h TOS as shown in Figure 2.16. However, carbon deposition increases with the APD of the fresh catalyst as shown in Figure 2.17 (Maity et al., 2003a). It is difficult to distinguish the deposited species by characterization of the textural properties, but it is confirmed by various researchers that initial decline in activity is attributed to coke formation (Yoshimura et al., 1991; Thakur and Thomas, 1984, 1985).

Apart from coke deposition, metals such as V, Ni, and Fe were deposited and modified the surface of the catalyst (CoMoS). Figure 2.18 shows that the presence of deposited Ni and V metals have been verified by their Kα X-ray emission (SEM-EDAX) line at 7.47 and 4.95 keV, respectively (Rana et al., 2005a). The deposited metals exist as sulfides of V_3S_4, N_3S_2, which is confirmed by using XRD

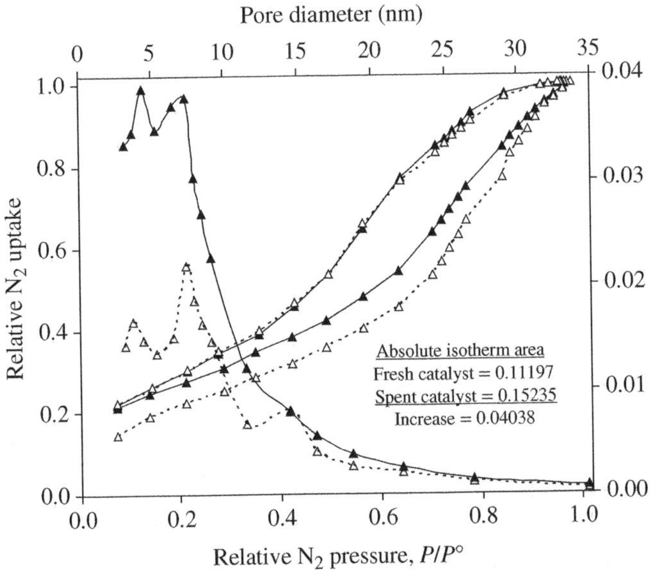

Figure 2.16 Pore size distributions and adsorption–desorption isotherms for $CoMo/Al_2O_3$. (▲) Fresh catalyst, (△) spent catalyst.

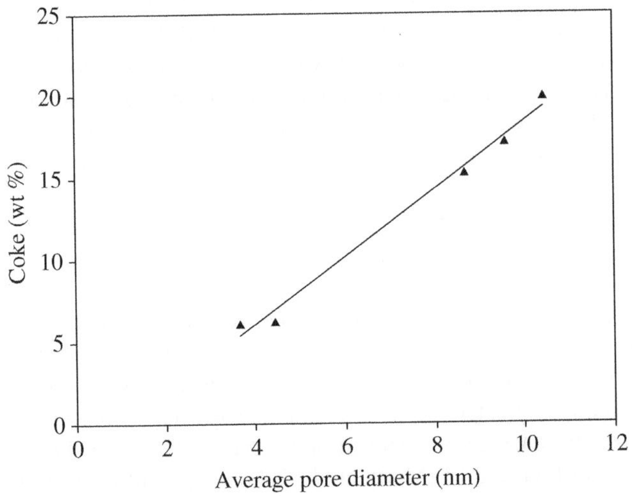

Figure 2.17 Average pore diameter of fresh catalysts versus coke formation at 60 h TOS during Maya crude hydrotreating (Adapted from Maity et al. (2003a)).

of the spent catalyst as shown in Figure 2.18 (Rana et al., 2004a). The XRD result indicates that the crystal size of deposited metal sulfide phases is larger than 4 nm. The deposition of Ni sulfide phase distributed homogeneously, inner and outer surfaces of the catalyst. The fact that this was not visible by XRD may be due to the small crystallite size (<4 nm). In contrast, V sulfides were distributed mostly at the

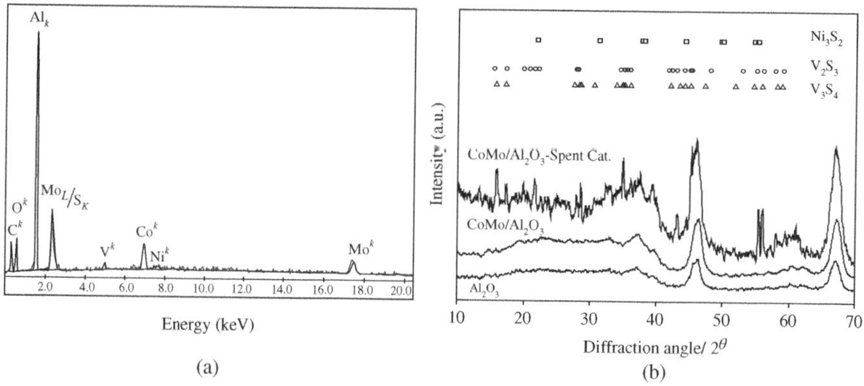

Figure 2.18 (a) SEM-EDAX microanalysis of spent $CoMo/Al_2O_3$ catalyst and (b) X-ray diffraction patterns for support, fresh, and spent $CoMo/Al_2O_3$ catalysts. Deposited Ni and V sulfides intensity is compared with JCPDS-ASTM data: (Δ) V_3S_4, (O) V_2S_3, (□) Ni_2S_3 (Adapted from Rana et al. (2005a)).

outer surface of the extrudate. This may be due to the formation of large crystallites, which were detected by XRD. Similar results were observed by other researchers (Kawada et al., 1975; Takeuchi et al., 1985; Toulhoat et al., 1987, 1990; Eijsbouts, 1999). They have reported that vanadium sulfide crystallite size ranged from 5 to 30 nm and that the crystallites grew up perpendicular to the support surface. Detailed studies on spent catalyst spectroscopic characterization were reported by Smith and Wei (1991a, 1991b, 1991c) using TEM, EDX, STEM, and XPS for model molecules such as nickel etioporphyrin and vanadyl etioporphyrin and concluded that nickel and vanadium sulfides deposited on HDM catalyst were in the form of relatively large spatially dispersed crystallites. The difference between Ni and V deposition could be due to the more refractory nature of the Ni molecule compared with the V molecule as shown in Figure 2.19 (Maity et al., 2003a). Since most of the metalloporphyrins are associated with the asphaltene molecules, it is plausible that some of the asphaltene molecules do not diffuse into the catalyst pore. Consequently, their conversion took place at the entrance of the pore mouth due to pore diffusion limitation. Furthermore, these results are confirmed by quantitative analysis of V metal as a function of extrudate radial distribution zones, showing an enrichment of V and carbon in the outer surface of the catalyst particle as illustrated in Figure 2.20 (Ancheyta et al., 2002, 2003).

2.5.8.2 ***Stability of Catalysts*** Catalyst stability is an important factor that has an immense impact on the catalyst development and is defined by the variation of conversion as a function of TOS. Heavy oil hydroprocessing catalysts finish operation either by pressure drop across the catalyst bed or when the operating temperature reaches its limiting value. The stability is usually affected by the pore diameter of the catalyst. For heavy feedstocks, it is reported that catalyst stability is directly proportional to the total pore volume and mean pores radius (Nomura et al., 1980). The

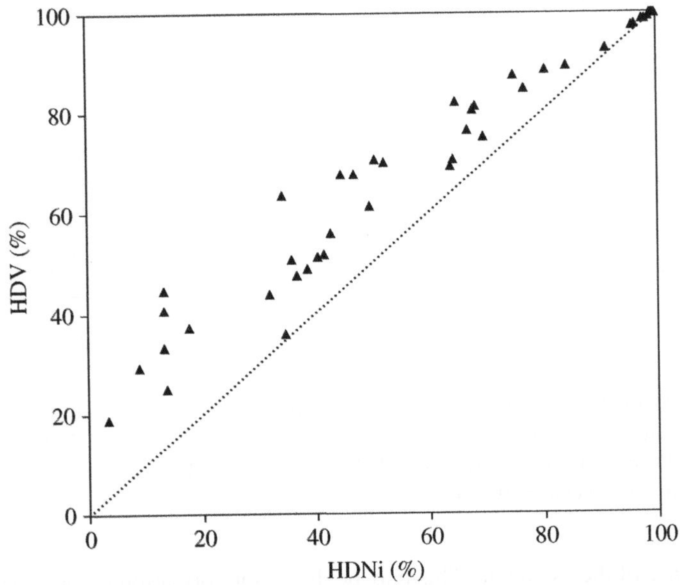

Figure 2.19 Removal of vanadium versus removal of nickel over $CoMo/TiO_2–Al_2O_3$ catalyst and Maya diluted with naphtha feedstock (Adapted from Maity et al. (2003a)).

smaller the pore diameter, the shorter the catalyst life. Thus, the catalyst stability can be drawn using the following equation:

$$s = a_1 Vr + b_1 \,(\text{cc/nm/g}) \tag{2.1}$$

where s is the catalyst stability, V the total pore volume, r the pore radius, and a_1 and b_1 are constants. This proportionality between the stability of catalyst has been confirmed by modeling with pore volume and pore radius. The stability phenomena indicate that catalysts with large pores and pore volume provide easy access and high capacity for inorganic metals and asphaltene molecules. Life of catalyst is closely dependent on the metal content in the feed, operating conditions, and textural properties of catalyst, since small pores, severe operating temperatures, and low pressure provoke stronger deactivations. The effect of feed metal level for residuum hydrotreating run length is shown in Figure 2.21 (Howell et al., 1985). Maya atmospheric residue has sulfur content similar to an Arabian Heavy, but the former has four times higher metal content. Higher metal content reduces the run length by a factor of 4 due to the pore plugging of the catalyst, which prevents reactive molecule diffusion to the catalytic sites. Another important parameter for the stability is temperature of the reactor. At low temperatures, the deactivation is slow; increasing temperature leads to sediment formation, which has an adverse effect on the catalyst life. Thus, for adequate conversion, textural properties, shape, and size of catalyst appear to be more important than chemical composition. The stability of catalyst is also sensitive

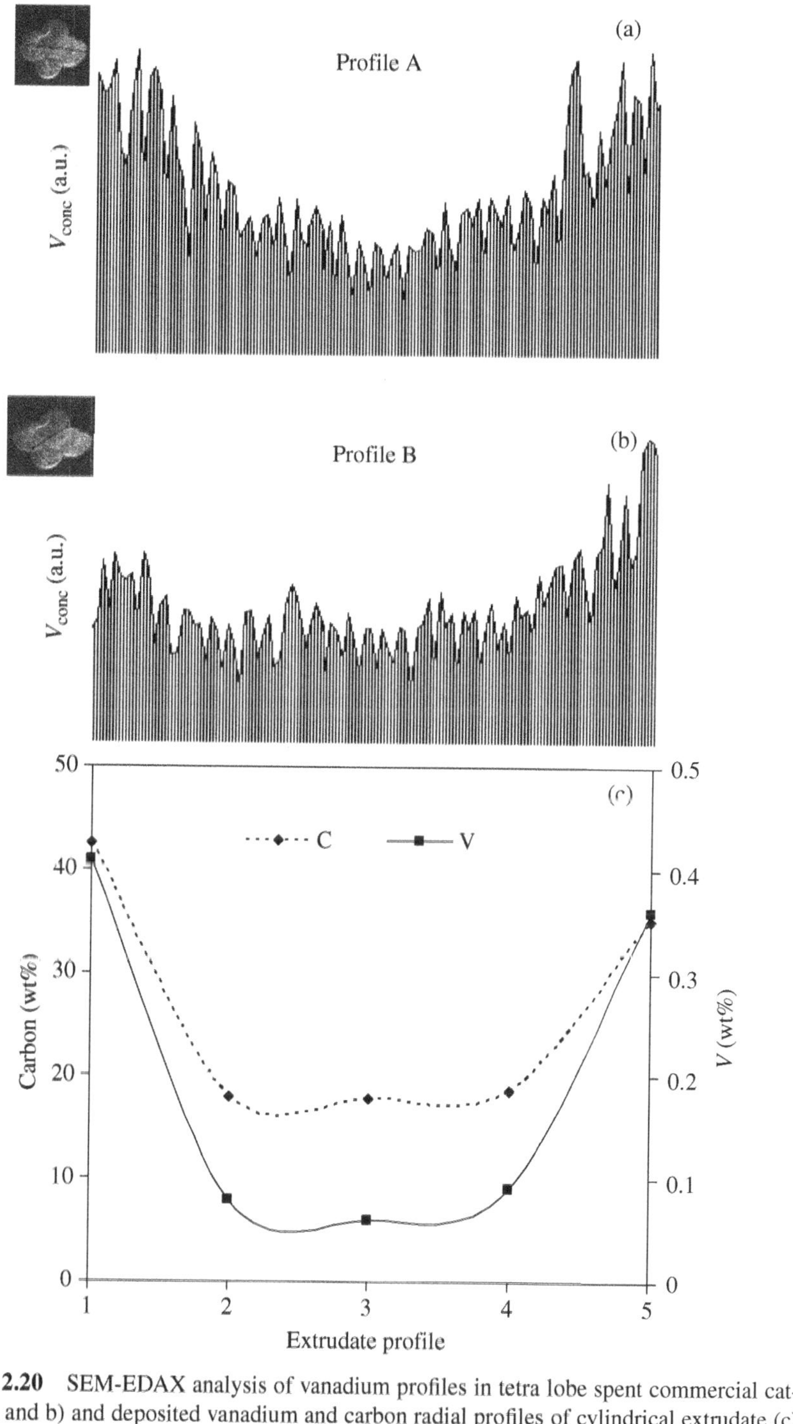

Figure 2.20 SEM-EDAX analysis of vanadium profiles in tetra lobe spent commercial catalyst (a and b) and deposited vanadium and carbon radial profiles of cylindrical extrudate (c) (Adapted from Ancheyta et al. (2002, 2003)).

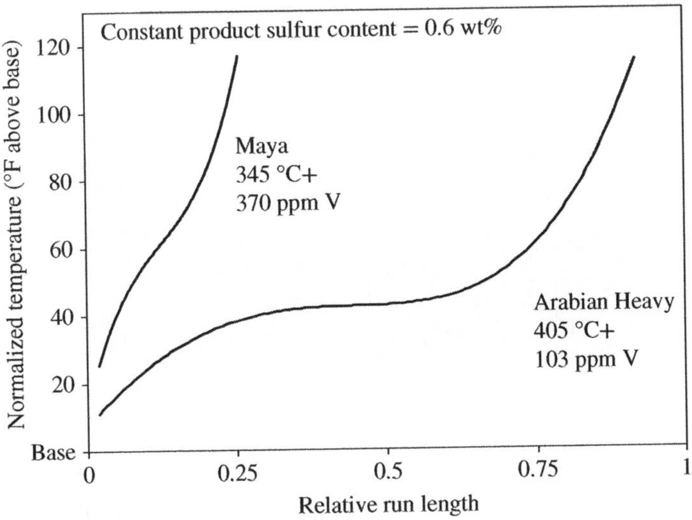

Figure 2.21 Effect of type of heavy feed on catalyst run length (Adapted from Howell et al. (1985)).

to the acidic function of catalyst, which is generally high for mixed-oxide-supported catalysts.

The activity of a catalyst having around 40% HDM conversion is not further decreasing with time (Figure 2.15), which indicates that deposited metal sulfides (Ni_xS_y and/or V_xS_y) may have some role in the catalytic sites, that is, autocatalysis. Generally, the metal loading of a heavy oil hydroprocessing catalyst is low and some parts of the support surface are not covered with the active metals (Mo, Co, or Ni). During HDM, metal sulfides coming from the feedstock are deposited on bare surface of the support as a part monolayer along MoS_2. In the case of regenerated catalysts, nickel and vanadium sulfides inhibit the HDS reaction, while they increase HDM activity (Kressmann et al., 1998). In this regard, an enhancement in HDM activity with increased H_2S partial pressure (Rana et al., 2007b) may also be attributed to the deposited metal sulfides, increasing the concentration of sulfhydryl groups, which further demonstrate high hydrogenation of porphyrin ring or HDM. Similarly, the sulfur content on spent catalysts slightly increases with APD; however, the molar ratio of S/Mo shows that the larger the APD the greater the S/Mo ratio, which again indicates that greater stability and metal retention capacity are obtained with larger APDs.

2.5.8.3 Pore Mouth Plugging (Semiquantitative Analysis of Pore Deactivation)

Nitrogen adsorption–desorption isotherms of fresh and spent catalysts are shown in Figures 2.16 and 2.22. An increase in the hysteresis loop of spent catalysts indicates that deactivation has occurred at the pore mouth. In fact, pores of fresh catalysts are considered cylindrical in shape; however, due to the metal and carbon deposition at the pore mouth, these cylindrical pores transform into ink-bottle type of pores

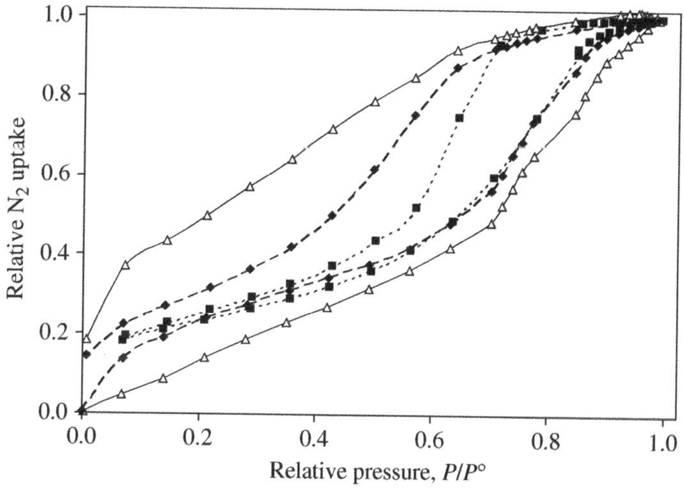

Figure 2.22 N_2 adsorption–desorption isotherms of NiMo/Al_2O_3–TiO_2 catalysts. (■) Fresh catalyst, (♦) spent catalyst (microflow reactor, 120 h TOS); (Δ) spent catalyst (bench-scale reactor, 200 h TOS) (Adapted from Rana et al. (2005a)).

in the spent catalyst. The deactivation at the pore mouth is also complementary to a decrease in the SSA and total pore volume. These deactivated catalyst isotherms display a large increase in the hysteresis. The isotherm absolute areas are shown in Table 2.12, which were calculated by using the trapezoidal rule integration method. The increase in the isotherm loop and the reduction in the total pore volume were considered representative of deactivation via pore mouth coking. These experimental results are also supported by demetalation diffusion-controlled reaction. The larger the pore diameter, the higher the metal retention capacity with even distribution of metals (V and Ni) on cylindrical-shaped catalysts (Tamm et al., 1981). It is also found that catalysts with large pores have a more even distribution of metals than small pore catalysts (Green and Broderick, 1981).

BJH (Barrett–Joyner–Halenda) adsorption and desorption surface areas were calculated to find another explanation for the formation of ink-bottle types of pores due to the deposition at the pore mouth. Table 2.12 is the compilation of adsorption–desorption area of the textural properties of fresh and spent catalysts. The adsorption data measure the true pore size distribution, which is not affected by the pore mouth deactivation (Fleisch et al., 1984). The constant ratios indicated that for fresh catalysts all pores are cylindrical in shape. On the other hand, desorption area/BET (Brunauer–Emmett–Teller)-SSA ratio of spent catalysts increases as the pore plugging increases. Hence, the following can be concluded:

- For D/A close to unity, the pores are cylindrical in nature.
- For D/BET greater than one, more deactivation occurs at the pore mouth.

TABLE 2.12 N_2 Adsorption–Desorption Measurements of Pore Mouth Plugging and N_2 Adsorption–Desorption Hysteresis Loop Area Analysis

Catalysts	Area Isotherm (Absolute Value)	D/A	D/BET	A/BET	Catalysts	Fresh Catalyst	Spent Catalyst	Increase, %
$CoMo/Al_2O_3$-acs	0.0274	1.0	1.1	1.1	$CoMo/Al_2O_3$-MFR	0.0865	0.101	
$CoMo/Al_2O_3$-acs-BU	—	1.3	1.2	0.9	$CoMo/Al_2O_3$-BSR		0.139	
$CoMo/Al_2O_3$-acs-U	0.0990	1.6	1.5	0.9	$NiMo/Ti/Al_2O_2$-MFR	0.0685	0.164	
$CoMo/Al_2O_3$-am	0.0564	1.1	1.1	1.0	$NiMo/Ti/Al_2O_2$-BSR		0.338	19.8
$CoMo/Al_2O_3$-am-U	0.4262	1.6	1.9	1.1				15.7
AT-1-WW		2.60	4.11	1.58			0.3173	6.2
AT-1		2.41	2.96	1.23		0.1191	0.2765	11.6
AT-1-R		1.30	1.4	1.07			0.1809	12.1
AT-2		1.37	1.78	1.31			0.2227	12.9
AT-3		1.33	1.69	1.27		0.1067	0.2358	13.1
AT-4		1.35	1.70	1.20		0.1144	0.2502	12.5
AT-5		1.23	1.52	1.24		0.1202	0.2558	19.8
AT-6		1.35	1.39	1.03		0.1248	0.2509	15.7

A, pore area calculated from BJH adsorption data; D, pore area calculated from desorption data; BET, total BET surface area; AT, Al_2O_3–TiO_2; U, spent catalyst (fixed bed, 60 h TOS); BU, spent catalyst (batch reactor, 6 h TOS); AT-1-WW (catalyst without washing); AT-1-R (catalyst regenerated) spent catalyst (120 h TOS), AT, Al_2O_3–TiO_2; MFR, microflow reactor; BSR, bench-scale reactor, AT-1-WW (catalyst without washing); AT-1-R (catalyst regenerated).

Source: Adapted from Rana et al. (2004a, 2005b, 2005d).

- The A/BET ratio should not be affected by the pore plugging and catalyst deactivation.

Therefore, the BJH desorption area is affected by the geometry of pores, while the adsorption area looks unaffected by the pore geometry and catalyst deactivation. The regenerated catalyst also followed the same principle of deactivation and indicates that around 60% of pore plugging can be recovered at the regeneration conditions used (Rana et al., 2005b).

Indeed, ink-bottle type of pores is expected to present difficulties to desorb until the relative pressure is reasonably low. It might be possible that adsorbed nitrogen in such a type of pores exists as liquid that is comparatively difficult to desorb and as a result the evaporation process takes place progressively rather than abruptly. It is relevant to mention that the same effect has been also predicted theoretically for tubular structures of low connectivity samples (Rojas et al., 2002). Therefore, the greater desorption/adsorption (BET-D/BET-A) area can be observed in the case of spent catalysts and this is more obvious in the case of $CoMo/Al_2O_3$-6.5, in which the APD is 6.5 nm. The difference in adsorption–desorption surface area is another reason to confirm pore mouth plugging. These results further indicate that true pore size distribution should be calculated using adsorption data; however, in the case of effective pore size distribution, desorption data are considered more reliable.

2.5.9 Final Comments

The petroleum refining industry has entered into a significant era and important changes have occurred over the past few years. The worldwide crude oil scenario is changing everyday with respect to crude oil price as well as origin; many of the world oil reserves are becoming heavy or extra heavy petroleum. At the same time, the demand for lighter products and middle distillates has grown steadily due to the demand of transportation fuels. The main goal of heavy crude oil hydroprocessing is to prepare upgraded crude oil with low impurities. The processes bring down the level of the impurities present in heavy crude, providing additional quantities of better quality feedstocks for fluid catalytic cracking and hydrocracking processes, which are the principal processes to convert the heavy fraction into the middle distillates or lighter fractions. The second target is to find out a catalyst that fulfills the stringent environmental regulations and that has sufficient stability with TOS.

Since catalyst function is crucial for these processes, for optimizing the properties a promising catalyst requires porous texture (bimodal porosity) specially designed for large molecules of sulfur, metals, asphaltene, and so on, along with optimum number of catalytic sites. The effect of supports on the different catalytic activities is an intriguing topic in heavy oil hydroprocessing. In spite of the intensive research knowledge about hydrotreating (model and/or middle distillate) catalysis, there are several questions that are not conceded for heavy oil processing. This is partly because of the complexity of the feedstock and the varied nature of supports, and limitations of the techniques employed to understand the role of metals and asphaltenes in various activities (hydrogenolysis, hydrogenation, and hydrocracking) that are induced by Mo-, CoMo-, and NiMo-supported systems. Asphaltenes, in which most of the metals tend to concentrate, require more attention with respect to the catalyst pore diameter and their diffusion to the active sites. The catalytic activity, without doubt, depends on the number of catalytic sites, but access to these sites is a matter of diffusing complex reactant molecules and adsorbed them on the catalytic site. Therefore, an accurate description of active sites under dynamic conditions is necessary.

Knowledge about the nature of active sites in spent catalysts will not only help know the role of the support but also will provide clues to prepare improved catalysts. The role of Ti containing support and the promotional effects are noticeable, but γ-Al_2O_3 remains as the main support for heavy oil hydroprocessing catalysts. To understand the deactivation behavior or catalyst stability is a subject of a detailed characterization of spent catalysts.

2.6 CONCLUDING REMARKS

Increased attention to the development of catalysts and processes for the hydroprocessing of heavy feeds will be necessary in view of the growing demand of distillates and the increasing need to process heavy feedstocks with high content of sulfur, nitrogen, metals, aromatics, and asphaltenes.

The conversion of heavy feeds to valuable products must be maximized by a proper match of feed, catalyst properties, reactor type, and operating conditions. It appears necessary to develop better analytical methods for the characterization of heavy feeds to be able to better predict the behavior of catalysts and catalytic reactors when processing such feeds.

The design of improved catalyst supports is necessary. Supports with adequate pore size distribution are necessary, which minimize diffusional limitations of large molecules to the pore interior while providing enhanced metal retention capacity and as large surface area as possible (without compromising the mechanical properties). For the hydroprocessing of heavy feeds, catalyst life can be further prolonged by applying balanced textural properties (high pore volume and enough surface area). Besides, the chemical composition of the support must be capable of providing enough acidity to perform the hydrocracking of large molecules but not so much as to lead to excessive coke formation. Studies of new catalyst formulations oriented to decrease catalyst deactivation, increase catalyst life and stability when processing heavy feeds are necessary.

The chemical composition of the active phase of the catalyst seems to depend on the processing scheme; the catalyst formulation will depend on the design of the catalytic bed system (single or multiple beds). For multiple bed reactors, the first catalyst bed must provide high HDM activity with high metal retention capacity. The second bed provides high HDS and some HDM, and the third bed must have adequate hydrocracking, HDS and HDN activities. The textural properties have to be adapted to the duty each catalyst bed performs. The shape and size of the catalyst particles have to be well designed to assure smooth operation of each catalyst bed. Attention must be paid to the use of optimized graded beds of the catalyst required to minimize the problem of increasing ΔP. Balanced catalyst formulations able to achieve maximum conversion with low formation of dry sludge for heavy feeds need to be developed. Thus, the chemical and physicochemical features of dry sludge formation and changes of asphaltenes structure during hydroprocessing must be studied in depth in order to construct mathematical models that will allow for the prediction of the conditions under which dry sludge is formed.

Knowledge obtained on the studies with model molecules and light feeds has to be integrated and applied to the processing of heavy feeds. However, any change or improvement in catalyst design or methodology of activation must take into consideration its feasibility for application in the refinery at industrial scale. Increasing catalyst stability and life is of utmost importance. In fixed-bed units, catalyst deactivation during the run is compensated by a progressive increase in bed temperature, up to a certain value dictated by metallurgical constraints or product quality. Moving- or ebullated-bed reactors with catalyst withdraw and replacement do not present the catalyst stability problem but create safety and environmental problems associated with catalyst disposal. Nevertheless, it appears that moving-bed or ebullated-bed processes employing suitable catalysts with high metal retention capacity represent the most efficient way of handling petroleum bottoms and other heavy feeds. The catalyst particles used in these reactors must be small enough to decrease the diffusional path of the molecules to the catalyst interior without compromising the mechanical and other properties necessary for the proper handling of the catalyst in the reactor.

REFERENCES

Absi-Halabi, M., Stanislaus, A., and Trimm, D.L. 1991. Coke formation on catalysts during the hydroprocessing of heavy oils. Appl. Catal. 72:193–215.

Absi-Halabi, M., Stanislaus, A., and Al-Zaid, H. 1993. Effect of acidic and basic vapors on pore size distribution of alumina under hydrothermal conditions. Appl. Catal., A 101:117–128.

Agudo, A.L., Llambias, F.J.G., Tascon, J.H.D., and Fierro, J.L.G. 1984. Characterization of sulfided NiMo/γ-Al_2O_3 catalysts by O_2 and NO chemisorption: influence of the method of preparation. Bull. Soc. Chim. Belg. 93:719–726.

Ancheyta, J., and Rana, M.S. 2005. Hydroprocessing of heavy oil fractions. Catal. Today 109:1–3.

Ancheyta, J., Betancourt, G., Centeno, G., Marroquín, G., Alonso, F., and Garciafigueroa, E. 2002. Catalyst deactivation during hydroprocessing of Maya heavy crude oil. I. Evaluation at constant operating conditions. Energy Fuels 16:1438–1443.

Ancheyta, J., Betancourt, G., Centeno, G., and Marroquín, G. 2003. Catalyst deactivation during hydroprocessing of Maya heavy crude oil. II. Effect of temperature during time-on-stream. Energy Fuels 17:462–467.

Ancheyta, J., Betancourt, G., Marroquín, G., Centeno, G., Alonso, F., and Muñoz, J.A. 2010. Process for the catalytic hydrotreatment of heavy hydrocarbons of petroleum. US Patent 7,651,604 B2.

Bartholdy, J., and Cooper, B.H. 1993. Metal and coke deactivation of resid hydroprocessing catalysts. Prepr. – Am. Chem. Soc., Div. Pet. Chem. 38:386–390.

Bonné, R.L.C., van Steenderen, P., and Moulijn, J.A. 2001. Hydrogenation of nickel and vanadyl tetraphenylporphyrin in absence of a catalyst: a kinetic study. Appl. Catal., A 206:171–181.

Boudart, M., Arrieta, J.S., and Dalla, B. 1983. Correlation between thiophene hydrodesulfurization activity and the number of first sulfur neighbors as determined by EXAFS in sulfided CoMo/γ-Al_2O_3. J. Am. Chem. Soc. 105:6501–6502.

Branco, V.A.M., Mansoori, G.A., de Almeida Xavier, L.C., Park, J.S., and Manafi, H. 2001. Asphaltene flocculation and collapse from petroleum fluids. J. Pet. Sci. Eng. 32:217–230.

Breysse, M., Furimsky, E., Kasztelan, S., Lacroix, M., and Perot, G. 2002. Hydrogen activation by transition metal sulfides. Catal. Rev.: Sci. Eng. 44:651–735.

Caloch, B., Rana, M.S., and Ancheyta, J. 2004. Improved hydrogenolysis (C–S, C–M) function with basic supported hydrodesulfurization catalysts. Catal. Today 98:91–98.

Carlsson, A., Brorson, M., and Topsøe, H. 2004. Morphology of WS_2 nanoclusters in WS_2/C hydrodesulfurization catalysts revealed by high-angle annular dark-field scanning transmission electron microscopy (HAADF-STEM) imaging. J. Catal. 227:530–536.

Choi, J., Maugé, F., Pichon, C., Olivier-Fourcade, J., Jumas, J., Petit-Clair, C., and Uzio, D. 2004. Alumina-supported cobalt-molybdenum sulfide modified by tin via surface organometallic chemistry: application to the simultaneous hydrodesulfurization of thiophenic compounds and the hydrogenation of olefins. Appl. Catal., A 267:203–216.

Colyar, J.J., and Wisdom, L.I. 1997. The H-oil process: a worldwide leader in vacuum residue processing, Proceedings of National Petroleum Refiners Association Annual Meeting, San Antonio, TX.

Cooper, B.H., Donnis, B.B.L., and Moyse, B.M. 1986. Technology: hydroprocessing conditions affect catalyst shape selection. Oil Gas J. 8:39–44.

Courty, Ph., Chamette, R., and Raimbault, C. 1999. Synthetic or reformulated fuels: a challenge for catalysis. Oil Gas Sci. Technol. 54(3):357–363.

Decroocq, D. 1997. Major scientific challenges about development of new processes in refining and petrochemistry. Rev. Inst. Fr. Pet. 52(5):469–489.

Dejonghe, S., Hubaut, R., Grimblot, J., Bonnelle, J.P., Des Courieres, T., and Faure, D. 1990. Hydrodemetallization of a vanadylporphyrin over sulfided NiMo/Al_2O_3, Mo/Al_2O_3, and Al_2O_3 catalysts – effect of the vanadium deposit on the toluene hydrogenation. Catal. Today 7(4):569–585.

Delannay, F. 1984. Characterization of heterogeneous catalysts, Marcel Dekker, Inc., New York.

Delannay, F. 1985. High resolution electron microscopy of hydrodesulfurization catalysts: a review. Appl. Catal. 16(2):i–ii.

Eijsbouts, S. 1999. Life cycle of hydroprocessing catalysts and total catalyst management. Stud. Surf. Sci. Catal. 127:21–36.

Eijsbouts, S., van den Oetelaar, L.C.A., and van Puijenbroek, R.R. 2005. MoS_2 morphology and promoter segregation in commercial Type 2 Ni–Mo/Al_2O_3 and Co–Mo/Al_2O_3 hydroprocessing catalysts. J. Catal. 229:352–364.

Falk, J.F. 1964. Porphyrin and metalloporphyrins, Elsevier Publishing Company, Amsterdam, p. 18.

Fleisch, T.H., Meyers, B.L., Hall, J.B., and Ott, G.L. 1984. Multitechnique analysis of a deactivated resid demetallation catalyst. J. Catal. 86:147–157.

Fleischer, E.B. 1963. The structure of Ni etioporphyrin: I. J. Am. Chem. Soc. 85:146–148.

Fukuyama, H., Terai, S., Uchida, M., Cano, J.L., and Ancheyta, J. 2004. Active carbon catalyst for heavy oil upgrading. Catal. Today 98:207–215.

Furimsky, E. 1998. Selection of catalysts and reactors for hydroprocessing, Appl. Catal., A 171:177–206.

Furimsky, E., and Massoth, F.E. 1993. Regeneration of hydroprocessing catalysts. Catal. Today 17:537–659.

Furimsky, E., and Massoth, F.E. 1999. Deactivation of hydroprocessing catalysts. Catal. Today 52:381–495.

Galarraga, C., and Ramírez de Agudelo, M.M. 1992. A stable catalyst for heavy oil processing. II. Preparation and characterization. J. Catal. 134(1):98–106.

Gardner, L.E., and Kukes S.G. 1989. Hydrofining employing a support material for fixed beds. US Patent 4,828,683, 9 May.

Gardner, L.E., Hogan, R.J., Sughrue, E.L., and Myers, J.W. 1987. Hydrotreating process employing a three stage catalyst system wherein a titanium compound is employed in the second stage. US Patent 4,657,663, 14 April.

Gorbaty, M.L., George, G.N., and Kelemen, S.R. 1990. Direct determination and quantification of sulphur forms in heavy petroleum and coals: 2. The sulphur K edge X-ray absorption spectroscopy approach. Fuel 69(8):945–949.

Gosselink, J.W. 1998. Sulfide catalysts in refineries. CATTECH 2(2):127–144.

Green, D.C., and Broderick, D.H. 1981. Residuum hydroprocessing in the 80s. Chem. Eng. Prog. 77(12):33–39.

Gregg, S.J., and Sing, K.S.W. 1982. Adsorption, surface area and porosity (2nd edition), Academic Press, New York.

Habib, M.M., Winslow, P.L., and Moore, R.O. 1995. Catalyst system for combined hydrotreating and hydrocracking and a process for upgrading hydrocarbonaceous feedstocks. US patents 5439,860, 8 August.

Habib, M.M., Winslow, P.L., and Moore, R.O. 1997. Catalyst system for combined hydrotreating and hydrocracking and a process for upgrading hydrocarbonaceous feedstocks. US patents 5,593,570, 14 January.

Hammer, G.P., and Clem, K.R. 1975. Ebullating bed process for hydrotreatment of heavy crudes and residua. US Patent 3,887,455, 3 June.

Hansford, R C. 1964. Silica–zirconia–titania hydrocracking catalyst. US Patent 3,159,588.

Harrison, G.E., McKinley, D.H., and Dennis, A.J. 1994. Multi-step hydrodesulphurizations process. US Patent 5,292,428, 8 March.

Hédoire, C., Louis, C., Davidson, A., Breysse, M., Maugé, F., and Vrinat, M. 2003. Support effect in hydrotreating catalysts: hydrogenation properties of molybdenum sulfide supported zeolites of various acidities. J. Catal. 220:433–441.

Hensen, E.J.M., Kooyman, P.J., van der Meer, Y., van der Kraan, A.M., de Beer, H.J., van Veen, J.A.R., and van Santen, R.A. 2001. The relation between morphology and hydrotreating activity for supported MoS_2 particles. J. Catal. 199:224–235.

Hensley, Jr A.L., and Quick, L.M. 1984. Three-catalyst process for the hydrotreating of heavy hydrocarbon streams. US Patent 4,431,525.

Hilbert, T.T., Mazzone, N.D., and Sarli, M.S. 1996. Gasoline upgrading process. US patent 5,510,016, 23 April.

Howell, J.A., and Wilson, M.E. 1987. Multi-stage hydrofining process. US patent 4,659,452.

Howell, R.L., Hung, C., Gibson, K.R., and Chen, H.C. 1985. Catalyst selection important for residuum hydroprocessing. Oil Gas J. 29:121–128.

Inoue, S., Asaoka, S., and Nakamura, M. 1998. Recent trends of industrial catalysts for resid hydroprocessing in Japan. Catal. Surv. Jpn. 2:87–97.

Jaffe, J. 1968. Coprecipitation method for making multi-component catalysts. US Patent 3,401,125.

Kabe, T., Ishihara, A., and Qian, W. 1999. Hydrodesulfurization and hydrodenitrogenation: chemistry and engineering, Wiley-VCH, New York.

Kaloidas, V., Thanos, A.M., Tsamatsoulis, D.C., and Papayannakos, N.G. 2000. Preparation of Al_2O_3 carriers in an integrated mini pilot unit. Chem. Eng. Process. 39:407–416.

Kawada, I., Onoda, M.N., Ishii, M., and Nakahira, M. 1975. Crystal structures of V_3S_4 and V_5S_8. J. Solid State Chem. 15:246–252.

Kelemen, S.R., George, G.N., and Gorbaty, M.L. 1990. Direct determination and quantification of sulphur forms in heavy petroleum and coals: 1. The X-ray photoelectron spectroscopy (XPS) approach. Fuel 69(8):939–944.

Kim, C.-S., and Massoth, F.E. 1993. Deactivation of a Ni/Mo hydrotreating catalyst by vanadium deposits. Fuel Process. Technol. 35:289–302.

Kressmann, S., Morel, F., Harlé, V., and Kasztelan, S. 1998. Recent developments in fixed-bed catalytic residue upgrading. Catal. Today 43:203–215.

Kunesh, J.G. 1982. Hydrocracking and hydrotreating shale oil in multiple catalytic reactors. US Patent 4,344,840, 17 August.

Kushiyama, S., Aizawa, R., Kobayashi, S., Koinuma, Y., and Uemasu, I. 1990. Effect of addition of sulphur and phosphorus on heavy oil hydrotreatment with dispersed molybdenum-based catalysts. Appl. Catal. 63:279–292.

Lauritsen, J.V., Bollinger, M., Lægsgaard, V.E., Jacobsen, K.W., Nørskov, J.K., Clausen, B.S., Topsøe, H., and Besenbacher, F. 2004. Atomic-scale insight into structure and morphology changes of MoS_2 nanoclusters in hydrotreating catalysts. J. Catal. 221:510–522.

Le Page, J.-F., Cosyns, J., Courty, P., Freund, E., Franck, J.-P., Jscquin, Y.J., Marcilly, B.C., Martino, G., Miquel, J., Montarnal, R., Sugier, A., and Landeghem van H. 1987. Applied heterogeneous catalysis, design, manufacture, use of solid catalysts, Eds. Technip, Paris.

Ledoux, M.J., and Hantzer, S. 1990. Hydrotreatment catalyst poisoning by vanadium and nickel porphyrin: ESR and NMR. Catal. Today 7(4):479–496.

Ledoux, M.J., Mchaux, O., Hantzer, P., Panissod, P.., Patit, P., André, J.J., and Callot, H.J. 1987. Hydrodesulfurization (HDS) poisoning by vanadium compounds: EPR and metal solid NMR analysis. J. Catal. 106:525–537.

Leofanti, G., Tozzola, G., Padovan, M., Petrini, G., Bordiga, S., and Zecchina, A. 1994. Catalyst characterization. in: D. Sanfilipo (Ed.) The catalytic process from laboratory to the industrial plant, Italian Chemical Society, Rimini.

Lowell, S. 1980. Continuous scan mercury porosimetry and the pore potential as a factor in porosimetry hysteresis. Powder Technol. 25(1):37–43.

Luck, F. 1991. A review of support effects on the activity and selectivity of hydrotreating catalysts. Bull. Soc. Chim. Belg. 100(11–12):781–800.

Maity, S.K., Ancheyta, J., Soberanis, L. Alonso, F., and Llanos, M.E. 2003a. Alumina–titania binary mixed oxide used as support of catalysts for hydrotreating of Maya heavy crude. Appl. Catal., A 244:141–153.

Maity, S.K., Ancheyta, J., Soberanis, L., and Alonso, F. 2003b. Alumina–silica binary mixed oxide used as support of catalysts for hydrotreating of Maya heavy crude. Appl. Catal., A 250:231–238.

Maity, S.K., Ancheyta, J., Soberanis, L., and Alonso, F. 2003c. Catalysts for hydroprocessing of Maya heavy crude. Appl. Catal., A 253:125–134.

Maity, S.K., Ancheyta, J., Alonso, F., and Rana, M.S. 2004. Preparation, characterization and evaluation of Maya crude hydroprocessing catalysts. Catal. Today 98:193–199.

Maity, S.K., Ancheyta, J., and Rana, M.S. 2005. Support effects on hydroprocessing of Maya heavy crude. Energy Fuels 19:343–347.

Maity, S.K., Ancheyta, J., Rana, M.S., and Rayo, P. 2006. Alumina–titania mixed oxide used as support for hydrotreating catalysts of Maya heavy crudes: effect of support preparation methods. Energy Fuels 20:427–431.

Mansoori, G.A. 1997. Modeling of asphaltene and other heavy organic depositions. J. Pet. Sci. Eng. 17:101–111.

Marroquín, G., Ancheyta, J., and Díaz, J.A.I. 2004. On the effect of reaction conditions on liquid phase sulfiding of a NiMo HDS catalyst. Catal. Today 98:75–81.

Massoth, F.E., and Muralidhar, G. 1982. Hydrodesulfurization catalysis, in: Proc. 4th Int. Conf. On Chemistry and uses of Molybdenum (H.F. Barry and P.C.H. Mitchell Eds), Climax Molybdenum Co., Ann Arbor, Michigan, 343–351.

McDaniel, M.P., and Hottovy, T.D. 1980. Total porosity of high pore-volume silicas by liquid adsorption. J. Colloid Interface Sci. 78:31–36.

Mitchell, P.C.H., and Valero, J.A. 1982. Hydrodemetallization and promoter effect of vanadium compounds on a Co–Mo/Al_2O_3 hydrodesulfurization. React. Kinet. Catal. Lett. 20:219–225.

Mitchell, P.C.H., Scott, C.S., Bonnelle, J.P., and Grimblot, J. 1985. Binding and decomposition of oxovanadium(IV) phthalocyanine, tetraphenylporphyrin and etioporphyrin on hydrotreating catalysts studied by X-ray photoelectron and ultraviolet–visible spectroscopies. J. Chem. Soc., Faraday Trans. 1 81:1047–1056.

Montarnal, R., and Le Page, J.F. 1967. La catalyse au laboratoire et dans l'industrie, Mason, Paris.

Morales, A., and Galiasso, R. 1982. Adsorption mechanism of Boscan porphyrins on MoO_3–Co_3O_4 and CoMo/Al_2O_3. Fuel 61:13–17.

Morel, F., Kressmann, S., Harlè, V., and Kasztelan, S. 1997. Process and catalyst for hydrocracking of heavy oil and residue. Stud. Surf. Sci. Catal. 106:1–16.

Moulijn, J.A., van Leeuwen, P.V.N.M., and van Santen, R.A. 1993. An integrated approach to homogeneous, heterogeneous and industrial catalysis, Elsevier, New York.

Moulijn, J.A., van Diepen, A.E., and Kapteijn, F. 2001. Catalyst deactivation: is it predictable? What to do? Appl. Catal., A 212:3–16.

Mounce, W. 1974. Hydrocracking and hydrodesulfurization process. US Patent 3,830,728, 20 August.

Muegge, B.D., and Massoth, F.E. 1991. Basic studies of deactivation of hydrotreating catalysts with anthracene. Fuel Process. Technol. 29:19–30.

Mullins, O.L. 1995. Sulfur and nitrogen molecular structures in asphaltenes and related materials quantified by XANES spectroscopy, in: Asphaltenes: fundamentals and applications (E.Y. Sheu and O.C. Mullins, Eds) Plenum Press, New York, Chapter II:53–96.

Nishijima, A., Shimada, H., Sato, T., Yoshimura, Y., and Hiraishi, J. 1986. Support effects on hydrocracking and hydrogenation activities of molybdenum catalysts used for upgrading coal-derived liquids. Polyhedron 5(1–2):243–247.

Nomura, H., Sekido, Y., and Oguchi, Y. 1980. Hydrodesulfurization reactions of residual oils (Part 3) Effects of various factors on catalyst life. Sekiyu Gakkaishi 23(5):321–327.

Ono, T., Ohguchi, Y., and Togari, O. 1983. Preparation of catalyst III (G. Poncelet, P. Grange and P. Jacob Eds), Elsevier, Amsterdam.

Oyama, S.T., Wang, X., Lee, Y.K., and Chun, W.J. 2004. Active phase of Ni_2P/SiO_2 in hydroprocessing reactions. J. Catal. 221:263–273.

Pecoraro, T.A., and Chianelli, R.R. 1981. Hydrodesulfurization catalysis by transition metal sulfides. J. Catal. 67:430–445.

Perego, C., and Villa, P. 1994. Catalyst preparation methods, in: The catalytic process from laboratory to the industrial plant (D. Sanfilipo Ed.) Italian Chemical Society, Rimini.

Peries, J.P., Billon, A., Hennico, A., Morrison, E., and Morel, F. 1986. Inst. Fr. du Petr. Rapport IFP No. 34297, June.

Portela, L., Grange, P., and Delmon, B. 1995. The adsorption of nitric oxide on supported Co–Mo hydrodesulfurization catalysts: a review. Catal. Rev.: Sci. Eng. 37(4):699–731.

Prins, R. 1992. Characterization of catalytic materials (I.E. Wachs and L.E. Fitzpatrick Eds), Chapter 6, Butterworth-Heinemann, USA.

Prins, R., de Beer, V.H.J., and Somorjai, G.A. 1989. Structure and function of the catalyst and the promoter in Co–Mo hydrodesulfurization catalyst. Catal. Rev.: Sci. Eng. 31:1–41.

Ramírez, J., and Gutiérrez-Alejandre, A. 1998. Relationship between hydrodesulfurization activity and morphological and structural changes in NiW hydrotreating catalysts supported on Al_2O_3–TiO_2 mixed oxides. Catal. Today 43:123–133.

Ramírez, J., Macías, G., and Cedeño, L., Gutiérrez-Alejandre, A., Cuevas, R., and Castillo, P. 2004. The role of titania in supported Mo, CoMo, NiMo and NiW hydrodesulfurization catalysts: analysis of past and new evidences. Catal. Today 98:19–30.

Ramírez, J., Rayo, P., Gutierrez–Alejandre, A., Ancheyta, J., and Rana, M.S. 2005. Analysis of the hydrotreatment of heavy Maya petroleum with NiMo catalysts supported on TiO_2–Al_2O_3 binary oxides. Effect of the incorporation method of Ti, Ni and Mo. Catal. Today 109:54–60.

Rana, M.S., Srinivas, B.N., Maity, S.K., Murali Dhar, G., and Prasada Rao, T.S.P.R. 2000. Origin of cracking functionality of sulfided (Ni) CoMo/SiO_2–ZrO_2 catalysts. J. Catal. 195:31–37.

Rana, M.S., Ancheyta, J., Rayo, P., and Maity, S.K. 2004a. Effect of alumina preparation on hydrodemetallization and hydrodesulfurization of Maya crude. Catal. Today 98:151–160.

Rana, M.S., Navarro, R., and Leglise, J. 2004b. Competitive effects of nitrogen and sulfur content on activity of hydrotreating CoMo/Al_2O_3 catalysts: a batch reactor study. Catal. Today 98:67–74.

Rana, M.S., Maity, S.K., Ancheyta, J., Murali Dhar, G., Prasada Rao, T.S.R. 2004c. Cumene cracking functionalities on sulfided Co(Ni)Mo/TiO_2–SiO_2 catalysts. Appl. Catal., A 258:215–225.

Rana, M.S., Ancheyta, J., and Rayo, P. 2005a. A comparative study for heavy oil hydroprocessing catalysts at microflow and bench-scale reactors. Catal. Today 109:24–32.

Rana, M.S., Ancheyta, J., Maity, S.K., and Rayo, P. 2005b. Maya crude hydrodemetallization and hydrodesulfurization catalysts: an effect of TiO_2 incorporation in Al_2O_3. Catal. Today 109:61–68.

Rana, M.S., Huidobro, M.L., Ancheyta, J., and Gómez, M.T. 2005c. Effect of support composition on hydrogenolysis of thiophene and Maya crude. Catal. Today 107–108:346–354.

Rana, M.S. Ancheyta, J., Maity, S.K., and Rayo, P. 2005d. Characteristics of Maya crude hydrodemetallization and hydrodesulfurization catalysts. Catal. Today 104:86–93.

Rana, M.S. Ancheyta, J., Rayo, P., and Maity, S.K. 2007a. Heavy oil hydroprocessing over supported NiMo sulfided catalyst: an inhibition effect by added H_2S. Fuel 86 (9): 1263–1269.

Rana, M.S. Ancheyta, J., Maity, S.K., and Rayo, P. 2007b. Hydrotreating of Maya crude oil: I. Effect of support composition and its pore-diameter on asphaltene conversion. Pet. Sci. Technol. 25:187–199.

Rankel, L.A. 1981. Reactions of metalloporphyrins and petroporphyrins with H_2S and H_2. Prepr. – Am. Chem. Soc., Div. Pet. Chem. 26:689–698.

Rayo, P. Ancheyta, J. Ramírez, J., and Gutierrez-Alejandre, A. 2004. Hydrotreating of diluted Maya crude with NiMo/Al_2O_3–TiO_2 catalysts: effect of diluent composition. Catal. Today 98:171–179.

Rayo P., Ancheyta, J., Ramírez, J., Maity, S.K., Rana, M.S., and Martínez, F.A. 2011. Catalyst, its preparation and use for hydrodesulfurization of residua and heavy crudes. US patent 7,968,069 B2.

Redey, A., Goldwasser, J., and Hall, W.K. 1988. The surface chemistry of molybdena-alumina catalysts reduced in H_2 at elevated temperatures. J. Catal. 113:82–95.

Richardson, J.T. 1989. Principles of catalyst development, Plenum Press, New York.

Ritter, H.L., and Drake, L.C. 1945. Pore-size distribution in porous materials. Pressure porosimeter and determination of complete macropore-size distribution. Ind. Eng. Chem., Anal. Ed. 17:782–786.

Rojas, F., Kornhauser, I., Felipe, C., Esparza, J.M., Cordero, S., Domínguez, A., and Ricardo, J.L. 2002. Capillary condensation in heterogeneous mesoporous networks consisting of variable connectivity and pore-size correlation. Phys. Chem. Chem. Phys. 4:2346–2355.

Ruiz, R.S., Alonso, F., and Ancheyta, J. 2004. Minimum fluidization velocity and bed expansion characteristics of hydrotreating catalysts in ebullated-bed systems. Energy Fuels 18(4):1149–1155.

Ruiz, R.S., Alonso, F., and Ancheyta, J. 2005. Pressure and temperature effects on the hydrodynamic characteristics of ebullated-bed systems. Catal. Today 109:205–213.

Scherzer, J. 1990. Octane-enhancing zeolite FCC catalysts, scientific and technical aspects, Marcel Dekker, Inc., New York and Basel.

Scott, A., Stevenson-Dumesic, J.A., and Baker, R.T.K. 1987. Metal support interaction in catalysis, sintering and redispersion, (E. Ruckenstein Ed.), Van Nostrand Reinhold Company, New York.

Shirota, Y., Fukui, Y., Ando, M., and Homma, Y. 1979. Two-step hydrodesulfurization of heavy hydrocarbon oil. US Patent 4,166,026, 28 August.

Smith, B.J., and Wei, J. 1991a. Deactivation in catalytic hydrodemetallation: I. Model compound kinetic studies. J. Catal. 132:1–20.

Smith, B.J., and Wei, J. 1991b. Deactivation in catalytic hydrodemetallation: II. Catalyst characterization. J. Catal. 132:21–40.

Smith, B.J., and Wei, J. 1991c. Deactivation in catalytic hydrodemetallation: III. Random-spheres catalyst models. J. Catal. 132:41–57.

Snel, R. 1984a. Control of the porous structure of amorphous silica–alumina: I. The effects of sodium ions and syneresis. Appl. Catal. 11:271–280.

Snel, R. 1984b. Control of the porous structure of amorphous silica–alumina: II. The effects of pH and reactant concentration. Appl. Catal. 12:189–200.

Snel, R. 1984c. Control of the porous structure of amorphous silica–alumina: III. The influence of pore regulating agents. Appl. Catal. 12:347–357.

Stanislaus, A., Abasi-Halabi, M., and Al-Dolama, K. 1988. Effect of phosphorus on the acidity of γ-alumina and on the thermal stability of γ-alumina supported nickel–molybdenum hydrotreating catalysts. Appl. Catal. 39:239–253.

Stiles, A.B. 1983. Catalyst manufacturer, Marcel Dekker, New York.

Takatsuka, T., Wada, Y., Hirohama, S., and Komatsu, S. 1988. More insights into the in situ liquid phase equilibrium in the ABC processes. AIChE Annual Meeting, Washington DC, Nov. 27–Dec. 2, 1988.

Takeuchi, C., Asaoka, S., Nakata, S.-i., and Shiroto, Y. 1985. Characteristics of residue hydrodemetallation catalyst. Prepr. – Am. Chem. Soc., Div. Pet. Chem. 30(1):96–107.

Tamm, P.W., Harnsberger, H.F., and Bridge, A.G. 1981. Effects of feed metals on catalyst aging in hydroprocessing residuum. Ind. Eng. Chem. Process Des. Dev. 20:262–273.

Tanaka, R., Hunt, J.E., Winans, R.E., Thiyagarajan, P., Sato, S., and Takanohashi, T. 2003. Aggregates structure analysis of petroleum asphaltenes with small-angle neutron scattering. Energy Fuels 17:127–134.

Thakur, D.S., and Thomas, M.G. 1984. Catalyst deactivation during direct coal liquefaction: a review. Ind. Eng. Chem. Prod. Res. Dev. 23:349–360.

Thakur, D.S., and Thomas, M.G. 1985. Catalyst deactivation in heavy petroleum and synthetic crude processing: a review. Appl. Catal. 15:197–225.

Topsøe, N.-Y. 2006. In situ FTIR: a versatile tool for the study of industrial catalysts. Catal. Today 113:58–64.

Topsøe, N.-Y., and Topsøe, H. 1993. FTIR studies of Mo/Al_2O_3-based catalysts: II. Evidence for the presence of SH groups and their role in acidity and activity. J. Catal. 139:641–651.

Topsøe, H.; Clausen, S., and Massoth, F.E. 1996. Hydrotreating catalysis science and technology, Springer-Verlag, New York.

Toulhoat, H., Plumail, J.C., Houpert, C., Szymanski, R., Bourseau, P., and Muratet, G. 1987. Modelling RDM catalyst deactivation by metal sulfides deposit: an original approach supported by HRTEM investigation and pilot test results. Prepr. – Am. Chem. Soc., Div. Pet. Chem. 32(2):463.

Toulhoat, H., Szymanski, R., and Plumail, J.C. 1990. Interrelations between initial pore structure, morphology and distribution of accumulated deposits, and lifetimes of hydrodemetallisation catalysts. Catal. Today 7:531–568.

Travert, A., Dujardin, C., Maugé, F., Cristol, S., Paul, J.F., Payen, E., and Bougeard, D. 2001. Parallel between infrared characterisation and ab initio calculations of CO adsorption on sulphided Mo catalysts. Catal. Today 70:255–269.

Trimm, D.L., and Stanislaus, A. 1986. The control of pore size in alumina catalyst supports: a review. Appl. Catal. 21:215–238.

van Veen, J.A.R. 2002. Hydrocracking, in: Zeolites for cleaner technologies, (M. Guisnet and J.-P. Gilson Eds), Imperial College Press, London.

Wachs, I.E. (Ed.). 1992. Characterization of catalytic materials, Butterworth-Heinemann, USA.

Wachs, I.E., Deo, G., Kim, D.S., Vuurman, M.A., and Hu, H. 1993. Molecular design of supported metal oxide catalysts, in: New frontiers in catalysis, Proc. 10th Int. Cong. on Catal. (L. Guczi, F. Solymosi, P. Tetenyi, Eds) Akademia Kiado, Budapest.

Walendziewski, J., and Trawczynski, J. 1993. Preparation of large pore alumina supports for hydrodesulfurization catalysts. Appl. Catal., A 96:163–174.

Walendziewski, J. and Trawczynski, J. 1994. Influence of the forming method on the pore structure of alumina supports. Appl. Catal., A 119:45–58.

Ware, R.A., and Wei, J. 1985. Catalytic hydrodemetallation of nickel porphyrins: II. Effects of pyridine and of sulfiding. J. Catal. 93:122–134.

Weissman, J.G., and Edwards, J.C. 1996. Characterization and aging of hydrotreating catalysts exposed to industrial processing conditions. Appl. Catal., A 142:289.

Weitkamp, J., Gerhardt, W., and Scholl, D. 1984. Hydrodemetalation of nickel porphyrins over sulfided and reduced $CoO–MoO_3/\gamma\text{-}Al_2O_3$. 8th Int. Cong. Catal., Berlin, II269–II280.

Wijngaarden, R.J., Kronberg, A., and Westerterp, K.R. 1998. Industrial catalysis, Wiley-VCH, New York.

Wolk, R.H., and Rovesti, W.C. 1974. Low sulfur fuel oil from high metals containing petroleum residuum. US patent 3,819509, 25 June.

Wolk, R.H., Nogbri, G., and Rovesti, W.C. 1975. Multi-zone method for demetallization and desulfurizing crude oil or atmospheric residual oil. US patent 3,901,792, 26 August.

Yoshimura, Y., Endo, S., Yoshitomi, S., Sato, T., Shimada, H., Matsubayashi, N., and Nishijima, A. 1991. Deactivation of hydrotreating molybdate catalysts by metal deposition. Fuel 70:733–739.

Zeuthen, P., Blom, P., and Massoth, F.E. 1991. Characterization of nitrogen on aged hydroprocessing catalysts by temperature-programmed oxidation. Appl. Catal. 78:265–276.

Zmierczak, W., Murali Dhar, G., and Massoth, F.E. 1982. Studies on molybdena catalysts: XI. Oxygen chemisorption on sulfided catalysts. J. Catal. 77:432–438.

3

DEACTIVATION OF HYDROPROCESSING CATALYSTS

3.1 INTRODUCTION

Deactivation of catalysts used for hydroprocessing of heavy oil is mainly due to the formation of carbonaceous deposits and metal deposition, which block the pore mouth leaving unutilized catalytic active sites. Coke is the carbonaceous deposit remaining after washing the spent catalyst with a solvent for removing light paraffins and adsorbed hydrocarbons. During hydroprocessing, asphaltenes are mainly responsible for coke formation. Adsorption of asphaltenes into the catalytic sites is due to the presence of nitrogen compounds originating condensation and dehydrogenation reactions of those compounds to finally form coke on the catalyst. Several authors have reported the presence of nitrogen in coke. Furimsky (1978) concluded that adsorption of nitrogen compounds is stronger compared with other hydrocarbons and adsorption is carried out by the interaction of the unpaired electrons of N with Lewis sites. As a consequence of adsorption of N-compound, coke is rich in nitrogen and the C/N atomic ratio increases in coke compared with the feedstock. Satterfield et al. (1978) stated that nitrogen-derivative compounds were present on the surface catalyst, and Choi and Gray (1988) also observed the presence of N-compounds (quinolones and amides) on the catalyst. According to Speight (1987), nitrogen remains in coke as nonvolatile compound after thermal reaction of asphaltenes, and Reynolds (1998) pointed out that the majority of sulfur, nitrogen, and metals, which are the most important asphaltenes constituents, are deposited on coke when thermal decomposition of asphaltenes is carried out. Dong et al. (1997) concluded that the presence of N-compounds (acridine and carbazole) produced a high content of coke on a $CoMo/Al_2O_3$ catalyst.

Deactivation of Heavy Oil Hydroprocessing Catalysts: Fundamentals and Modeling, First Edition. Jorge Ancheyta.

The deposition of carbonaceous material can be classified as specific and nonspecific depending on the location of carbon deposits with respect to the catalytic active sites. Van Doorn and Moulijn (1990) reported the formation of filamentary carbon on Co and Ni supported on Al_2O_3 catalysts and stated that coke deposition on spent hydroprocessing catalysts is of a three-dimensional carbonaceous structure. Traditionally, it is considered that coke is primarily responsible for initial deactivation of the catalyst, that is, during the first hours-on-stream, while continued deactivation has been attributed to metal deposition. Nielsen et al. (1981) have stated that distribution of coke and metals throughout the pore system is the main factor to determine the deactivation rate. According to these authors, there are optimal conditions; that is, partial pressure of hydrogen, catalyst particle size, and quality of the feedstock, for maximum utilization of catalysts. However, the optimum catalytic performance is not depending on only one type of catalyst. It is desirable to have different types of catalysts during hydroprocessing of heavy oil, where a special catalyst designed for metal removal is placed at the top of the bed or in the first reactors and good catalysts for HDS/hydrocracking are situated at the bottom of the bed or in the last reactors.

3.2 HYDROPROCESSING OF HEAVY OILS

3.2.1 General Aspects

Catalytic hydroprocessing is extensively applied in petroleum refining industry to remove impurities (sulfur, nitrogen, oxygen, polynuclear aromatics) and metal-containing compounds (V and Ni, mainly). The concentration of these impurities increases as the boiling point of the petroleum fraction also does. High-molecular-weight feedstocks, that is, heavy oils, contain the same impurities in higher concentrations as well as complex V- and Ni-containing compounds and asphaltenes.

Severity of reaction conditions depends on the type of feed and the desired quality of the product. In general, the higher the feed boiling point, the higher the reaction severity.

To carry out the hydroprocessing reaction, there are different types of reactors, which are similar in terms of principle of operation, but they differ in some technical details.

- Fixed-bed reactor (FBR)
- Moving-bed reactor (MBR)
- Expanded- or ebullated-bed reactor (EBR)
- Slurry-phase reactor (SPR).

FBR is more utilized for processing of light feeds, such as naphtha and middle distillates, but under certain conditions and process configuration, it is also used for hydroprocessing of heavier oils. It must be highlighted that when the feed contains high amount of metals and asphaltenes, the use of FBR has to be carefully examined according to catalyst deactivation and cycle life. Alternatively, MBR and EBR have demonstrated reliable operation with difficult feeds. Life of catalyst is crucial to retain its activity and selectivity for some time. Depending on the feed, the catalyst life may vary in the order of months or a year. It is then clear that the timescale of deactivation influences the choice of reactor (Moulijn et al., 2001).

Hydroprocessing of heavy oils is strongly influenced by the method of feed introduction, arrangement of the catalytic beds, and mode of operation of the reactors. Hence, proper selection and design of reactors are very important for this.

In FBR, the liquid hydrocarbon trickles down through the fixed catalyst bed from top to bottom of the reactor. Hydrogen gas passes cocurrently through the bed. A single tailor-made optimum catalyst having high hydrodesulfurization activity and low metal tolerance can handle feedstocks having less than 25 wppm of metal for a cycle length of 1 year. For feedstocks containing metals in the range of 25–50 wppm, a dual-catalyst system is more effective. In such a system, one catalyst having a higher metal tolerance is placed in the front of the reactor, whereas the second catalyst located in the tail end of the reactor is generally of higher desulfurization activity (Scheffer et al., 1988; Kressmann et al., 1998). A triple catalyst system consisting of an HDM, a balanced HDM/HDS, and a refining catalyst is generally used to handle feedstocks having metal content in the range of 100–150 wppm for an average cycle length of 1 year. For feeds with metal content higher than this range, the HDM catalyst gets quickly exhausted, and for achieving a cycle length of 1 year, a new swing FBR concept has been introduced by IFP (Institute Français du Petrole). The IFP process includes two swing fixed-bed HDM reactors that operate in a switchable mode and the catalyst can be unloaded/reloaded without disturbing the operation of the system. The swing reactors are generally followed by various FBR in series containing HDS and other hydrofining catalysts.

To overcome the problems of handling feeds having metal content higher than 150 wppm, in addition to the concept of swing FBR, an MBR involving continuous withdrawal of deactivated catalyst and simultaneous addition of fresh catalyst has been introduced (Topsøe et al., 1996). The development of Shell bunker reactor technology is an example of such a system, in which the fresh catalyst and reacting fluids (both hydrocarbon and hydrogen) move in a cocurrent downflow mode and a special arrangement of internals inside the reactor allows withdrawal of the spent catalyst from the bottom of the reactor (Kressmann et al., 1998). In another design, as developed by Chevron in their Onstream Catalyst Replacement (OCR) technology (Reynolds et al., 1992), the catalyst moves in a downflow mode, while the reacting fluid moves countercurrently from the bottom to top of the reactor. A similar system has been designed by ASVAHL group (Agrawal and Wei, 1984). In all these cases, a combination of MBR followed by FBR is used. The MBR containing the HDM catalyst is placed first, which is followed by other FBR containing HDS and hydrofining catalysts.

Another alternative for handling metal-rich feedstocks, besides the use of swing guard FBR and MBR, is EBR. Reactants are introduced at the bottom and the product is taken out from the top. The catalyst bed is kept expanded by the upward flow of reactants. The spent catalyst is continuously withdrawn and the fresh catalyst is added without disturbing the operation and thereby providing a high onstream factor. SPR can also be used for hydroprocessing of heavy oil, but it has limited commercial experience.

3.2.2 Reactors for Hydroprocessing

3.2.2.1 ***Fixed-Bed Reactor*** FBR is the most used reactor system in commercial hydroprocessing. Except for naphtha, for which the reaction is carried out in two phases (gas–solid) because at the reaction conditions naphtha is completely vaporized, for heavier feeds three phases are commonly found (hydrogen, a liquid–gas mixture of the partially vaporized feed, and the solid catalyst). This latter system is called trickle-bed reactor (TBR), a reactor in which a liquid phase and a gas phase flow cocurrently downward through a fixed bed of catalyst particles while reactions take place (Rodriguez and Ancheyta, 2004). The name "trickle-bed reactor" originates then from this flow pattern. The gas is the continuous phase and the liquid is the disperse phase (Quann et al., 1988). A schematic representation of the phenomena occurring in a TBR based on the three-film theory is presented in Figure 3.1 (Korsten and Hoffmann, 1996; Bhaskar et al., 2004).

When hydroprocessing heavy oils, multiple catalyst beds with cooling in between are used (Robinson and Dolbear, 2006). Multibed configuration with a hydrogen quench system is usually employed. Figure 3.2 shows an example of an FBR with three catalytic beds and hydrogen quenches, in which feed distributor, quench zones, catalytic beds, and catalyst support are clearly indicated. Because hydroprocessing reactions are exothermic, controlling reaction temperature becomes very important in order to prevent carbon deposition on the catalyst and preserve the quality of products at desired levels. The common method of controlling the reaction temperature is by combining the hot process fluids from the preceding bed with relatively cold hydrogen-rich gas before the mixture passes into the next bed, that is, between the catalytic beds, the so-called "quench" (Robinson and Dolbear, 2006). The problem when quenching with hydrogen is its availability in refineries. Hydrogen is also needed to keep the hydrogen-to-oil ratio (H_2/oil) along the reactors. Quenching with hydrogen takes place in some part of the reactor length and has two main functions: control of reaction temperature and improvement of flow distribution in the reactor bed. The quench zone is commonly a mixing chamber, where the bed effluent is mixed with the hydrogen recycle stream (Muñoz et al., 2005). On the other hand, it is well known that a poor distribution design impacts on the catalyst utilization efficiency and to minimize this problem a liquid distributor should have high distribution element density, low sensitivity to tray out-of-levelness, low pressure drop, large spray angle, turndown flexibility, and be easy for cleaning (Chou, 2004).

It should be mentioned that for gas-phase reactors distribution trays are not necessary, but are important for TBR . The general benefits of using reactor internals are (Sarli et al., 1993; Ouwerkerk et al., 1999; Gorshteyn et al., 2002) to provide optimal conditions for catalyst performance in terms of liquid/gas distribution, inter- and intraphase mixing, quenching performance, and fouling resistance; to be space efficient and easy to maintain; equal distribution of liquid over the catalyst to achieve

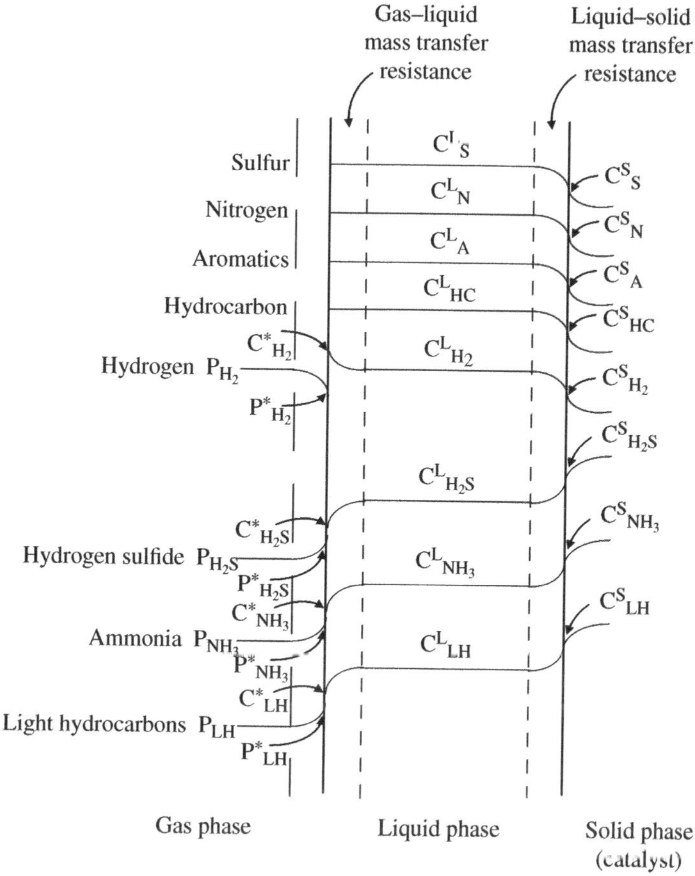

Figure 3.1 Profiles of concentration in a hydroprocessing TBR (Adapted from Korsten and Hoffmann (1996) and Bhaskar et al. (2004)).

maximum utilization of the inventory; to avoid potential risks formed by stagnant zones and channeling; to extend the cycle length by reducing the average bed temperature and/or deactivation rate; improved product quality by reducing the risk of dry spots in the catalyst bed; enhanced reactant distribution and uniform radial temperatures; improved temperature control; and design of large diameter reactors.

During FBR hydroprocessing of heavy oils (with either single catalyst systems or multiple catalyst systems, in one or more reactors in series, and sometimes several parallel trains of several reactors in series), it is common to use guard materials at the top of the reactor or in a separate vessel before the main hydrotreating reactors to catch foulants. Other particulates present in the feed can also settle in the catalysts, causing catalytic bed plugging. As a result, pressure drop increases and reactor performance declines, which eventually cause shutdown of hydroprocessing plants. Pressure drop in hydroprocessing units is usually the result of accumulation of a dense layer of particulates or formation of gum from reactive species in the feed. In

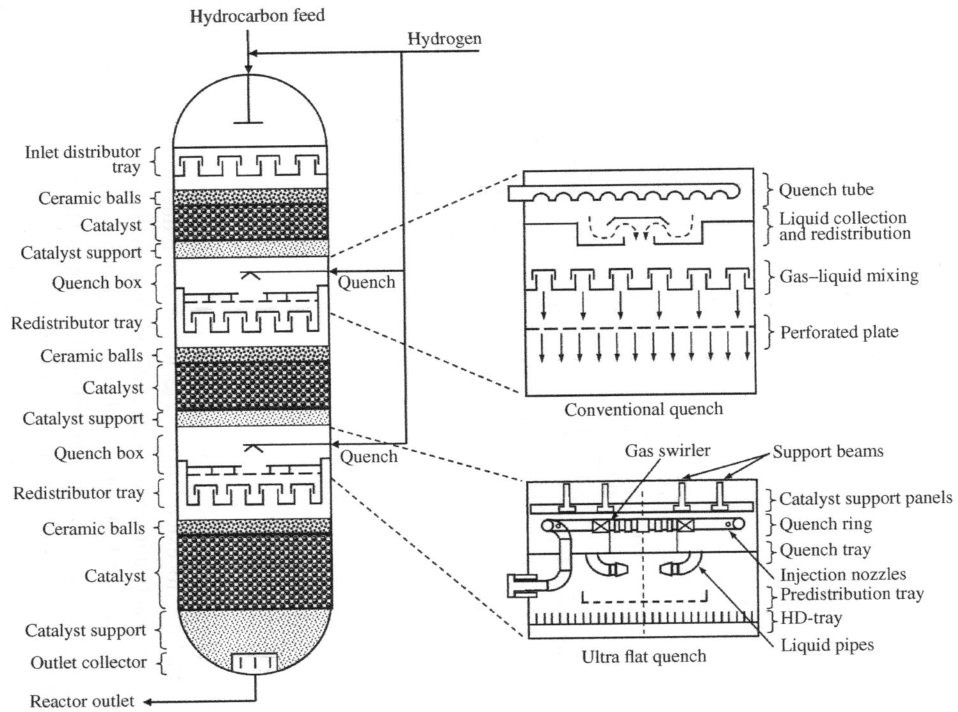

Figure 3.2 FBR with multicatalytic bed and different quench technologies (Adapted from Ouwerkerk et al. (1999), Minderhoud et al. (1999), and Robinson and Dolbear (2006)).

other words, ΔP develops when the particulates layer reduces the void fraction in the catalyst bed during time. For heavy oils, proper bed grading can be used to increase the cycle length and prevent premature shutdown. The control of particulates is done by using large diameter and large void fraction catalysts to spread out the zone of deposition of particulates. Each layer of catalyst, with particular shape and size, collects particulates within a certain size range and prevents formation of a dense layer of particulate accumulation so that pressure drop due to gum formation is prevented. Macroporous materials and bed grading substantially reduce the pressure drop problems, and as consequence extend the life of the catalyst bed and eliminate the need for frequent shutdown to skim the head part of the bed to remove crust, plugs, and agglomerates of catalyst particles, and replace fouled catalyst. The main disadvantage of using bed grading is the loss of reactor volume for loading the catalyst since a part of the grading material can be inert. However, the use of materials with some catalytic activity mitigates these effects. Hence, finding a right balance between fouling prevention and preserving a sufficient overall catalyst activity is the key issue in optimizing bed grading (Minderhoud et al., 1999).

There are other technologies based on FBR, such as quick catalyst replacement (QCR) reactor (Van Zijll Langhout et al., 1980), radial-flow FBR, and upflow reactor (UFR; Speight, 2000). However, among the different reactors based on fixed-bed system, TBR downflow FBR is the most used approach for hydroprocessing, whose advantages and disadvantages are as follows (Gianetto and Specchia, 1992; Al-Dahhan et al., 1997):

- *Advantages*. Liquid flow behaves mainly as piston flow, very little backmixing; low catalyst loss, no catalyst attrition; no moving parts; flexibility of operation at high pressure and temperature; larger reactor sizes; low investment and operating costs; high catalyst loading per unit volume of the liquid; and low energy dissipation rate.
- *Disadvantages*. Lower catalyst effectiveness because of the use of large catalyst particle size; long diffusion distance; high pressure drop; inappropriate catalyst wetting with low liquid flow rates; possibility of liquid maldistribution, which may give rise to hot spots and reactor runaway; short catalyst cycle life due to impracticality for reactions with rapid deactivation catalysts. This can be overcome by using a guard-bed reactor in order to minimize the metal and coke deposition on the downstream reactors.

The problem of short catalyst cycle life, which is the main reason for declining the use of FBR when feeds have high amount of metals, can be solved by the association of adequate HDM and HDS catalysts as well as appropriate selection of reaction conditions and feed preparation (Ancheyta et al., 2010).

3.2.2.2 Moving-Bed Reactor In MBR, the catalyst goes in downflow through the reactor by gravitational forces. The fresh catalyst enters at the top of the reactor and the deactivated catalyst leaves the reactor at the bottom, while the hydrocarbon goes either in counter- or cocurrent flow through the reactor. With this moving-bed system, the catalyst can be replaced either continuously or discontinuously (Gosselink, 1998).

The main differences and advantages of MBR over FBR are as follows: in MBR the catalysts are homogeneously loaded by metal sulfides, while in FBR a profile exists over the catalyst bed; higher catalyst utilization in MBR; when using more than one reactor, for example, MBR combined with FBR, this configuration enables onstream replacement of one catalyst system by another, thus increasing the flexibility of the unit; although catalysts used in MBR are chemically quite similar to those of FBR, their mechanical strengths and shapes (i.e., strong, attrition-resistant catalyst) should meet the more demanding situation, for example, severe grinding and abrasion; in MBR, the top layer of the moving bed consists of fresh catalyst, and metals deposited on the top move downward with the catalyst and are released at the bottom; tolerance to metals and other contaminants is much greater in MBR compared with FBR.

One example of MBR is the bunker reactor used in HYCON process, which enables easy catalyst replacement without interrupting operation by means of valves of lock hoppers. The catalyst and heavy oil are fed in cocurrent flow; the fresh catalyst enters at the top of the reactor, while the deactivated catalyst is removed from the bottom (Figure 3.3). A special design of the reactor internals ensures that the catalyst can move down in plug flow. Special screens separate the catalysts from the process fluids before leaving the reactor. Catalyst replacement frequency depends on the deactivation rate and is done at a rate to ensure a total plant run time of at least one year, which depends mainly on the metal content in the feed. For feeds containing high amount of metals, metal sulfides can be better accommodated onto the catalyst in bunker-flow reactor compared with other reactors. Operating conditions and catalyst addition/withdrawal rate can be also adjusted to ensure that the catalyst taken out is completely spent, while still retaining an acceptable average activity in the reactor (Sie, 2001).

OCR process is another option for hydroprocessing heavy oils with significant amount of metals. OCR is an MBR operating in a countercurrent mode at high temperature and pressure. Fresh catalyst is added at the top of the reactor and the feed into the bottom, and both move through the reactor in a countercurrent flow, causing the dirtiest feed to contact the oldest catalyst first. The fresh catalyst can be added at the top of the reactor and the spent catalyst removed from the OCR reactor while the unit is onstream (Figure 3.3). Its main application is before FBR so that heavier feeds with higher levels of impurities can be processed, while maintaining the product quality and economical FBR run lengths constant (Scheuerman et al., 1993).

Another alternative of moving systems, with feed and catalyst flowing countercurrent, is the MBR used in Hyvahl-M process (Euzen, 1991; Billon et al., 1991), which requires special equipment and procedures for safe and effective catalyst transfer into and out of the high pressure unit, similar to the OCR reactor. Catalyst is taken at atmospheric pressure and transferred to a reactor operating under hydrogen pressure, and then the catalyst is taken from the reactor at the high operating conditions and discharged to the atmosphere.

Catalyst replacement in MBR is usually a batch operation, which is done typically once or twice a week. Adding and removing the catalyst is the most critical section.

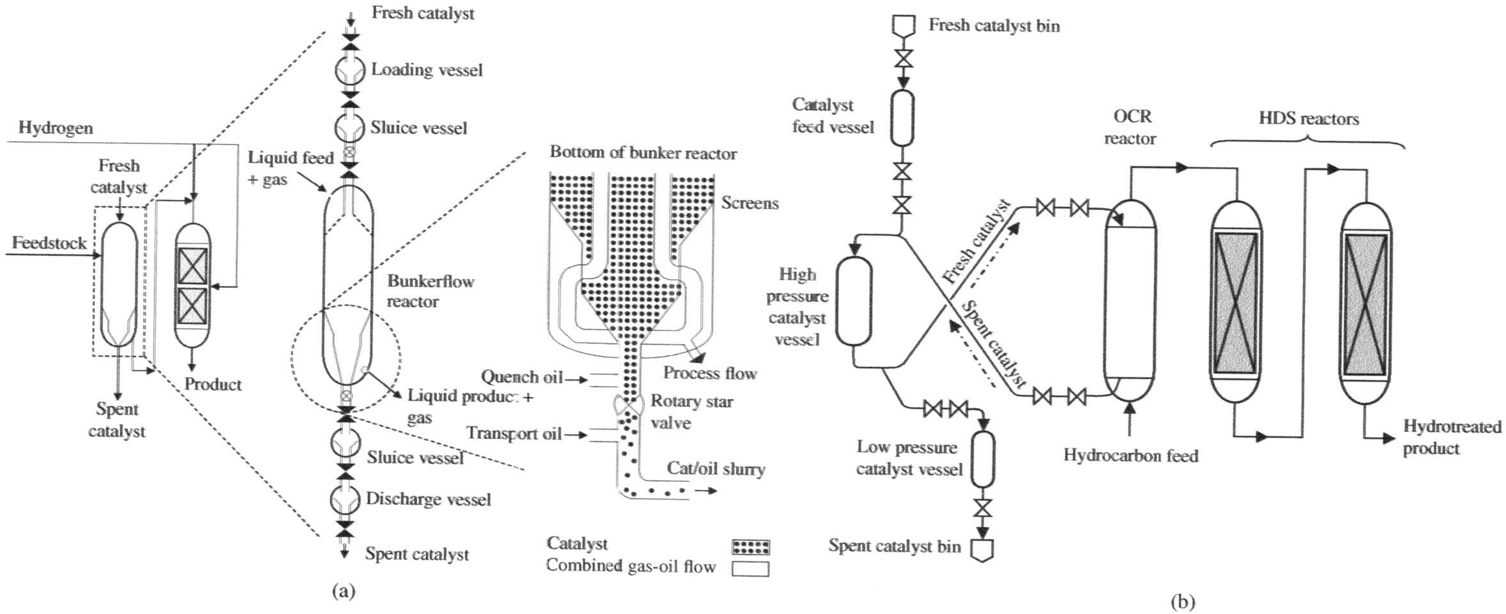

Figure 3.3 Example of MBR for hydroprocessing: (a) OCR reactor and (b) bunker reactor (Adapted from Scheuerman et al. (1993)).

Countercurrent mode of operation of MBR seems to be the best configuration since the spent catalyst contacts the fresh feed at the MBR bottom while the fresh catalyst reacts with an almost already hydrodemetallized feed at the MBR top, resulting in lower catalyst consumption (Morel et al., 1997).

3.2.2.3 ***Ebullated-Bed Reactor*** EBR is used to overcome some of the deficiencies of FBR. H-Oil, T-Star (an extension of H-Oil), and LC-Fining processes are the commercialized technologies that use EBR. These hydroprocessing technologies possess very similar features (process parameters and reactor design) but are different in mechanical details. Figure 3.4 shows both EBRs employed in H-Oil and LC-Fining technologies (Daniel et al., 1988).

In EBR, hydrocarbon feed and hydrogen are fed upflow through a catalyst bed, expanding and backmixing the bed, minimizing bed plugging, and consequently reducing pressure drop problems. The mixture of the gas (make-up and recycle hydrogen) and liquid (feed and recycle oils) reactants enters the reactor plenum chamber and is mixed well through the specially designed gas/liquid mixer, spargers, and catalyst support grid plates. An homogeneous environment is created to hydrotreat and hydrocrack the heavy feedstocks (Kam et al., 1999). Product quality is constantly maintained at a high level by intermittent catalyst addition and withdrawal. Reactor features include onstream catalyst addition and withdrawal, thereby eliminating the need to shut down for catalyst replacement.

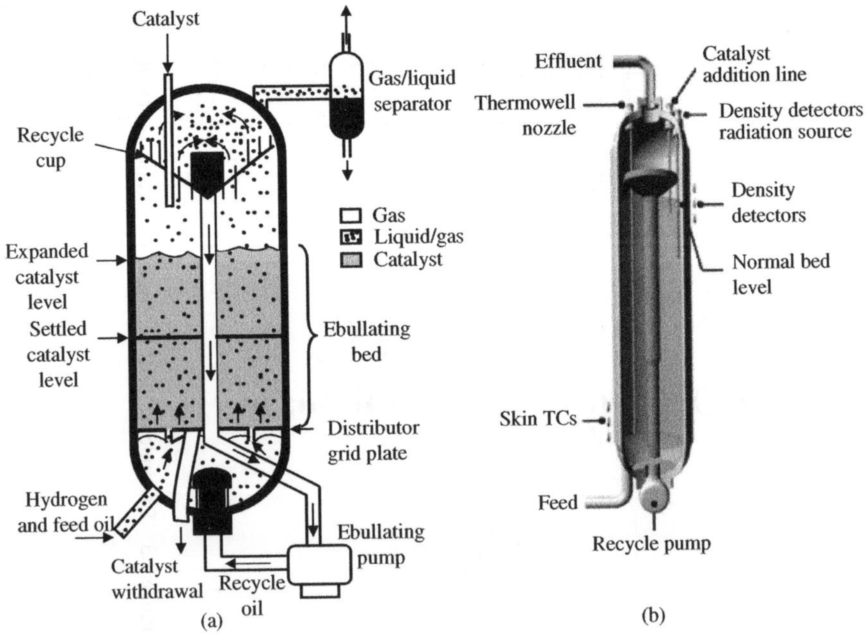

Figure 3.4 Examples of EBR for hydroprocessing: (a) H-oil process and (b) LC-fining process (Adapted from Daniel et al. (1988)).

The main advantages and disadvantages of EBR are as follows (Quann et al., 1988; Furimsky, 1998; Gosselink, 1998; Sie, 2001; Babich and Moulijn, 2003; Ruiz et al., 2004a, 2004b, 2005):

- *Advantages.* Very flexible operation (high and low conversion modes); capable of periodical withdrawal and/or addition of the catalyst to the reactor without interrupting the operation to maintain the necessary catalyst activity; this characteristic increases the operating factor of the unit and decreases refinery maintenance costs associated with catalyst dumping and reloading compared with FBR; its design ensures ample free space between particles, allowing entrained solids to pass through the bed without accumulation, plugging, or increased pressure drop; increased reaction rate by the use of smaller diameter catalyst particles due to significantly diminished diffusion limitations, that is, higher effectiveness factors; this smaller size of the catalyst also makes it less susceptible to pore mouth plugging by metal deposits; good heat transfer so that overheating of the catalyst bed is minimized and less coke forms; the averaged activity of the catalyst inventory reaches a steady state, which results in constant product properties during the cycle length, which is a major difference compared with FBR in which reaction selectivities are changing during the run in response to increases in operating temperature necessary to compensate declining catalyst activity; operation is nearly isothermal since the EBR behaves as well-mixed reactor, which easily dissipates the exothermic heat of reaction and enables higher operating temperatures and thus higher conversion levels.
- *Disadvantages.* Absence of plug-flow regime that is kinetically more favorable than well-mixed regime, which can be partially improved by setting several EBR in series; catalyst attrition and erosion, which means that the catalyst has to be mechanically stable and resist attrition; the smaller catalyst particle size and the lower catalyst holdup (higher void fraction) require a higher reactor volume than FBR and MBR; catalyst consumption rate is quite high; stagnant zones may be developed and careful monitoring is required to prevent the growth of such zones, which can lead to unstable and runaway conditions; sediment formation; scale-up and design is more difficult than for other reactors since it requires much more information, for example, data of feedstock composition, catalyst properties, catalysis and chemical kinetics aspects, hydrodynamic phenomena, heat transfer at the catalyst pellet and bed scales, and so on.

When operating hydroprocessing EBR at high level of conversion (>50%), the main problem is sediment formation, which leads to deposits on the reactor internal parts and downstream vessels (heat exchangers, separators, and fractionating towers) and transfer lines. Sedimentation also causes EBR operability problems, catalyst deactivation, increase of catalyst consumption, and decrease in residual fuel stability, which eventually cause shutdown of commercial units (Morel et al., 1997). For this reason, refiners commonly limit sediment formation at 0.8–1.0 wt% to warranty continuous operation of commercial hydrocracking units.

3.2.2.4 Slurry-Phase Reactor SPR can also be used for hydroprocessing of feeds with very high metal content to obtain lower boiling products using a single reactor. SPR achieves a similar intimate contacting of oil and catalyst and may operate with lower degree of backmixing than EBR. Differently to FBR and EBR, in SPR small amount of finely divided powder is used (typically from 0.1 to 3.0 wt%), which can be an additive or catalyst (or catalyst precursors). The catalyst is mixed with the feed (heavy oil) and both are fed upward with hydrogen through an empty reactor vessel. The SPR is free of internal equipment and operates in a three-phase mode. The solid additive particles are suspended in the primary liquid hydrocarbon phase through which the hydrogen and product gases flow rapidly in bubble form. Since the oil and catalyst flow cocurrently, the mixture approaches plug-flow behavior (Quann et al., 1988; Speight, 2000).

In an SPR, the fresh catalyst is slurried with heavy oil prior to entering the reactor, and when the reaction finishes, the spent catalyst leaves the SPR together with the heavy fraction and remains in the unconverted residue in a benign form (Furimsky, 1998). Inside the reactor, the liquid/powder mixture behaves as a single phase (homogeneous phase) due to the small size of the catalyst/additive particles. Figure 3.5 shows a typical simplified process scheme for catalyst (or additive) addition to an SPR. It has been reported that the powder mostly serves as a site where the small amount of coke can deposit, so as to keep walls, valves, and heat exchangers clean and thus maintain good operability (Schulman and Dickenson,

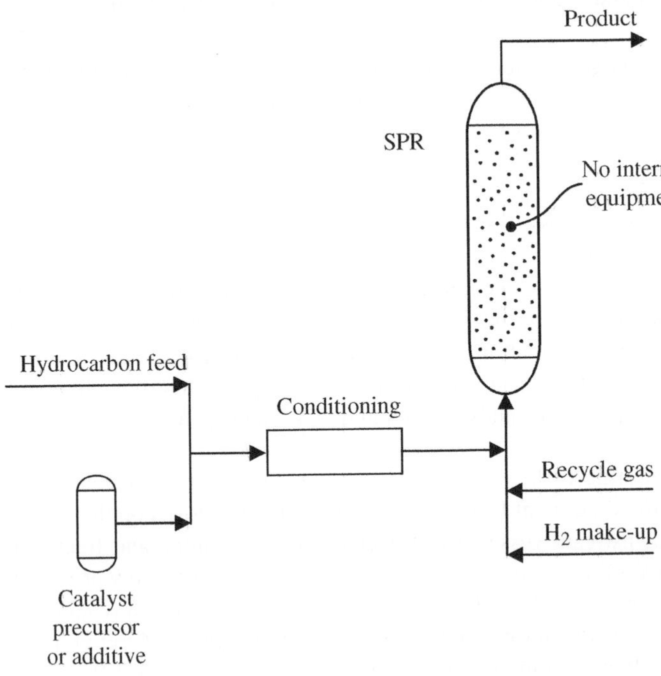

Figure 3.5 Example of SPR for hydroprocessing.

1991). In other words, the use of a selected catalyst dispersed into the feed inhibits coke formation.

The following are the main advantages and disadvantages of SPR (Morel et al., 1997; Furimsky, 1998; Speight, 2000; Robinson and Dolbear, 2006):

- *Advantages.* Use of disposable catalysts with low cost, which possess high surface area and micron-scale particle size, that means high catalyst utilization; the catalyst or additive used in SPR has more number of external pores on the surface than the extruded catalyst used in FBR, EBR, and MBR, which makes slurry catalyst less likely to have metal contaminants plugging up these pores because of their size; spent hydroprocessing catalysts from SPR are not hazardous wastes and do not require special handling procedures such as those recovered from FBR, MBR, or EBR; SPR can be used for hydroprocessing of very poor quality feedstocks; good external mass transfer; the smaller size of catalysts used in SPR compared with catalysts employed in other reactors results in smaller distances between particles and less time for a reactant molecule to find an active catalyst site; the effectiveness of the dual-role additive (hydrogenation and suppression of coke formation) used in SPR allows for the use of operating temperatures that give high conversion in a single-stage reactor; residue conversion higher than 90% can be obtained at space velocities much higher than those used for FBR, MBR, or EBR; because of its particular features (empty reactor without internal equipment) the volume of the SPR is maximized; the homogeneous phase operation of SPR provides a thermally stable operation (uniform temperature) with no possibility of temperature runaway.
- *Disadvantages.* Product quality is very poor; the poor quality of the unconverted pitch makes it almost impossible to be used as a fuel unless it is blended with other streams; catalyst particle size must be strictly designed to achieve excellent dispersion conditions in order to obtain high level of conversion; SPR has also to be carefully designed to maintain a mixed three-phase slurry of heavy oil, fine power catalyst, and hydrogen, and to promote effective contact; the effectiveness of the hydroprocessing SPR is highly dependent upon catalyst selection, since a slurry catalyst with poor activity will result in coke formation and a biphasic, incompatible product, which means that before commercial use a slurry catalyst has to be subjected to different coke suppression and product compatibility tests, among others; the inexpensive additives or carbonaceous material to control coke formation show very low activity toward hydrogenation reactions compared with transition metals; these low-cost materials can be used in a once-through mode; however, high hydrogen pressures are necessary to compensate the scarce catalytic performance; these restrictions have limited commercialization of hydroprocessing SPR due to high investment, operating costs, and the severity of the process; SPR-based hydroprocessing technologies are still at semicommercial scale and the system has to be demonstrated at a larger scale for refiners to consider this as a commercially competitive alternative.

3.2.3 Process Variables

The four most important process variables in hydroprocessing are total pressure and hydrogen partial pressure, reaction temperature, H_2/oil ratio and recycle gas rate, and space velocity and fresh feed rate.

3.2.3.1 Total Pressure and Hydrogen Partial Pressure The performance of any hydroprocessing reactor is limited by the hydrogen partial pressure at the inlet to the reactor. The higher the hydrogen partial pressure, the better the hydroprocessing reactor performance. The overall effect of increasing the partial pressure of hydrogen is to increase the extent of the conversion (Speight, 2000). The presence of heteroatom compounds with different reactivities in a hydroprocessing feed makes, for instance, hydrodesulfurization of refractory multiring sulfur compounds very difficult and a high demanding hydrogen process, pathway of which goes through prehydrogenation of one of the aromatic rings. Therefore, high hydrogen partial pressure is required; otherwise (Ho, 2003), HDN rate may be so slow that nitrogen compounds block off virtually all active sites available for HDS; HDS rate of refractory sulfur compounds may be limited by thermodynamically mandated low hydrogenation rate; the catalyst surface may be starved or adsorbed hydrogen.

Total pressure of a hydroprocessing reactor is fixed by the design and is controlled by the pressure that is maintained at the high pressure separator (HPS), while hydrogen partial pressure is determined by multiplying total pressure (at the inlet of the reactor) by the hydrogen purity of the recycle gas. Hydrogen partial pressure and other variables are determined by the type of feedstock, feed quality, and process objectives. Operating the reactor at higher hydrogen partial pressures yields the following main benefits (Mehra and Al-Abdulal, 2005; Gruia, 2006): longer catalyst cycle life, capability for processing heavier feeds, higher throughput capability, higher conversion capability, better product quality, purge gas elimination.

A hydroprocessing reactor must be operated at hydrogen partial pressure close to the design value; otherwise, catalyst deactivation rate will substantially be increased and consequently catalyst cycle life will be reduced due to excessive condensation reactions that form coke. Catalyst activity will also be negatively affected by operating the hydroprocessing reactor at low hydrogen partial pressure. However, equipment limitations (reactors, heaters, exchangers, and vessels) restrict the operation at a pressure close to or a little bit higher than the design value, and the unique option to increase hydrogen partial pressure is by increasing purity of the recycle gas (Gruia, 2006). In commercial operations, hydrogen partial pressure is mainly obtained by feeding the proper amount of make-up gas.

The increase of catalyst activity for achieving higher impurity removal and conversion rates would require significant modifications in hydroprocessing reactor operation, primarily through the use of higher pressure, and also by increasing hydrogen rate and purity, reducing space velocity, and proper selection of catalyst. Pressure requirements would depend, of course, on the feedstock quality and on the product quality target of each refinery. At higher hydrogen partial pressures, sulfur (and other impurities) removal is easier; however, reactors are more expensive. Also, at higher pressures, hydrogen consumption of the unit increases, which can become a significant cost factor for the refinery. The minimum pressure required typically goes up with the required severity of the unit, that is, the heavier the feedstock, the lower levels of sulfur in product required. For new units, higher hydrogen partial pressure can be achieved by higher total pressure plant design, but the increase is not as pronounced as that achieved by increasing H_2 purity, since with the former approach H_2S partial pressure is also increased, whereas an increase in the recycle gas purity does not affect (or slightly reduces) the H_2S partial pressure (Bingham and Christensen, 2000).

3.2.3.2 ***Reaction Temperature*** In combination with hydrogen partial pressure, the reactor temperature generally determines the type of components that can be removed from the hydrocarbon feed and also establishes the working life of the catalyst. In general, an increase in reaction temperature will substantially increase reaction rates and consequently removal of impurities (Takahashi et al., 2005). However, at temperatures above 410 °C, thermal cracking of the hydrocarbon constituents becomes more prominent, which can lead to formation of considerable amounts of low-molecular-weight hydrocarbon liquids and gases and also to catalyst deactivation much more quickly than at lower temperatures. Thermal cracking also produces olefins, which when hydrogenated release heat, increasing the temperatures further and thermal cracking rates increase (hot spots). Finally, this condition inside the reactor provokes temperatures higher than the safe upper limits for the reactor walls (Speight, 2000).

Since hydroprocessing reactions are mostly exothermic, the reactor temperature will increase as the reactants proceed through the catalyst beds. This means that outlet reactor temperature will be higher than inlet reactor temperature. This increase in reactor temperature is not linear (Rodriguez and Ancheyta, 2004; Mederos et al., 2006). In other words, for determining the average temperature of the reactor, an arithmetic average of temperatures will not be the more accurate representative reactor temperature. To properly monitor the reactor temperature, it is necessary to observe the temperature at the inlet and outlet and axially (and radially) throughout the catalyst bed. A temperature profile can be constructed by plotting the catalyst temperature versus weight percent of the catalyst. For this reason, the weight average bed temperature (WABT) is typically used as a measure of the reactor temperature. Commercial reactors commonly have various temperature indicators (TIs) located in different zones of catalytic beds. Most of the times TIs are placed at the inlet and outlet of the beds; then for determining the global WABT, average WABT_i of each catalytic bed is first determined as follows (Stefanidis et al., 2005):

$$\text{WABT}_i = \frac{1}{3}T_i^{\text{in}} + \frac{2}{3}T_i^{\text{out}} = \frac{T_i^{\text{in}} + 2T_i^{\text{out}}}{3} \tag{3.1}$$

and the global WABT is calculated with the following equation:

$$\text{WABT} = \sum_{i=1}^{N} (\text{WABT}_i)(\text{Wc}_i\%) \tag{3.2}$$

where N is the number of catalyst beds, T_i^{in} and T_i^{out} are the inlet and outlet temperatures in each catalytic bed, respectively, and $Wc_i\%$ is the weight percent of catalyst in each bed with respect to the total.

During hydroprocessing of heavy oils, temperature has to be increased to compensate for catalyst deactivation in order to maintain the desired conversion level. The initial temperature is known as start-of-run (SOR) temperature ($WABT_{SOR}$) and the final temperature as end-of-run (EOR) temperature ($WABT_{EOR}$). Values of these two temperatures depend mainly on the feed type. Typically, ($WABT_{EOR} - WABT_{SOR}$) = 30 °C. The temperature increase in many cases is very small to prevent catalyst coking, about 1–2 °C per month. For feed containing relatively high amount of metals, this increase in temperature is required more frequently. When WABT reaches a value close to the designed maximum, the catalyst has to be replaced.

3.2.3.3 H_2-to-Oil Ratio and Recycle Gas Rate The choice of gas rate is governed by economic considerations. Recycle is used to maintain the hydrogen partial pressure and the physical contact of the hydrogen with the catalyst and hydrocarbon to ensure adequate conversion and impurities removal while minimizing carbon deposition. Increasing the hydrogen partial pressure reduces the reactor SOR temperature and also reduces the rate of catalyst deactivation. Above certain gas rate, the change in hydrogen partial pressure will be relatively small. In general, higher gas rates than necessary incur extra heating and cooling rates, which may outweigh other advantages.

Increasing the recycle-gas rate increases the H_2/oil ratio in the reactor. This increased ratio acts in much the same manner as increased hydrogen partial pressure. H_2/oil ratio is determined by

$$H_2/\text{oil} = \frac{\text{Total gas to the reactor, SCF/day}}{\text{Total feed to the reactor, Bbl/day}} [=] \frac{\text{SCF}}{\text{Bbl}} \tag{3.3}$$

In addition to affecting the hydrogen partial pressure, the gas rate is important as it acts to strip volatile products from the reactor liquids and thus affects the concentration of various components in the reactive liquid phase. Similarly to hydrogen partial pressure, H_2/oil ratio must be maintained at the design value and any reduction below the design minimum will have adverse effects on the catalyst life.

In the case of EBR, excessive gas rates affect the nature of the ebullating bed, result in catalyst carryover into the downcomer feeding the ebullating pumps, and increase the concentration of heavy materials in the reactor liquid phase by stripping volatile products, which tends to increase the conversion level but can lead to loss of liquid in the reactor or to an excessive concentration of very heavy materials.

With high-sulfur feedstocks, the level of H_2S can build up to high values reducing purity of the recycle gas stream and also the hydrogen partial pressure and inhibiting the HDS reaction. Thus, the recycle gas from the HPS is generally water-washed to remove ammonia, preventing the formation of ammonium sulfide that might form blockages in the reactor effluent cooler, and then sent to the sour water plant to remove H_2S. If a recycle gas scrubbed is not available, the reactor temperature must be increased to reduce the inhibition effect of H_2S, which effect is greater at higher total reactor pressure.

Make-up hydrogen is very important to compensate hydrogen consumption during hydroprocessing reactions, which is recognized as an important operation factor in refineries from the viewpoint of hydrogen supply investment and energy cost reduction.

3.2.3.4 Space Velocity and Fresh Feed Rate Space velocity is a variable normally used to relate the amount of catalyst to the amount of feed and is calculated on volume basis (LHSV: liquid hourly space velocity) or weight basis (WHSV: weight hourly space velocity), the former being the most commonly used in hydroprocessing. LHSV and WHSV are calculated as follows:

$$\text{LHSV} = \frac{\text{Total volumetric feed flowrate to the reactor}}{\text{Total catalyst volume}} [=]\ \text{h}^{-1} \tag{3.4}$$

$$\text{WHSV} = \frac{\text{Total mass feed flowrate to the reactor}}{\text{Total catalyst weight}} [=]\ \text{h}^{-1} \tag{3.5}$$

Increasing the space velocity (higher feed rate for a given amount of catalyst) requires a higher reactor temperature to maintain the same impurity removal and conversion levels and results in an increase in the deactivation rate, thus reducing catalyst life.

With a fixed reactor size, the residence time is inversely proportional to the fresh feed rate and to the space velocity as well. When the feed rate exceeds the design value, the conversion level will often have to be lower than the design level

3.2.4 Effect of Reaction Conditions on Catalyst Deactivation

The influence of process variables in coke and metal deposition on catalyst is explained in terms of temperature, pressure of hydrogen, and type of feedstock.

3.2.4.1 Influence of Temperature Increasing the temperature is common practice for compensating activity with time-on-stream of a catalyst and keeping constant a level of conversion. During this process, increase in temperature induces the reactant molecule and provokes metal and carbon deposition on the catalyst surface. Thus, it is observed that more vanadium and carbon deposited occur on the edge of the pellet because the reaction takes place at the surface of the catalyst and not on the active sites. Dong et al. (1997) studied the influence of N-compounds in VGO (vacuum gas oil) along with the increase in temperature and observed that the higher the temperature, the higher the coke deposition. However, the presence of carbazole and acridine in VGO considerably increased the tendency to form coke on a $CoMo/Al_2O_3$ catalyst. Working with FBRs, Takahashi et al. (2005) concluded that the increase in temperature resulted in an accelerated decomposition rate of maltenes and asphaltenes with the consequent deposition of metals and coke in the guard reactor, which was responsible for the rapid catalyst deactivation. The transformation of maltenes is the main factor that renders asphaltenes insoluble in the medium. In addition, rupture of alkyl carbons of asphaltenes makes them less soluble and aromatic cores are very prone to form coke. These reactions are increased at higher temperatures, and it is to be expected that higher amount of coke is formed at higher temperatures.

3.2.4.2 Influence of Hydrogen Pressure Experiments for studying the influence of pressure have been carried out by Tamm et al. (1981). They studied the effect of two partial pressures (12.5 and 18.6 MPa) keeping temperature constant at 370 °C using Arabian Heavy Atmospheric Residue as feedstock. When increasing partial pressure, more vanadium was deposited at the edge of the pellet and less deposition at the center was observed. The opposite trend was observed at the lowest pressure, that is, a slight increase of vanadium deposit was found at the center of the catalyst pellet. With respect to coke deposition, it was well documented that initial coke deposition onto the catalyst surface diminished slightly with increasing hydrogen pressure.

3.2.4.3 Influence of the Type of Feedstock Metal deposition is highly dependent on the nature of feedstock. The deposition profile of vanadium is more prominent with feedstocks containing more asphaltenes and Ni+V. In this case, the deposition in the edge is more obvious than light feedstock (Tamm et al., 1981). It has been stated that the kinetics of hydrodemetallization of heavy feedstocks is not fully represented by a first-order power law. Instead, demetallization approaches to second-order kinetics because not all molecules have the same reactivity during the reaction. A simple kinetic representation of the reaction does not match the metal deposition profile in integral reactor operated as fixed bed and a better representation of the reaction could include parallel reactions. It is known that catalyst deactivation rate is more than five times faster for a high-metal-content feedstock (e.g., Boscan feed with 1308 wppm Ni+V) than for a conventional Kuwaiti straight-run atmospheric residue of Kuwait Export Crude (84 wppm Ni+V).

3.3 MECHANISMS OF CATALYST DEACTIVATION

Hydroprocessing catalysts are active for a number of important reactions, all of which involve hydrogenolysis of the heteroatom–carbon bond and breaking of C–C bonds. Hydroprocessing reactions take place on the active sites of the catalyst, which are believed to be located on the sulfur vacancies located on the edges of the MoS_2 crystallites. The vacancies associated with the Ni or Co promoters are considerably more active than those associated with Mo only. Since the sulfur vacancies have a Lewis acid character, poisoning of hydroprocessing reactions will occur when a strongly adsorbed molecule such as a nitrogen compound, a coke molecule, a metal, or any other basic molecule occupies an active vacancy. Basically, poisoning of hydroprocessing catalysts occurs by single or multiple causes such as active site blockage by strongly adsorbed species, active site coverage by deposits of coke or metals, pore mouth constriction/blockage, and sintering of the active phase (Dautzenberg et al., 1978; Furimsky and Massoth, 1999).

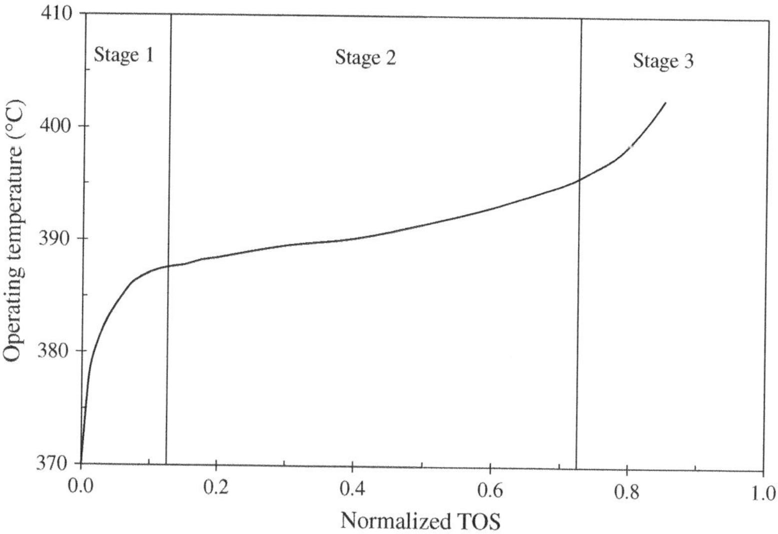

Figure 3.6 Typical profile of catalyst deactivation. Stage 1: deactivation by coke, stage 2: deactivation by metals, and stage 3: pore blockage.

The degree and type of deactivation will depend on the feed properties and will be evidenced by an S-shape curve of temperature versus time-on-stream as shown in Figure 3.6. Initially, coke deposition causes a rapid deactivation that in several hours attains a pseudo-steady state. Then, deactivation by metal deposits is observed during a longer period of time until a drastic deactivation caused by pore restriction and blockage is observed.

A poison is a substance that competes for adsorption on the active sites. The adsorption of the poison can be reversible or irreversible (Figure 3.7). The effect of a

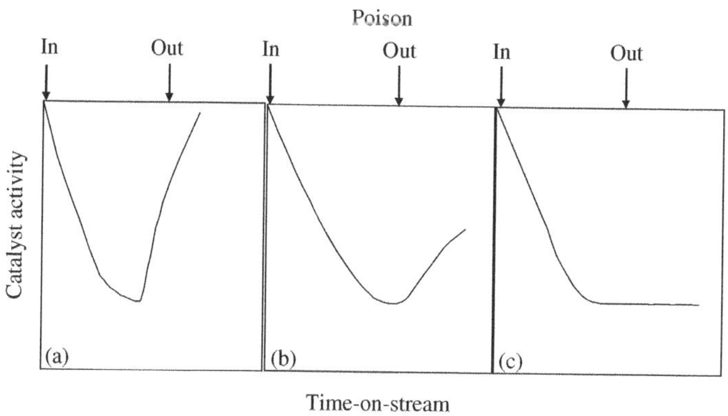

Figure 3.7 Profiles of catalyst activity versus time-on-stream for (a) reversible, (b) quasi-irreversible, (c) irreversible poisoning (Adapted from Furimsky and Massoth (1999)).

poison (p) on the reaction rate, r_A, can be taken into account in the denominator term of a Langmuir–Hinshelwood mechanism:

$$r_A = \frac{k_A f(C_A)}{\left(1 + \sum K_i C_i + K_p C_p\right)^n} \tag{3.6}$$

where k_A is the rate constant, C_A the concentration of reactant A, $f(C_A)$ its particular rate form, K_p and K_i are the adsorption constants of the poison and the adsorbing species associated with the reaction, respectively, n is the power of the inhibition term and its value is associated with the number of active sites involved in the reaction mechanism. Equation 3.6 shows that the extent of inhibition will be related to the magnitude of the poison adsorption constant and concentration with respect to those of the species associated with the main reaction.

Deactivation by coke and metals occur to a greater extent in heavy oils and follows a trend different from the one observed for light oils. For heavy feeds, coke deposits rapidly during the earlier reaction stages and then reaches a steady state while the deposition of metals increases continuously with time in a more or less linear manner. Figure 3.8 shows the time evolution of coke and metals for catalyst at different locations in the catalytic bed.

Coke deposition increases with the boiling range of the feed. However, for feeds with a similar boiling range, the one with higher concentration of coke precursors, such as aromatics and heterocyclic compounds, will cause a more severe poisoning.

Metals and coke affect to a different extent the various catalyst functionalities. Cable et al. (1981) compared the activities of fresh and regenerated catalysts and found that metals had a moderate effect on HDS and a weak effect on hydrogenation and cracking, while coke had a moderate effect on HDS but a strong effect on hydrogenation and cracking. These results are summarized in Table 3.1.

The composition of the catalytic surface is of major importance to control the extent of coke deposition. Coke precursors such as alkenes, aromatics, and heterocyclics can transform into higher-molecular-weight species if sufficient active hydrogen is not available to prevent it.

In the case of heavy oils, precipitation of asphaltenes from the feed is another important source of coke formation. Asphaltenes are stabilized by the presence of resins. Therefore, if during hydroprocessing, resins are eliminated at a faster rate than asphaltenes, precipitation of the latter occurs on the surface of the catalyst, leading to the possibility of greater coke formation (Dautzenberg and de Deken, 1985). It is believed that asphaltenes have a greater tendency than resins to coke formation (Banerjee et al., 1986).

In addition to feed composition and catalyst properties, hydrogen partial pressure and operating temperature are important for controlling the extent of coke formation. The presence of high concentration of active hydrogen on the surface can convert coke precursors into stable products. However, at high operating temperatures, the probability of this conversion decreases because of thermodynamical limitations.

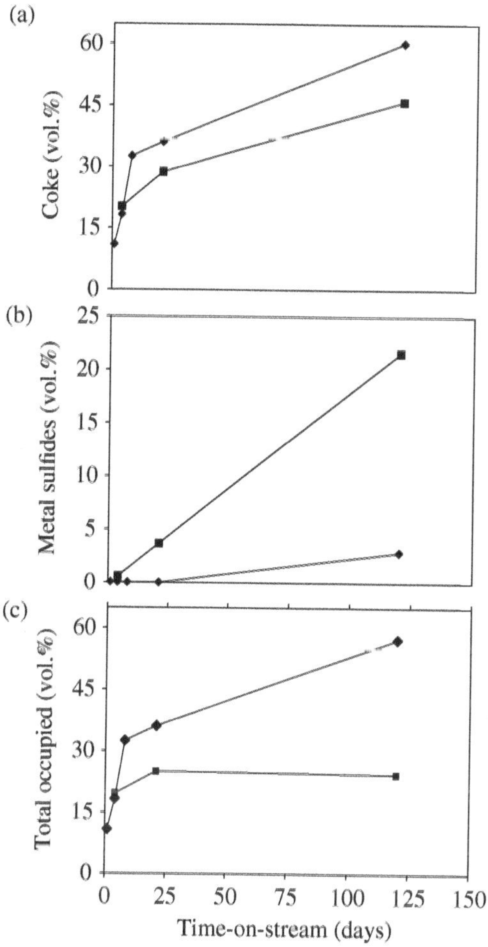

Figure 3.8 Evolution of (a) coke, (b) metals, and (c) total occupied volumes as a function of time-on-stream (♦) bottom of catalytic bed, (■) top of catalytic bed (Adapted from Fonseca et al. (1996)).

TABLE 3.1 Relative Effect of Deactivation by Metals and Coke on Catalyst Functionalities

	Hydrodesulfurization	Hydrogenation	Hydrocracking
Metals	Moderate	Weak	Weak
Coke	Moderate	Strong	Very strong

Source: Adapted from Cable et al. (1981).

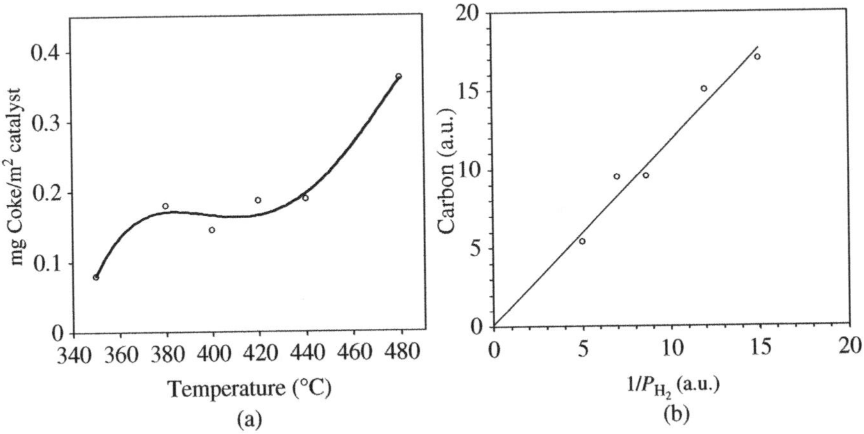

Figure 3.9 Effect of (a) temperature on catalyst coke deposits and (b) coke steady-state level as a function of hydrogen pressure (Adapted from Ternan et al. (1979) and Oelderik et al. (1989)).

Figure 3.9 suggests the existence of an optimal combination of catalyst properties and process conditions, which ensures the best performance of the catalyst (Ternan et al., 1979).

At low temperatures, coke formation increases because the conversion of heavy species to lighter fractions is slow and the lifetime of heavy molecules on the surface is prolonged increasing the possibility of polymerization reactions. Another reason for this observation could be that resins are converted faster than asphaltenes causing the precipitation of the latter. Above 375 °C, the transformation of asphaltenes may increase not only due to the higher temperature but also because the hydrogenation of coke precursors competes successfully with their polymerization. The drastic increase in coke formation above 440 °C is possibly due to an increase in dehydrogenation followed by polycondensation reactions. The shape of the curve in Figure 3.9 can be altered by changing the partial pressure of hydrogen. The results shown in Figure 3.9 also indicate that the steady-state level of carbon on the catalyst can be significantly altered by changing the hydrogen partial pressure. Different aspects of the poisoning by coke in hydroprocessing catalysts have been addressed in Furimsky and Massoth (1999).

Pore volume and pore size distribution are important to maximize catalyst utilization when coke deposition is a problem. Figure 3.10 compares the nitrogen physisorption isotherms of fresh and spent catalysts with small (6 nm) and large (17 nm) diameter pores. The variations in the shape of the hysteresis loops were considered as representative of pore mouth coking (Rana et al., 2005).

As shown in Figure 3.6, catalyst life is greatly affected by the content of metals in the feed. For heavy oils, which contain high concentration of metals, the catalyst life can be several times lower than for light oils. Deactivation by metals occurs along with the deactivation by coke deposition. However, while coke reaches pseudo-steady-state equilibrium after an initial rapid buildup, the deposition of

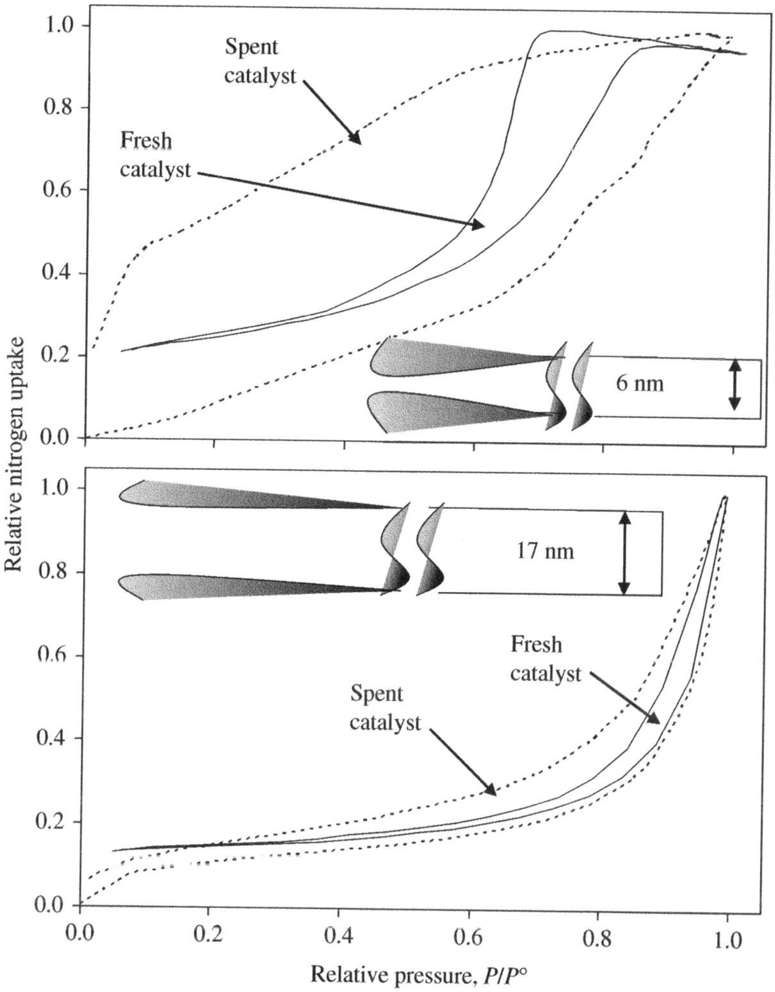

Figure 3.10 Adsorption–desorption isotherms for fresh and spent $CoMo/Al_2O_3$ catalysts with different average pore diameters. (Adapted from Rana et al. (2005)).

metals continuously increases with time-on-stream. One of the reasons for the continuous increase in catalyst metal content is that during hydroprocessing, metals, mostly V and Ni, are converted to sulfides, which autocatalyze the metal deposition reaction (Sie, 1980). Hence, the hydrodemetallization reaction can continue on the newly created sites after the original active sites of the catalyst have been covered by metal deposits. Permanent deactivation by metal deposition causes an important problem to refiners. Because of this, hydroprocessing schemes often use HDM catalysts to selectively remove metals and protect downstream catalysts.

The S-shaped catalyst deactivation profile observed for heavy oils (Figure 3.6) is then related to coke formation and metal deposits on the catalyst. The initial period of rapid deactivation is generally attributed to the buildup of an equilibrium coke load on the catalyst surface (Sie, 1980; Pazos et al., 1983) but has also been ascribed in part to the buildup of monolayer or submonolayer metal deposits (Tamm et al., 1981; Johnson et al., 1986). The almost constant deactivation rate observed in the intermediate period is ascribed to increasing diffusional resistance caused by accumulation of metals on the pore walls. The final rapid deactivation that causes the end of catalyst life is due to pore plugging caused by metal sulfide deposits. However, studies performed by Smith and Wei (1991) on deactivation of HDM catalysts with model molecules have shown evidence that metal deposits have a crystalline form that is inconsistent with the general idea of uniform metal deposits. It was also observed by the same authors that both a sulfided $CoMo/Al_2O_3$ catalyst and a low promoter alumina carrier neither deactivated nor acquired catalytic activity as nickel sulfide deposits accumulated. These observations suggested that part of the active components of the catalyst remains active for HDM even in the presence of high levels of deposited nickel sulfide and are inconsistent with a contribution of the nickel deposits to the rapid initial deactivation of the HDM catalyst. In contrast, HDM experiments carried out with vanadyl etioporphyrin (VO-EP) in squalane were diffusion limited due to the low effective diffusion of VO-EP. The catalyst deactivation was consistent with the general idea that accumulation of metal sulfide deposits are responsible for the relatively slow steady deactivation that characterizes the intermediate period of the life of a hydroprocessing catalyst. In a study of deactivation of residue HDM catalysts, it was shown that different results are obtained by carrying out the HDM experiments in batch or continuous flow reactors. The batch reactor yielded carbon deposits with almost no metals, whereas the continuous reactor provided used catalysts containing significant amounts of both carbon and metal. It was also found that a small amount of well-dispersed vanadium inside the catalyst grain was more deactivating than a large amount of carbon (Gualda and Kasztelan, 1996).

About one half of the Ni and V present in heavy feeds are present as porphyrins, the other part of metals are in less characterized forms that include bonds with nitrogen, oxygen, and sulfur in the defect centers of asphaltene sheets (Mitchell, 1990; Beaton and Bertolacini, 1991). Due to the bulkiness of the porphyrins and asphaltene molecules, the demetallization reaction is affected by pore diffusion limitations. Consequently, metals are generally deposited on the outer zone of the catalyst particle. However, the characteristics of the catalyst can alter the radial profile of metal deposits making it more shallow or broad. Figure 3.11 shows the radial profiles of vanadium obtained by electron microprobe analysis along the diameter of catalysts used in the hydroprocessing of residua.

The pore size of the catalyst determines, to a greater extent, the type of metal penetration profile. For microporous catalysts, the metal deposits are restricted to the outer shell of the catalyst particle entailing pore mouth plugging. Catalysts with wide or intermediate pore diameters lead to deep or intermediate metal penetration and high and medium metal storage capacities, respectively. So, catalyst life will be determined to a great extent by the pore volume available for storing metal sulfide deposits. However, it must be considered that the increase in pore diameter and volume brings about a decrease in surface area and, therefore, lower catalytic activity

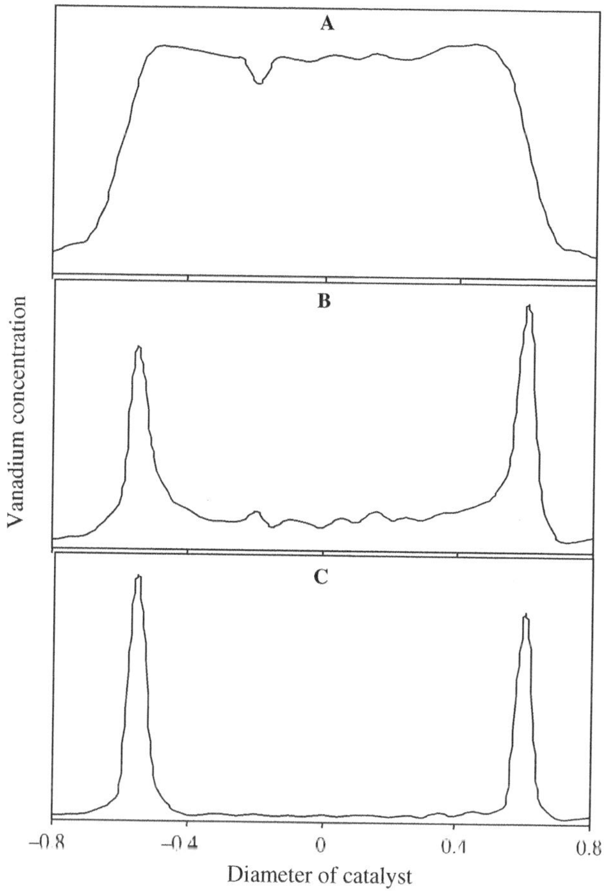

Catalyst	A	B	C
Pore size	Wide	Intermediate	Narrow
Metals penetration	Deep	Intermediate	Shallow
Metal storage capacity	High	Medium	Low
Stability	Very good	Fair	Poor
HC/HDS activity	Low	Fair	High
Application	HDM catalyst	HC/HDS catalyst	HC/HDS catalyst
Location in reactor train	Front end	Middle	Tail end

HC, hydroconversion.

Figure 3.11 Radial concentration profiles of vanadium determined by electron microprobe scans along the diameter of catalysts used in hydroprocessing of residue (Adapted from Sie (2001)).

for other hydrotreatment reactions. Hence, porosity and relative activity in hydroprocessing must be balanced so as to obtain the best catalyst performance. The change in activity with pore diameter for the different reactions (HDS, HDM, and HDAs) indicates that the optimum pore diameter is larger for hydrodemetallization and deasphaltenization than for hydrodesulfurization.

The opposite effects of pore size variation on catalyst stability and activity are illustrated in Figure 3.12. It is clear that catalyst deactivation has important consequences for process design and operation. In general, the design of the catalyst has to reach a compromise between catalyst life and activity.

3.4 ASPHALTENES AND THEIR EFFECT ON CATALYST DEACTIVATION

The complex structures of asphaltenes are responsible for many reactions when upgrading heavy oils. It is to be remembered that heavy oils contain asphaltenes that can undergo thermal or catalytic reactions.

3.4.1 Thermal Reaction

For heavy oil hydroprocessing, it is usually accepted that coke formation is associated with asphaltene cracking. Coke formation is a thermal process due to dehydrogenation of paraffins and naphthenes into olefins and cycloolefins, respectively. These

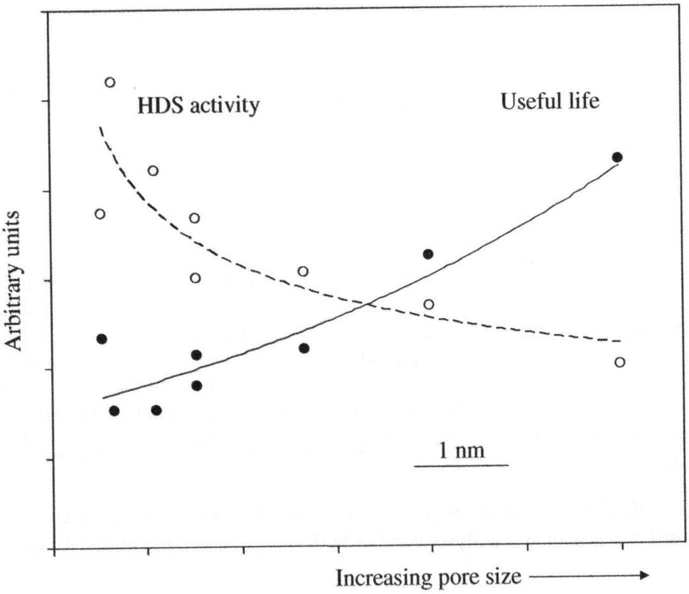

Figure 3.12 Effect of pore size on useful life and hydrodesulfurization activity for catalysts with narrow monomodal pore size distributions tested under standard conditions (Adapted from Sie (2001)).

olefins further condense and form coke, which is time dependent as reported by Del Bianco et al. (1993). Some kinetic studies to analyze the direct influence of asphaltenes toward coke formation have been carried out by Banerjee et al. (1986), with fractionated Athabasca bitumen, Topped Athabasca bitumen, Light Arabian vacuum bottoms, and Cold Lake heavy oil into asphaltenes, hard and soft resins, aromatics, and saturates. Soft and hard resins were obtained by column fractionation with methyl ethyl ketone (MEK) and tetrahydrofuran, respectively. It was observed that aromatic fractions produced coke but at a lower velocity than asphaltenes did. The rate of coke formation from saturates is low because saturates need to be converted into several products (olefins) before producing coke. The reaction order was supposed to be unity and activation energy of coke formation for different fractions is as follows:

Asphaltenes: 41–47 kJ/mol
Soft resins: 36–40 kJ/mol
Hard resins: 34–40 kJ/mol
Aromatics: 52–58 kJ/mol.

The influence of temperature is one of the most important parameters in coke formation. Regarding this, studies involving residue with high content of asphaltenes have been carried out by Zhang et al. (2006), who established a link between coking, colloidal stability, and molecular properties of petroleum residue during thermal reactions. A parameter called "colloidal stability" was defined based on the mass of added *n*-heptane-to-mass of residue; it was observed that during the induction period (time necessary to coke appears as a separated phase) for different residue samples analyzed coke formation was not observed, but it was formed very fast after reaching the induction period. At the same time, the colloidal stability decreased during the induction period and almost remained unchanged after it. The main observations were as follows:

- Asphaltenes concentration in residue increases during the induction period reaching a maximum at the end of coke induction period and after that there is a smooth decrease as time progresses indicating that aggregation of asphaltenes occurs and they turn into coke. At the same time, maltenes amount decreases sharply during the induction period and after that the decrease is very smooth.
- The diminution of the amount of maltenes reduces the resins content along with colloidal stability of asphaltenes during the induction period, which leads to saturation of asphaltenes in the colloidal medium making asphaltenes prone to aggregate and transform in coke.
- Molecular weight of maltenes diminishes as thermal reaction progresses. Molecular weight of asphaltenes increases during the induction period having a maximum due to aggregation reactions but after that period, asphaltenes undergo a reduction of their molecular weight.
- Dipole moments of residue increases as thermal reactions progress and its maximum value is reached at the end of the induction period and then decreased.

This means that asphaltenes with high dipole moment tend to aggregate during the induction period and after that they form coke.

As thermal reactions advance, asphaltenes and resin ratio will tend to vary with temperature and accelerate the deterioration of colloidal stability. Thus, asphaltenes having large conglomerating parameters prefer to aggregate into coke. It is well known that the resin/asphaltene ratio of less than one favors sediment formation.

The induction period has been also referred as the time necessary for the mesophase to become visible in the microscope having bright under cross-polarized light. Mesophase appears when precursors, such as asphaltenes, exceed their solubility limit and depends on the solvent power of maltenes present in heavy crude, which stabilize asphaltenes; however, maltenes change their solvent power due to the extent of thermal reactions that modify colloidal stability of the medium (Gentzis et al., 2001). Conversion of maltenes during reaction leaves asphaltenes unprotected and more likely to agglomerate and precipitate. At one point, the physical appearance of the medium changes as a consequence of the progress of thermal reactions provoking that the initial appearance, that is, asphaltenes in maltenes, changes to another phase, maltenes in asphaltenes, as reported by Gimaev et al. (1980). This phenomenon is known as phase inversion and was observed earlier by Brooks and Taylor (1965). Gimaev et al. (1980) concluded as follows:

- As thermal reactions proceed, free radicals are formed and condensation reactions make that products accumulate in the system with a consequent phase transition in the liquid medium.
- Asphaltenes experience a gradual separation of phases because of intermolecular attractions and by attraction of unpaired electrons of free radicals originating the formation of mesophase considered as anisotropic.
- Accumulation of aromatic structures produces a second liquid phase and van der Waals and dipole–dipole forces do this phase orient the molecules of the mesophase sphere.
- When condensation occurs within mesophase, then solid crystallites of coke appear.
- Phase separation is mainly promoted by incompatibility of the medium with asphaltenes due to their shortening of alkyl chains when processed at high temperatures. Asphaltenes and resins are not thermally stable and they begin to suffer changes when temperature increases.

Hence, the main reasons for incompatibility of asphaltenes and the medium are cracking and polymerization reactions as stated by Wiehe (1993), who also reported that coke formation is dependent on the behavior of asphaltenes. As conversion increases, asphaltenes core becomes more aromatic until the solubility limit is exceeded with the consequent phase separation forming a second liquid phase that has less hydrogen to extract and stabilize free radicals. This makes that recombination of asphaltene cores be feasible and can give higher yields of coke. In this respect,

Wiehe (1994) created the concept of pendant-core building model for explaining that macromolecules can be approached by a combination of two blocks linked by weak bonds. The pendant unit is a volatile unit, whereas the core is nonvolatile. Cracking of heavy fractions leaves the pendant units free and cores tend to concentrate. According to this model, all fractions can be obtained by a combination of pendants and cores units. Further cracking permits only recombination of cores and apparition of coke is imminent, whereas all distillates are formed by pendant units.

Rahmani et al. (2003) observed phase inversion in some parts of coke produced from asphaltenes. During phase inversion, coke is the major phase and oil is dispersed in it, which was shown by SEM. Coke is separated as spheres or as coalesced structure from the continuous oil phase, whereas coke is the continuous phase. Le Page et al. (1987) have also explained the mesophase formation during thermal treatment and proposed that at the beginning, asphaltenes present some void spaces in their structures; however, heating produces cleavage of alkyl chains of asphaltenes and they turn into more compact structures, which cannot be dissolved even in high medium of resin. The core aggregation indicated that asphaltenes have a tridimensional structure prior to apparition of the mesophase. At this point, a bigger rearrangement occurs and mesophase appears with a disklike shape. Moreover, Storm et al. (1996) explained rheological changes of petroleum as a consequence of the formation of dispersed droplets of liquid phase. It can be drawn that there is a change that depends on temperature directly leading asphaltenes to reorganize the asphaltic micelles during coke induction period in which large two-dimensional particles are formed. Induction period is the main cause of changes of asphaltenes, that is, heptane-soluble material is turned into pentane insolubles, and at the same time pentane insolubles are converted into heptane insolubles. These changes are responsible for macroscopic phase separation, which has been supported by SAXS (small-angle X-ray scattering) analysis. It has been stated that the size of original micelles is 0.004 μm, and after reorganization of particles their characteristic dimension is in between 0.02 and 0.03 μm (Storm et al., 1998).

3.4.2 Catalytic Reaction

To define the best method for conversion of bottoms of barrel into valuable products, a detailed feedstock analysis is required. One of the most important components of the feedstock for this selection is asphaltenes. This macromolecule is closely related to other crude oil properties, such as viscosity and API gravity, among others. Asphaltenes are commonly defined by their solubility in toluene and insolubility in *n*-heptane and are a highly aromatic, polydisperse mixture consisting of the heaviest and most polar fraction of crude oil. It is well known that asphaltenes contain in their molecule most of metals, sulfur, and nitrogen present in crude oils.

To design a catalyst for hydroprocessing of heavy oils, and particularly to maximize asphaltenes conversion, two distinct properties are required: the number of acid sites and large pore diameter, which allow large molecules for easy diffusion into the pores (or catalytic sites). Acid sites are essential to enhance hydrocracking selectivity (Delmon, 1990); however, coke formation is also favored by these catalytic acid sites, consequently, a balance between acidity and textural properties of catalysts is necessary. One portion of asphaltenes is adsorbed on the catalyst surface creating a steady state, which will lead the adsorbed asphaltenes onto the catalyst surface to react to produce coke.

The function of the catalyst is to remove contaminants such as sulfur, nitrogen, and metals present in the feedstock. The catalyst also hydrogenates cracked fragments and increases the H/C atomic ratio of products. However, the catalyst suffers from deactivation mainly due to coke and metal deposition (Ni+V). Catalysts with narrower pores and high surface area possess high hydrotreating activity, but they are very fast deactivated (Bartholdy and Cooper, 1993; Absi-Halabi et al., 1991; Thakur and Thomas, 1985; Rana et al., 2004). Since asphaltenes are aggregate big molecules present in crude, their molecular size makes them difficult to enter inside the pores. Increasing the pore size of catalyst improves their diffusion; however, diffusivity is also controlled by catalyst particle and pore structure. That is why it would be desirable that the catalyst has the maximum possible surface area to minimize diffusion restrictions. Metals are also linked to asphaltenes as metalloporphyrins. Large unimodal pore catalysts with pores ranging from 100 to 300 nm exhibited the highest activity for vanadium removal for Kuwait vacuum residue as reported by Absi-Halabi et al. (1995). Nitrogen removal is achieved mainly in meso- and macropores because their structures have been related to porphyrinic compounds as well.

Since deactivation through pore diffusion will depend on the catalyst pore diameter as well as on the composition of the feedstock, the external surface of catalyst particles and diffusion into the catalyst interior become rate-limiting factors especially for the molecules of complex metals (Ni, V) and asphaltene. This can be followed by the textural properties and HDS, HDM, and HDAs conversions as shown in Figure 3.13. This behavior is observed for different average pore diameter (APD) of alumina-supported CoMo catalysts, which contain similar amounts of Co and Mo. In general, HDM and HDAs activities increase with pore diameter, while HDS activity decreases. These results further confirm diffusion limitations against large molecules into pores or catalytic sites, that is, CUS.

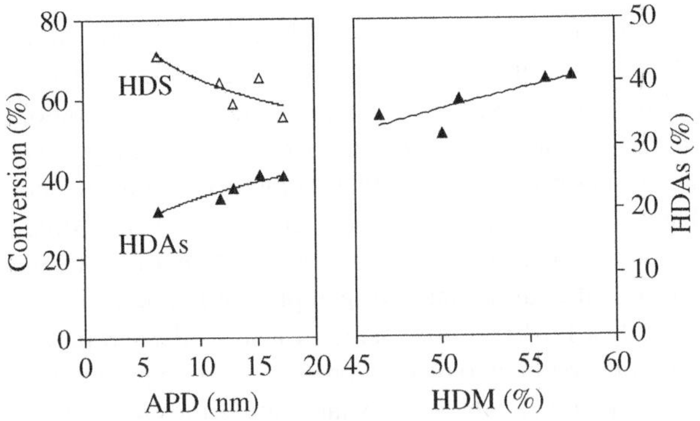

Figure 3.13 Effect of pore diameter on the HDS and HDAs, and HDAs versus HDM at 120 h time-on-stream (Adapted from Rana et al. (2007)).

Coke deposition over the catalyst was related to pore size and it was concluded that big molecules such as asphaltenes, which are prone to decompose and form coke, spend more time in narrower pores making the reaction diffusion-limited (Ancheyta et al., 2005). Asphaltenes that can go inside small pores undergo overcracking in the pores because they move slowly and for this reason more carbon is generated. Catalysts with more number of macropores favor that asphaltenes enter deeper inside and go out easily spending less time and producing less carbon. Deactivation can also be carried out by metals, for example, metalloporphyrins cannot enter very deep inside small pores by which molecules have restricted access to internal surface causing demetallization at the mouth of the pores.

When upgrading heavy crude oils, heavy and light fractions are converted at different degree. Lighter fractions have smaller size, and lower aromaticity and polarity than heavy fractions, which can turn these molecules into pentane soluble during catalytic hydrocracking. However, the highly condensed aromatic core is a potential precursor of coke (Zhao et al., 2001). In order to prevent coke formation, different types of treatments are available; one of them is deasphalting with solvents. Since asphaltenes are coke precursors, if they are removed, then the tendency to form coke is reduced. Other possibility is to increase partial pressure of hydrogen during the reaction. Capping of radicals is achieved by hydrogen and free radical intermediates are removed before polymerization, condensation, or coking take place. The hydrodemetallization of feedstock is also recommended in a guard reactor. The catalyst employed for HDM must possess bigger pores for capturing metals, which usually depend on the combination of textural properties of the catalyst. However, apart from the reaction conditions, the deposition of metals also depends on the metal content in the feedstock. This indicates that metal deposition affects more or less at a similar magnitude the conversion of HDAs and HDM, while the HDS activity is less affected. A plausible explanation for this behavior is smaller molecules of sulfur compound compared with those of the asphaltene, as well as Ni and V porphyrin molecules that can be easily diffused into the pores. The activity results are well supported by the textural properties of spent catalysts in which pore volume and specific surface area are reduced by 40% and 25%, respectively, at 120 h TOS (Rana et al., 2007).

The presence of hydrogen during thermal reactions initially limits the hydrogen donation ability of asphaltenes preserving their structures, especially the naphthenic and aliphatic bonds, which can be further cracked. Since the content of hydrogen in asphaltenes is still high at the beginning of thermal reactions, asphaltenes conversion is slower and coke is also slowly formed. In the absence of an effective hydrogen donor, the rate of coke formation is faster; however, by using a good hydrogen donor, the reaction proceeds slowly but increasing the selectivity toward maltenes. The presence of a catalyst keeps high selectivity to maltenes at intermediate reaction rate. The catalyst uses efficiently the hydrogen dissolved in the organic medium and hydrogenates part of residual asphaltenes (Savage and Klein, 1988). When hydrogen is not present in the reaction, a deficiency of hydrogen in asphaltenes is created originating free radicals that can condensate and/or polymerize and, in consequence, coke is readily formed. By using moderate hydroprocessing conditions, asphaltenes catalytic conversion is limited (<50%).

During hydroprocessing in a fixed-bed reactor, Guibard et al. (2005) have pointed out that the presence of asphaltenes in the HDM section limits the HDS of resins and aromatics.

Asphaltene aggregates that go to the HDS section proceeding from the HDM section are still large so that they cannot easily enter inside the pores, thereby limiting the HDS reaction. To establish the influence that asphaltenes have during HDS, Le Lannic et al. (2007) carried out experiments in which demetallized and deasphalted feedstocks were submitted to HDS. Resins and aromatics were desulfurized at around 100% when asphaltenes were previously removed. Compared with asphaltenic feedstocks, HDS was slower due to the presence of asphaltenes that block and retard the access to the active sites. Thus, it is important to remember that asphaltenes, resins, aromatics, and saturates are present in residue in a delicate balance. However, during hydroprocessing, destabilization of asphaltenes occurs when resins are converted, leaving asphaltenes unprotected to the point in which they are no longer soluble, causing instability and incompatibility. Higher temperatures during hydroprocessing convert the medium into more incompatible due to competition of thermal and cracking reactions and the difference in reactivity of distinct components. Aromatics and resins are less complex structures than asphaltenes and react more easily but due to resins under hydrogenating conditions, they modify their structure and at the same time asphaltenes undergo cracking of alkyl chains that leaves the aromatic core with a minimum number of branched chains. Asphaltenes are not soluble in hydrogenated resins and precipitate, enhancing coke formation on the catalyst as reported by Matsushita et al. (2004).

During catalytic processes and at initial stages, thermal reactions take place quickly and the catalyst does not play a role in that moment and consequently thermolysis of asphaltenes in residue occurs. Thermal and catalytic coke depends on the time in which they are formed. The influence of temperature is notorious in coke generation even though catalysts are used when processing heavy fractions. When using a catalyst for cracking, for example, zeolite-based catalyst, which is commonly used to upgrade residue, fast-growing coke produced by the influence of temperature is capable of damaging zeolite-supported catalysts by blockage or encapsulation before zeolite develops its cracking capability. The cleavage of alkyl chains will favor the generation of free radicals, which in turn will combine together to condensate and create bigger molecules that precipitate from the crude and deposit as coke. As reaction progresses, a second type of coke is observed on the catalyst, which is formed by associated molecules of asphaltenes and maltenes. In the case of asphaltenes, the presence of nitrogen in their structure makes them basic in nature being preferentially adsorbed even in weakly acidic surfaces. Desorption is not as easy to carry out and molecules tend to cross-linking giving a product with higher molecular weight. As the time goes on, more cross-linking is achieved and formation of coke is imminent but in this case having a catalytic influence. When using zeolite-supported catalysts, not all molecules are prone to be adsorbed onto the pores because large molecules still remain. Big molecules such as asphaltenes will deposit at the entrance of the pores, blocking the access to its interior and forming coke. Only when molecules have smaller size than pores, the catalyst will be really useful and will exert its cracking activity, but it has to be considered that at this stage many pores could be blocked by coke as reported by Ho (1992).

de Jong (1994) has also stated that coke formation is a parallel reaction that involves thermal and catalytic reactions. Thermal reactions enhance the formation of free radicals especially in big molecules such as asphaltenes; at this point, two cases are possible:

- Free radicals recombine to form a condensate molecule, which eventually grows up to form coke.
- Free radicals could be neutralized by the action of activated hydrogen (hydrogen radicals).

Activated hydrogen is formed by catalytic action, and in the case of hydroprocessing catalysts it is due to the presence of molybdenum even at low concentrations. However, at high temperatures, dehydrogenation of naphthenes is thermodynamically favored by the catalyst, which could imply that naphthenes linked to aromatic rings dehydrogenate increasing the aromaticity of molecules. The continuously repeated dehydrogenation would be responsible of forming more polynuclear aromatic rings.

Images of asphaltenes obtained by SEM magnified at around 2000 times are reported in Figure 3.14. Fractionated asphaltenes from hydrotreated crude at reaction temperatures of 400 and 420 °C along with asphaltenes from virgin Maya crude oil are shown in this figure with two different magnifications for each sample for better comparison. Asphaltenes were separated by using *n*-heptane and washed

Figure 3.14 SEM images of purified asphaltenes: (a) from Maya crude oil, (b) from hydrotreated Maya crude at 100 kg/cm^2, 400 °C and LHSV = 1 h^{-1}, (c) from hydrotreated Maya crude at 100 kg/cm^2, 420 °C and LHSV = 1 h^{-1}.

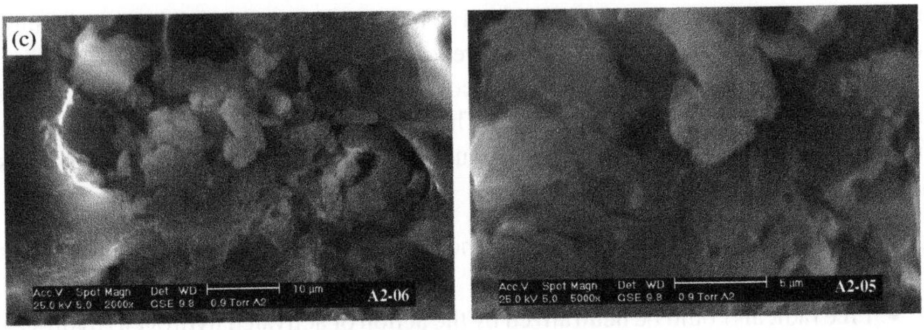

Figure 3.14 (*Continued*)

with a mixture of solvents (67/33 toluene/n-C_7). SEM images clearly indicate the aggregates of asphaltenes at the two reaction temperatures. This change in morphology can be attributed to the temperature at which reaction was carried out and significant differences in precipitated asphaltenes are observed, particularly when the reaction temperature changes in this range (from 400 to 420 °C). It appears from the figure that the virgin asphaltenes have small porosity and structure as fragile aggregates, while in the case of hydrotreated asphaltenes structure became stronger or defined, which is further magnified with the HDT reaction temperature of 420 °C. This structural information also indicated that with increasing reaction temperature only side and fragile structures are converted into more solid and highly aromatic in nature, which is more difficult to crack catalytically. Similar types of results have been reported by others, indicating that variation of structure of asphaltenes just depends on temperature.

REFERENCES

Absi-Halabi, M., Stanislaus, A., Al-Mughni, T., Khan, S., and Qamra, A. 1995. Hydroprocessing of vacuum residues: relation between catalyst activity, deactivation and pore size distribution. Fuel 74(8):1211–1215.

Absi-Halabi, M., Stanislaus, A., and Trimm, D.L. 1991. Coke formation on catalysts during the hydroprocessing of heavy oils. Appl. Catal., A 72(2):193–215.

Al-Dahhan, M.H., Karachi, F., Dudukovic, M.P., and Laurent, A. 1997. High pressure trickle-bed reactors: a review. Ind. Eng. Chem. Res. 36:3292–3314.

Agrawal, R.M., and Wei, J. 1984. Hydrodemetallation of nickel and vanadium porphyrins. 1. Intrinsic kinetics. Ind. Eng. Chem. Process Des. Dev. 23:505–514.

Ancheyta, J., Rana, M.S., and Furimsky, E. 2005. Hydroprocessing of heavy petroleum feeds: tutorial. Catal. Today 109:3–15.

Ancheyta, J., Betancourt, G., Marroquín, G., Centeno, G., Alonso, F., and Muñoz, J.A. 2010. Process for the catalytic hydrotreatment of heavy hydrocarbons of petroleum. US Patent 7,651,604 B2.

Babich, I.V., and Moulijn, J.A. 2003. Science and technology of novel processes for deep desulfurization of oil refinery streams: a review. Fuel 82:607–631.

Banerjee, D.K., Laidler, K.J., Nandi, B.N., and Patmore, D.J. 1986. Kinetic studies of coke formation in hydrocarbon fractions of heavy crudes. Fuel 65(4):480–484.

Bartholdy, J. and Cooper, B.H. 1993. Metal and coke deactivation of resid hydroprocessing catalysts. Prepr. – Am. Chem. Soc., Div. Pet. Chem.. 38(2), 386–390.

Beaton, W.I., and Bertolacini, R.J. 1991. Resid hydroprocessing at Amoco. Catal. Rev. Sci. Eng. 33:281–317.

Bhaskar, M., Valavarasu, G., Sairam, B., Balaraman, K.S., and Balu, K. 2004. Three-phase reactor model to simulate the performance of pilot-plant and industrial trickle-bed reactors sustaining hydrotreating reactors. Ind. Eng. Chem. Res., 43:6654–6669.

Billon, A., Peries, J.P., Espeillan, M., and des Courieres, T. 1991. Hyvahl F versus Hyvahl M, swing reactor or moving bed. Paper presented at the 1991 NPRA Annual Meeting, March 17–19, 1991, San Antonio, Texas.

Bingham, F.E., and Christensen, P. 2000. Revamping HDS units to meet high quality diesel specifications. Presented at the Asian Pacific Refining Technology Conference, March 8–10, 2000, Kuala Lumpur, Malaysia.

Brooks, J.D. and Taylor, G.H. 1965. The formation of graphitizing carbons from the liquid phase. Carbon 3(2):185–186.

Cable, T.L., Massoth, F.E. and Thomas, M.G. 1981. Studies on an aged H-coal catalyst. Fuel Process. Technol. 4:265–275.

Choi, J.H.K. and Gray, M.R. 1988. Structural analysis of extracts from spent hydroprocessing catalysts. Ind. Eng. Chem. Res. 27(9):1587–1595.

Chou, T. 2004. Causes of fouling in hydroprocessing units. Pet. Technol. Quart. 3:1–5.

Daniel, M., Lerman, D.B., and Peck, L.B. 1988. Amocos LC-fining residue hydrocracker yield and performance correlations from a commercial unit. Paper presented at the 1988 NPRA annual meeting.

Dautzenberg, F.M., van Klinken, J., Pronk, K.M.A., Sie, S.T., and Wijffels, J.B. 1978. Catalyst deactivation through pore mouth plugging during residue desulfurization. ACS Symp. Ser. 65:254–267.

Dautzenberg, F.M., and de Deken, J.C. 1985. Modes of operation in hydrodemetallization. Prepr. – Am. Chem. Soc., Div. Pet. Chem. 30:8–20.

de Jong, K.P. 1994. Effects of vapor–liquid equilibria on coke deposition in trickle-bed reactors during heavy oil processing. 1. Experimental results. Ind. Eng. Chem. Res. 33(4):821–824.

Del Bianco, A., Panariti, N., Anelli, M., Beltrame, P.L., and Carniti, P. 1993. Thermal cracking of petroleum residues. 1. Kinetic analysis of the reaction. Fuel 72(1):75–85.

Delmon, B. 1990. Advances in hydropurification catalysts and catalysis, in: D.L. Trimm, S. Akashah, M. Absi-Halabi, and H. Bishara (Eds.), Catalysts in petroleum refining, Elsevier, The Netherlands, 1–40, p. 1989.

Dong, D., Jeong, S., and Massoth, F.E. 1997. Effect of nitrogen compounds on deactivation of hydrotreating catalysts by coke. Catal. Today 37(3):267–275.

Euzen, J.P. 1991. Moving-bed process for residue hydrotreating. Rev. Inst. Fr. Pet. 46:517–527.

Fonseca, A., Zeuthen, P., and Nagy, J.B. 1996. ^{13}C NMR quantitative analysis of catalyst carbon deposits. Fuel 75:1363–1376.

Furimsky, E. 1978. Chemical origin of coke deposited on catalyst surface. Ind. Eng. Chem. Prod. Res. Dev. 17(4):329–331.

Furimsky, E. 1998. Selection of catalysts and reactors for hydroprocessing. Appl. Catal., A 171:177–206.

Furimsky, E., and Massoth, F.E. 1999. Deactivation of hydroprocessing catalysts. Catal. Today 52:381–495.

Gentzis, T., Rahimi, P., Malhotra, R., and Hirschon, A.S. 2001. The effect of carbon additives on the mesophase induction period of Athabasca bitumen. Fuel Process. Technol. 69(3):191–203.

Gianetto, A., and Specchia, V. 1992. Trickle-bed reactors: state of art and perspectives. Chem. Eng. Sci. 47(13–14):3197–3213.

Gimaev, R.N., Gubaidullin, V.Z., Rogacheva, O.V., Davydov, G.F., and Danil'yan, T.D. 1980. Mechanism of coke formation from liquid phase. Chem. Tech. Fuels Oils 16(3):196–199.

Gorshteyn, A., Kamiesnky, P., Davis, T., Novak, W., and Lee, M. 2002. ExxonMobil catalytic dewaxing – a commercial proven technology. 2nd Russian Refining Technology Conference, Sept. 26–27, 2002, Moscow.

Gosselink, J.W. 1998. Sulfide catalysts in refineries. CATTECH 2(2):127–144.

Gruia, A. 2006. Recent advances in hydrocracking, Chapter 7, in: Ch.H. Hsu and P.R. Robinson (Eds), Practical advances in petroleum processing, Springer, New York, USA.

Gualda, G., and Kasztelan, S. 1996. Initial deactivation of residue hydrodemetallization catalysts. J. Catal. 161(1):319–337.

Guibard, I., Merdrignac, I., and Kressmann, S. 2005. Characterization of refractory sulfur compounds in petroleum residue. ACS Symp. Ser. 895, 51–64.

Ho, T.C. 1992. Study of coke formation in resid catalytic cracking. Ind. Eng. Chem. Res. 31(10):2281–2286.

Ho, T.C. 2003. Hydrodesulfurization with RuS_2 at low hydrogen pressures. Catal. Lett. 89(1–2):21–25.

Johnson, B.J., Massoth, F.E. and Bartholdy, J. 1986. Diffusion and catalytic activity studies on resid-deactivated HDS catalysts. AIChE J. 32:1980–1987.

Kam, E.K.T., Al-Mashan, M.-H., and Al-Zami, H. 1999. The mixing aspects of NiMo and CoMo hydrotreating catalysts in ebullated-bed reactors. Catal. Today 48:229–236.

Korsten, H., and Hoffmann, U. 1996. Three-phase reactor model for hydrotreating in pilot trickle-bed reactors. AIChE J. 42(5):1350–1357.

Kressmann, S., Morel, F., Harlé, V., and Kasztelan, S. 1998. Recent developments in fixed-bed catalytic residue upgrading. Catal. Today 43:203–215.

Le Lannic, K., Guibard, I., and Merdrignac, I. 2007. Behavior and role of asphaltenes in a two-stage fixed bed hydrotreating process. Pet. Sci. Technol. 25(1–2):169–186.

Le Page, J.F., Morel, F., Trassard, A.M., and Bousquet, J. 1987. Thermal cracking under hydrogen pressure: preliminary step to the conversion of heavy oils and residues. Prepr. – Am. Chem. Soc., Div. Pet. Chem.. 32(2):470–476.

Matsushita, K., Marafi, A., Hauser, A., and Stanislaus, A. 2004. Relation between relative solubility of asphaltenes in the product oil and coke deposition in residue hydroprocessing. Fuel 83(11–12):1669–1674.

Mederos, F.S., Rodriguez, M.A., Ancheyta, J., and Arce, E. 2006. Dynamic modeling and simulation of catalytic hydrotreating reactors. Energy Fuels 20:936–945.

Mehra, Y.R., and Al-Abdulal, A.H. 2005. Hydrogen purification in hydroprocessing (HPH^{SM} technology). Paper presented at the 103rd NPRA annual meeting, March 13–15, 2005, San Francisco, California, USA.

Minderhoud, J.K., van Veen, J.A.R., and Hagan, A.P. 1999. Hydrocracking in the year 2000: a strong interaction between technology development and market requirements, Stud. Surf. Sci. Catal. 127:3–20.

Mitchell, P.C.H., 1990. Hydrodemetallisation of crude petroleum: fundamental studies. Catal. Today 7:439–445.

Morel, F., Kressmann, S., Harlé, V., and Kasztelan, S. 1997. Processes and catalysts for hydrocracking of heavy oil and residues, in: G.F. Froment, B. Delmon, and P. Grange (Eds.) Hydrotreatment and hydrocracking of oil fractions, Elsevier, Amsterdam.

Moulijn, J.A., van Diepen, A.E., and Kapteijn, F. 2001. Catalyst deactivation: is it predictable? What to do? Appl. Catal., A 212:3–16.

Muñoz, J.A.D., Alvarez, A., Ancheyta, J., Rodríguez, M.A., and Marroquín, G. 2005. Process heat integration of a heavy crude hydrotreatment plant. Catal. Today 109:214–218.

Nielsen, A., Cooper, B.H., and Jacobsen, A.C. 1981. Composite catalyst beds for hydroprocessing of heavy residua. Prepr. – Am. Chem. Soc., Div. Pet. Chem. 26(2):440–455.

Oelderik, J.M., Sie, S.T., and Bode, D. 1989. Progress in the catalysis of the upgrading of petroleum residue: a review of 25 years of R&D on Shell's residue hydroconversion technology. Appl. Catal. 47:1–24.

Ouwerkerk, C.E.D., Bratland, E.S., Hagan, A.P., Kikkert, B.L.J.P., and Zonnevylle, M.C. 1999. Performance optimisation of fixed bed processes. Pet. Technol. Quart. 4(1):21–30.

Pazos, J.M., Gonzalez, J.C., and Salazar-Guillen, A.J. 1983. Effect of catalyst properties and operating conditions on hydroprocessing high metals feeds. Ind. Eng. Chem. Process Des. Dev. 22:653–659.

Quann, R.J., Ware, R.A., Hung, Ch., and Wei, J. 1988. Catalytic hydrodemetallation of petroleum. Adv. Chem. Eng. 14:95–259.

Rahmani, S., McCaffrey, W., Elliott, J.A.W., and Gray, M.R. 2003. Liquid-phase behavior during the cracking of asphaltenes. Ind. Eng. Chem. Res. 42(17):4101–4108.

Rana, M., Ancheyta, J., Rayo, P., and Maity, S.K. 2004. Effect of alumina preparation on hydrodemetallization and hydrodesulfurization of Maya crude. Catal. Today 98(1–2):151–160.

Rana, M., Ancheyta, J., Maity, S.K., and Rayo, P. 2005. Characteristics of Maya crude hydrodemetallization and hydrodesulfurization. Catal. Today 104(1):86–93.

Reynolds, B.E., Bachtel, R.W., and Yagi, K. 1992. Chevron's onstream catalyst replacement (OCR^{TM}) provides enhanced flexibility to residue hydrotreaters. Paper presented at the 1992 NPRA annual meeting, March 22–24, 1992, New Orleans, Louisiana, USA.

Reynolds, J.G. 1998. Metals and heteroelements, in heavy crude oils, in: J.G. Speight, (Ed.), Petroleum chemistry and refining, Taylor & Francis, Washington, DC, Chapter 3.

Robinson, P.R., and Dolbear, G.E. 2006. Hydrotreating and hydrocracking: fundamentals, in: Ch.H. Hsu and P.R. Robinson (Eds), Practical advances in petroleum processing, Springer, New York, USA, Chapter 7.

Rodriguez, M.A., and Ancheyta, J. 2004. Modeling of hydrodesulfurization (HDS), hydrodenitrogenation (HDN), and the hydrogenation of aromatics (HDA) in a vacuum gas oil hydrotreater. Energy Fuels 18:789–794.

Ruiz, R.S., Alonso, F., and Ancheyta, J. 2004a. Minimum fluidization velocity and bed expansion characteristics of hydrotreating catalysts in ebullated-bed systems. Energy Fuels 18:1149–1155.

Ruiz, R.S., Alonso, F., and Ancheyta, J. 2004b. Effect of high pressure operation on overall phase holdups in ebullated-bed reactors. Catal. Today 98:265–271.

Ruiz, R.S., Alonso, F., and Ancheyta, J. 2005. Pressure and temperature effects on the hydrodynamic characteristics of ebullated-bed systems. Catal. Today 109:205–213.

Sarli, M.S., McGovern, S.J., Lewis, D.W., and Snider, P.W. 1993. Improved hydrocracker temperature control: Mobil quench zone technology. Paper presented at the 1993 NPRA annual meeting, March 21–23, Houston, TX.

Satterfield C.N., Modell, M., Hites, R.A., and Declerck, C.J. 1978. Intermediate reactions in the catalytic hydrodenitrogenation of quinoline. Ind. Eng. Chem. Prod. Res. Dev. 17(2):141–148.

Savage, P.E. and Klein, M.T. 1988. Asphaltene reaction pathways. 3. Effect of reaction environment. Energy Fuels 2(5), 619–628.

Scheffer, B., van Koten, M.A., Robschlager, K.W., and de Boks, F.C. 1988. The shell residue hydroconversion process: development and achievements. Catal. Today 43:217–224.

Scheuerman, G.L., Johnson, D.R., Reynolds, B.E., Bachtel, R.W., and Threlkel, R.S. 1993. Advances in Chevron RDS technology for heavy oil upgrading flexibility. Fuel. Process. Technol. 35:39–54.

Schulman, B.L., and Dickenson, R.L. 1991. Upgrading heavy crudes: a wide range of excellent technologies now available. UNITAR 5th International Conference, Caracas, Ven. Aug. 105–113.

Sie, S.T. 1980. Catalyst deactivation by pore plugging in petroleum processing, in: B. Delmon and G.F. Froment (Eds), Catalyst deactivation, Elsevier, Amsterdam, 545–569.

Sie, S.T. 2001. Consequences of catalyst deactivation for process design and operation. Appl. Catal., A 212:129–151.

Smith, B.J. and Wei, J. 1991. Deactivation in catalytic hydrodemetallation. III. Random-spheres catalyst models. J. Catal. 132:41–57.

Speight, J.G. 1987. Initial reactions in the coking of residua. Prepr. – Am. Chem. Soc., Div. Pet. Chem. 32(2):413–418.

Speight, J.G. 2000. The desulfurization of heavy oils and residua. 2nd edition. Marcel Dekker, New York.

Stefanidis, G.D., Bellos, G.D., and Papayannakos, N.G. 2005. An improved weighted average reactor temperature estimation for simulation of adiabatic industrial hydrotreaters. Fuel Process. Technol. 86:1761–1775.

Storm, D.A., Barresi, R.J., and Sheu, E.Y. 1996. Flocculation of asphaltenes in heavy oil at elevated temperatures. Pet. Sci. Technol. 14(1–2):243–260.

Storm, D.A., Barresi, R.J., Sheu, E.Y., Bhattacharya, A.K., and DeRosa, T.F. 1998. Microphase behavior of asphaltic micelles during catalytic and thermal upgrading. Energy Fuels 12(1):120–128.

Takahashi, T., Higashi, H., and Kai, T. 2005. Development of a new hydrodemetallization catalyst for deep desulfurization of atmospheric residue and the effect of reaction temperature on catalyst deactivation. Catal. Today 104:76–85.

Tamm, P.W., Harnsberger, H.F., and Bridge, A.G. 1981. Effects of feed metals on catalyst aging in hydroprocessing residuum. Ind. Eng. Chem. Proc. Des. Dev. 20(2):262–273.

Ternan, M., Furimsky, E., and Parsons, B.I. 1979. Coke formation on hydrodesulphurization catalysts. Fuel Process. Technol. 2:45–55.

Thakur, D.S. and Thomas, M.G. 1985. Catalyst deactivation during coal liquefaction: a review. Appl. Catal., A 45(2):197–225.

Topsøe, H., Clausen, S., and Massoth, F.E. 1996. Hydrotreating catalysis science and technology, Springer-Verlag, Berlin.

Van Doorn, J. and Moulijn, J.A. 1990. Carbon deposition on catalysts. Catal. Today 7(2):257–266.

Van Zijll Langhout, W.C., Ouwerkerk, C., and Pronk, K.M.A. 1980. New process hydrotreats metal-rich feedstocks. Oil Gas J. 1:120–126.

Wiehe, I.A. 1993. A phase-separation kinetic model for coke formation. Ind. Eng. Chem. Res. 32(11):2447–2454.

Wiehe, I.A. 1994. The pendant-core building block model of petroleum residua. Energy Fuels 8(3):536–544.

Zhang, L., Yang, G., Que, G., Zhang, Q., and Yang, P. 2006. Colloidal stability variation of petroleum residue during thermal reaction. Energy Fuels 20(5):2008–2012.

Zhao, S., Kotlyar, L.S., Woods, J.R., Sparks, B.D., Hardacre, K., and Chung, K.H. 2001. Molecular transformation of Athabasca bitumen end-cuts during coking and hydrocracking. Fuel 80(8):1155–1163.

4

CHARACTERIZATION OF SPENT HYDROPROCESSING CATALYST

4.1 INTRODUCTION

Heavy oils contain high amount of metals, sulfur, nitrogen and have lower API gravity than light crude oils. To improve these properties, they require to be passed through an upgrading process, such as hydroprocessing. However, hydroprocessing catalysts decrease their activity with time-on-stream (TOS), the so-called catalyst deactivation or catalyst activity decay. Catalyst deactivation is an inevitable phenomenon that involves both chemical and physical changes in the catalyst nature along with the catalytic reaction.

In fixed-bed hydroprocessing units, catalyst deactivation is due to three main causes:

- *Step I.* Carbon lay down
- *Step II.* Slow metal deposition on active phase
- *Step III.* Diffusion limitation into the pores.

The coke and metal deposition is associated with the loss of activity due to the decrease in catalytic sites as well as textural properties.

Hydroprocessing of heavy oils is often accomplished in multistage, that is, a series of reactor and/or several catalytic beds aiming at protecting the tail-end catalysts from premature deactivation. Through the different stages, one or several specific goals are aimed: diminish metal content, transform asphaltenes and resins, extinguish metal-containing molecules, lessen sulfur and nitrogen content, and achieve ultradeep

Deactivation of Heavy Oil Hydroprocessing Catalysts: Fundamentals and Modeling, First Edition. Jorge Ancheyta.

sulfur levels. Therefore, the mechanism and extent of catalytic deactivation depend on the hydroprocessing stage.

To maintain the desired activity, the catalyst deactivation is compensated by continuously raising the temperature. After certain time, the deactivated catalyst needs to be replaced by new one. Depending on the nature of crude oils and requirement, the life of hydrotreating catalyst for heavy oil varies from 6 months to 1 year.

In general, deactivation on supported catalysts involves complex physical and chemical occurrences. The deposited coke and metals are due to different reasons because deposition patterns are different. Thus, it is important to study the deactivation behavior of heavy crude oil hydroprocessing catalyst by investigating deposited species, state and distribution of coke and metals on catalyst.

4.2 CHARACTERIZATION TECHNIQUES

Prior to the characterization of spent catalysts, it is necessary to wash them by using a Soxhlet equipment in order to remove the physically adsorbed hydrocarbon on the catalyst. Figure 4.1 shows a typical Soxhlet apparatus.

In some cases, Soxhlet washing of spent catalysts with toluene followed by LGO (light gas oil) and again toluene washing is applied (Saito and Kumata, 1998). The use of LGO allows for removing some carbonaceous compounds (25–45%) deposited on the catalyst surface. During this treatment, the coke scrub down is known as "soft coke," while the remaining coke is called "hard coke." It was found that hard coke was mainly deposited in larger pores. Soft coke can be considered a temporal poison that exists on the catalyst surface. Thus, the removal of light carbonaceous compounds by LGO reflux permits to characterize spent catalysts in a better way. Spent hydroprocessing catalysts can be characterized by a number of routine techniques (atomic absorption, textural properties, etc.) as well as by more specialized methods. Some of the main techniques are described in the following sections.

4.2.1 Temperature Programmed Oxidation (TPO)

Characterization of spent catalyst by temperature programmed oxidation (TPO) is used to obtain information for oxidation of coke at different temperatures along with the oxidation of deposited S and N species that are present in coke. Since sulfur can be released from the metal sulfides (Mo, Co, Ni, V, Fe, etc.), so it is difficult to quantify the total sulfur obtained as SO_2. Al-Dalama and Stanislaus (2006) carried out TPO analyses for CO_2, SO_2, and NO_2 on spent catalysts obtained from different beds in fixed-bed reactor. In the case of TPO for CO_2 analysis, it was observed that in the first reactors, whose catalysts are more poisoned with vanadium, the peak for CO_2 was from 200 to 450 °C. On the other hand, in the last reactors, whose catalysts have less vanadium deposited, the peak for CO_2 began around 250 °C and it was extended up to 550 °C. Spent catalysts from the first reactors exhibited a lower temperature range for coke oxidation, which can be due to the presence of vanadium that may catalyze the oxidation as reported by Massoth (1981). The TPO profiles for SO_2 exhibited three

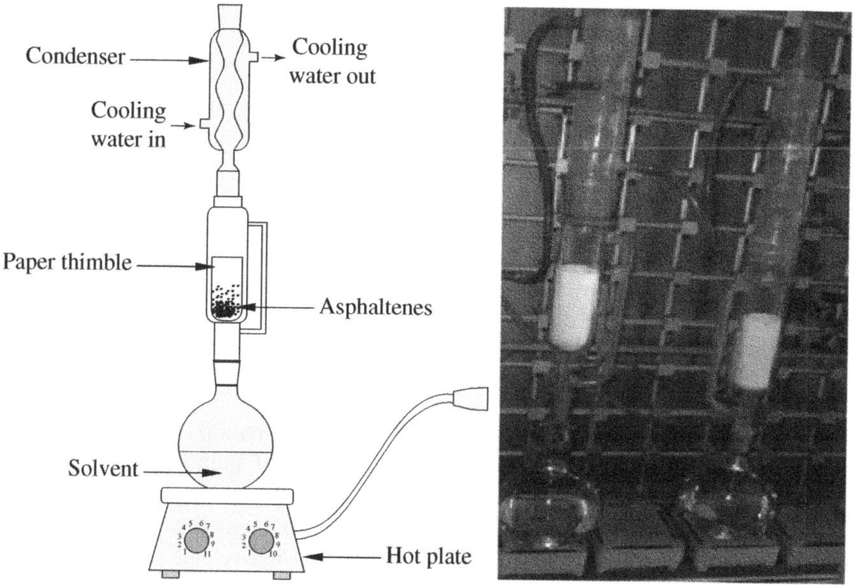

Figure 4.1 Typical Soxhlet apparatus.

peaks. The first peak between 200 and 250 °C corresponds to the oxidation of metal sulfides (Yoshimura et al., 1991). The second peak is broader, about 300–500 °C, which may correspond to CO_2 profiles and can be due to sulfur in coke is oxidized along with carbon. The third peak is located in the interval of 600–700 °C and could be assigned to decomposition of sulfates into SO_2 (Clark et al., 1993). TPO of NO_2 profiles showed three peaks: the first peak coincides with that of the SO_2 and could be due to adsorption of basic nitrogen in the sulfide phase by which nitrogen species could be oxidized along with sulfur species; the second peak coincides with CO_2 at around 380 °C, which could imply a relationship between C and N in coke, some authors have reported the presence of N in coke (Speight, 1987); and the third peak around 700 °C is not fully identified; however, it could be due to the presence of metal nitrates along with metal sulfates. Figure 4.2 shows typical TPO profiles obtained by Al-Dalama and Stanislaus (2006). In order to perceive soft coke and hard coke by TPO, soft coke burns at lower temperatures (low-temperature CO_2 peak) while hard coke at higher temperatures (high-temperature CO_2 peak) particularly coke formed due to the precipitation of asphaltenes or due to sediment formation.

4.2.2 Nuclear Magnetic Resonance

Generally, coke is either surrounded by carbon atom or by very few hydrogen atoms. In order to identify the nature of carbon and its surrounding atoms connectivity ^{13}C NMR is the best technique to determine its structure. Usually, coke deposition increases along the reactor and its nature also becomes more aromatic at the end

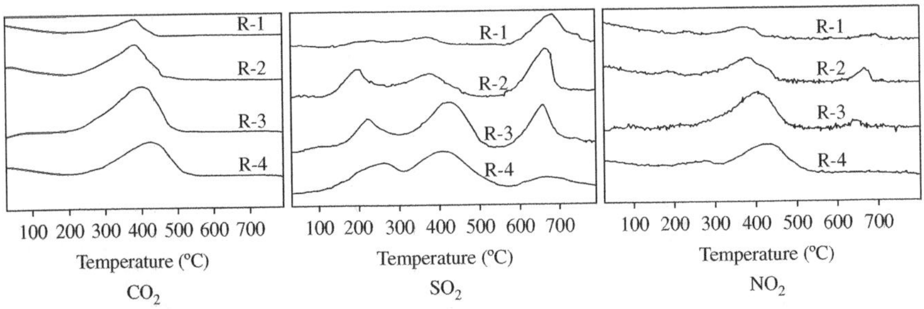

Figure 4.2 TPO profiles of spent catalysts from four reactors in a series (R-1 to R-4) (Adapted from Al-Dalama and Stanislaus (2006)).

of the reactor, which has less alkyl chains linked. Structural parameters can be identified easily by using well-established techniques of high power decoupling, magic-angle spinning (MAS) to remove chemical shift anisotropy (~100 ppm for aromatic carbons), and cross-polarization (CP) in which magnetization is transferred from abundant ^{1}H to dilute ^{13}C spins to improve sensitivity and so avoiding long relaxation delays. By using solid-state ^{13}C NMR, it has been reported that coke represents different types of structures (tertiary and quaternary carbons, other carbon types), carbon groups (aliphatic or aromatic), which are usually affected by the nature of feedstock, catalyst type, and reaction conditions (Al-Dalama and Stanislaus, 2006). Egiebor et al. (1989) have characterized organic residues on spent hydroprocessing catalysts by NMR and confirmed that coke contained a significant amount of aliphatic, hydroaromatic, and aromatic compounds. Snape et al. (1999) have defined the existence of two types of coke depending on the percent of deposited carbon on the catalyst and type of solvent used to extract coke. The soft coke is chloroform-extractable and accounts for about 30% of the total carbonaceous deposit on the catalyst being highly polar with significant aliphatic nature. Hard coke is not solvent-extractable (5–7 wt% of carbon content) and has an aromaticity of around 50%. When longer periods of deactivation by coke are present, carbon content is around 15 wt% or higher and is highly aromatic (90% or even more). ^{13}C NMR analysis indicated that hard coke is highly aromatic and soft coke is having aliphatic peak and for this reason this type of coke is chloroform-extractable. For deep analysis of hard coke, different experiments have been carried out in order to identify clearly the contribution of different groups such as ^{13}C SPE/MAS (single-pulse excitation/magic angle spinning) and CP/MAS (cross-polarization/magic angle spinning). NMR experiments were employed to quantitatively measure insoluble coke deposits (Sahoo et al., 2004). The authors have also tried to elucidate coke composition from MAS NMR along with other analytical techniques (soluble and insoluble components) that are extracted from spent hydrocracking catalyst. It was also reported that soft coke is more related to the feedstock composition, whereas the insoluble coke or hard coke formation may depend on the nature of catalyst and its functional reactions, which polycondensed to aromatics.

4.2.3 Raman Spectrometry

Characterization of spent catalysts by Raman spectrometry has been used by several authors in order to elucidate coke structure (Digne et al., 2007; Amemiya et al., 2003; Chua and Stair, 2003; Espinat et al., 1985). Digne et al. (2007) have characterized a $NiMo/Al_2O_3$ spent catalyst from a hydrodesulfurization unit operating for 2 years. Classical characteristic bands of MoS_2 phase at 384 and 406 cm^{-1} were identified, which were also reported by Brown et al. (1977). Apart from MoS_2, two more peaks were detected, first one between 1550 and 1650 cm^{-1} and the second one around 1250 and 1450 cm^{-1}, which in appearance has similarities to graphite spectrum, which exhibits two peaks at 1585 cm^{-1} (G mode) and 1355 cm^{-1} (D mode) that correspond to the ring stretching of polynuclear aromatic rings. However, when using different feedstocks, such as standard gas oil, vacuum distillate, and vacuum gas oil over the same catalyst in a pilot plant (PP) test, supplementary differences were found. The signal at around 1600–1605 cm^{-1} remained almost without changes, but when vacuum distillate was used, two peaks at 1585 and 1618 cm^{-1} appeared indicating the presence of intermediate species. The peak around 1250 and 1450 cm^{-1} presents some displacements, and in the case vacuum distillate and vacuum gas oil, it is broader. Amemiya et al. (2003) have also reported the presence of peaks at 1600 and 1350 cm^{-1} when characterizing spent catalyst for vacuum gas oil hydrotreating operating unit at 8 MPa and 360–400 °C for 1 year in a fixed-bed reactor. These two peaks are stronger as the depth of catalytic bed increases. Carbon deposition is slightly more remarkable on the edge and less at the center of the catalyst particle. Both signals at 1600 and 1350 cm^{-1} showed this trend; that is, higher intensity near the edge, intermediate intensity in the middle, and lower intensity at the center of the catalyst particle. Calcination of spent catalyst increased the intensity of the bands mainly in the catalyst located at the top of the catalytic bed. The catalyst located in the lower section of the catalytic bed exhibited partially graphitized carbon. Interpretation of Raman spectra of coke is not easy; however, this technique is useful for characterizing not only fresh catalysts but also deposited coke.

4.2.4 SEM-EDX Analysis

The traces of deposited species can be identified by using different characterization techniques such as energy-dispersive XRD. With this technique, different metals such as Ni, V, Fe, and Na deposited on the catalyst surface have been detected. Since most of the metalloporphyrins are associated with asphaltene molecules, it is likely that these complex molecules of asphaltene do not diffuse into the catalyst pore. Consequently, their conversion takes place at the entrance of the pore mouth due to pore diffusion limitation. Thus, normally the distribution of vanadium and carbon deposition as a function of catalyst particle (extrudate) adopts a U-shape pattern with higher deposition near the edge of the particle as shown in Figure 4.3. Al-Dalama and Stanislaus (2006) characterized vanadium deposits across the pellet in a system of four fixed-bed reactors in series by electron microprobe analysis. Profiles of vanadium deposition changed from the first to the fourth reactor. Vanadium concentration

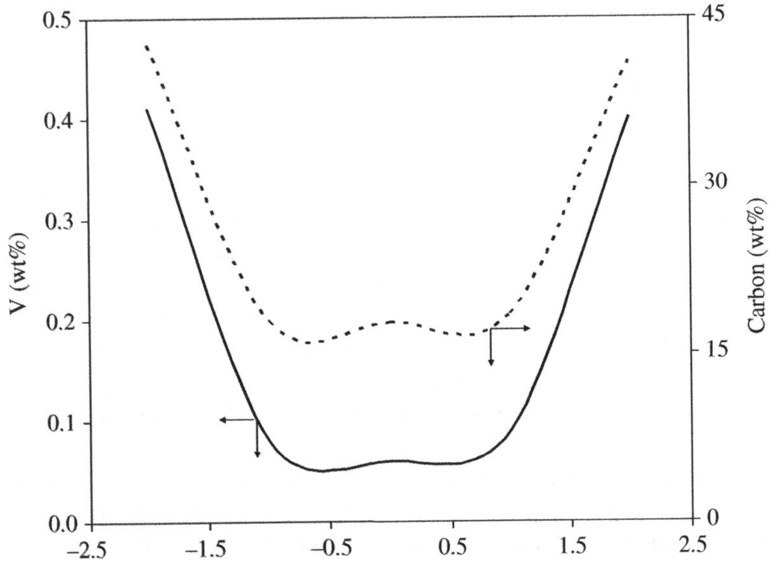

Figure 4.3 Deposited vanadium and carbon concentrations profiles obtained in radial distribution of cylindrical extrudate (Adapted from Rana et al. (2007)).

diminished from the edge of the pellet to the center. Usually, heavy oil hydrotreating catalysts are loaded as graded bed approach, that is, a catalyst with big pores for high metal retention and low activity is loaded at the top bed, where the rate of diffusion of molecules containing metals is high. Metals are deeper transported inside the pores and its deposition is homogeneous across the pellet. On the other hand, catalysts used in the last reactors are more active having smaller pores by which diffusion rate will be low. Molecules have a tendency to react as they enter inside the pores and metal deposition will occur near the pore mouth. In consequence, nonhomogeneous deposition of vanadium will be obtained in the pellets of the last reactors having maximum concentrations in the edges.

Ancheyta et al. (2003) performed hydrotreating of Maya crude in a fixed-bed reactor using a NiMo/Al_2O_3 tetralobe commercial catalyst and found by SEM analysis that deposition profiles of V along the longest and the shortest distance through the lobes were almost similar. TOS was 1120 h, and temperature was increased (400, 420, and 430 °C) for compensating catalyst deactivation. The deposition of V metal as a function of extrudate radial distribution zones is shown in Figure 4.4, which shows an enrichment of V in the superficial region of the catalyst particle and its radial analyses. These patterns differ from those obtained at lower TOS (490 h) and at constant temperature (400 °C) reported elsewhere (Ancheyta et al., 2002), where the longest distance through the lobes presented more deposited vanadium. Differences between both cases are due to the TOS and the increase in temperature. Qualitative deposition of Ni and V on spent catalysts can be evaluated also by scanning transmission

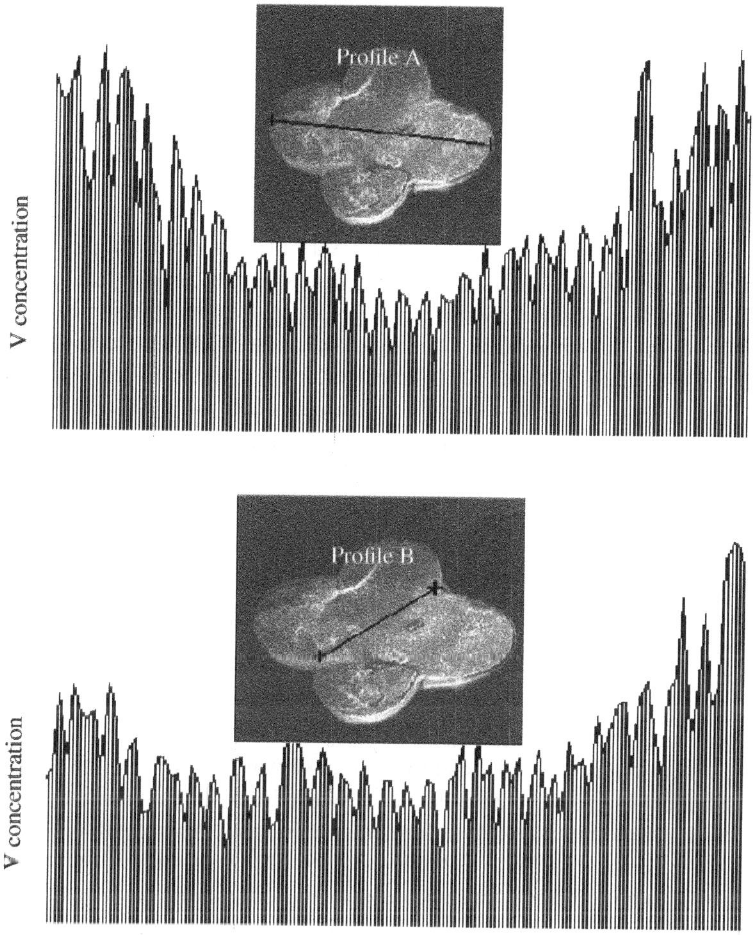

Figure 4.4 Intraparticle SEM-EDAX analysis of V profiles in tetralobe spent commercial catalyst (Adapted from Ancheyta et al. (2002, 2003)).

electron microscopy. Trejo et al. (2008) have shown that the distribution of Ni and V over CoMo/Al_2O_3–MgO spent catalysts after 204 h of TOS exhibited a homogeneous deposition of metals on catalyst particle. However, the catalyst with larger pore diameters (CoMo/Al–Mg-0.1) has a higher amount of metals due to its wider pores. On the other hand, smaller-pore-diameter catalysts have more coke deposition. The less coke formation could be due to the higher MgO content in the catalyst that may prevent coke formation due to its basic nature. The distribution of vanadium is in the outer surface of the catalyst while Ni distribution is more uniform throughout the catalyst particle. Diffusion of nickel porphyrin reactant molecule inside the pores is higher and for this reason nickel is uniformly deposited across the pellet. It has been established that vanadium is more reactive than nickel due to the presence of a double bond in vanadium, which makes it more prone to be attacked. Some authors have reported that reactivity of vanadium compounds is 3–5 times higher than that of nickel. However, the effective diffusivity of nickel is 10 times higher than that of vanadium (Sato et al., 1971).

4.2.5 Thermogravimetric Analysis (TGA)

Thermogravimetric analyses (TGA) of coke not only identify the type of coke but also are a useful tool to estimate kinetic parameters, which can be obtained from the pyrolysis of asphaltenes. Reaction order in coking is frequently assumed to be one. A number of kinetic studies have been reported using TGA data during the coking reactions (Cotte and Calderon, 1981; Schucker, 1983; Jiménez-Mateos et al., 1996; Li and Yue, 2003). This nonisothermal technique is helpful because a short time is required and problems due to initial heat-up period that accompanies the isothermal experiments are avoided (Shih and Sohn, 1980).

4.2.5.1 TGA Experiments The influence of asphaltenes in coke formation in different atmospheric residues was analyzed by Gonçalves et al. (2007). They observed that not only asphaltenes are responsible for coke formation when thermal cracking is used for upgrading residues but also other heavy components are coke precursors such as polar aromatics. In one test, atmospheric residue was thermally cracked and the amount of coke was measured. In another test, atmospheric residue without asphaltenes, which were precipitated from residue by using the IP-143 standard method, was also cracked and the amount of coke was compared with that obtained when asphaltenes were present in residue. In spite of removing asphaltenes by precipitation, a small amount of coke was formed in each case. When SARA fractionation and TGA are combined, it is possible to establish the propensity of each fraction to form coke as reported by Jiménez-Mateos et al. (1996) who fractionated different heavy products into saturates, aromatics, resins, and asphaltenes, and each fraction along with the whole bitumen was analyzed by TGA to show quantitatively that asphaltenes are responsible for 50 wt% of carbonaceous residue at least during pyrolysis, followed by resins that are capable to form up to 20 wt%. The whole bitumen as such is responsible for around 20% of coke during pyrolysis, whereas the least fractions prone to form coke are aromatics and saturates. In all cases, the loss of weight was almost constant beyond 500 °C.

4.2.5.2 Kinetics of Coking The use of nonisothermal kinetics when applying TGA has been reviewed to great extent by Flynn and Wall (1966). The variation of heating rate was also discussed and reliable measurements were obtained. In this case, the method was applied to polymers, but the mathematical treatment was not exclusive to these materials and thermogravimetry of heavy fractions of petroleum and residue can be analyzed in the same way. Different methods of analysis have been reported, which include (1) integral method, (2) differential method, (3) difference-differential method, (4) initial reaction rates, and (5) nonlinear or cyclic heating rate methods.

Trejo et al. (2008) used a nonisothermal technique with various heating rates applied to the determination of kinetics of coking of asphaltenes from a 13°API crude that were purified under Soxhlet reflux with n-C_7 overnight in order to remove adsorbed resins. Approximately 10 mg of purified sample was analyzed at four different heating rates: 4, 8, 12, and 16 °C/min, ranging from room temperature up to 900 °C with a nitrogen flow of 50 mL/min. TGA data were analyzed by using Friedman's procedure (Friedman, 1964) and kinetics of coking was well fitted to first order. It was assumed that asphaltenes are decomposed thermally according to the following equation (Schucker, 1983):

$$\text{Asp} \rightarrow aC + (1 - a)V \tag{4.1}$$

where Asp is asphaltene, C is the coke product, V is the volatile fraction, and a is the stoichiometric coefficient. From this, it is possible to have a kinetic expression for volatiles formation according to

$$\frac{1}{Vo}\frac{dV}{dt} = k_0 e^{-E_A/RT}\left(1 - \frac{V}{Vo}\right) \tag{4.2}$$

where Vo is the total amount of volatilized material and t is time. Taking the logarithm of both sides and assuming that $V/Vo = x$, the following expression is obtained (Li and Yue, 2003):

$$\ln\left[\frac{dx}{dt}\right] = \ln[k_0(1 - x)] - \frac{E_A}{RT} \tag{4.3}$$

By applying a linear regression to this equation, a series of activation energies (E_A) and preexponential factors (k_0) are obtained as function of volatilized asphaltenes (x). The left-hand side of the equation is the rate of change of weight that is directly obtained from the time derivative on the TGA.

Analyzed asphaltenes are responsible for around 47 wt% of coke generated when suffering thermal decomposition in this case. Thermograms for asphaltenes are shown in Figure 4.5 at four different heating rates. It is observed that a more prominent weight loss is obtained at higher velocities. According to pyrolysis, the obtained results can be divided into three stages when analyzing asphaltenes:

- *First stage* corresponds to a minimal loss of weight, indicating that chemical reactions are negligible at this point, where temperature increases from ambient to around 350 °C.
- *Second stage* corresponds to an increase in temperature from 350 to 450 °C approximately. In this stage there is a loss of weight due to volatiles; however, evolved gas is not high. The weight is reduced because intermolecular associations and weaker chemical bonds are destroyed accompanied by rupture of alkyl chains to generate small gaseous molecules.
- *Third stage* is normally located above 450 °C, where more volatiles are produced constantly until no more gases are released. In this point, the stronger

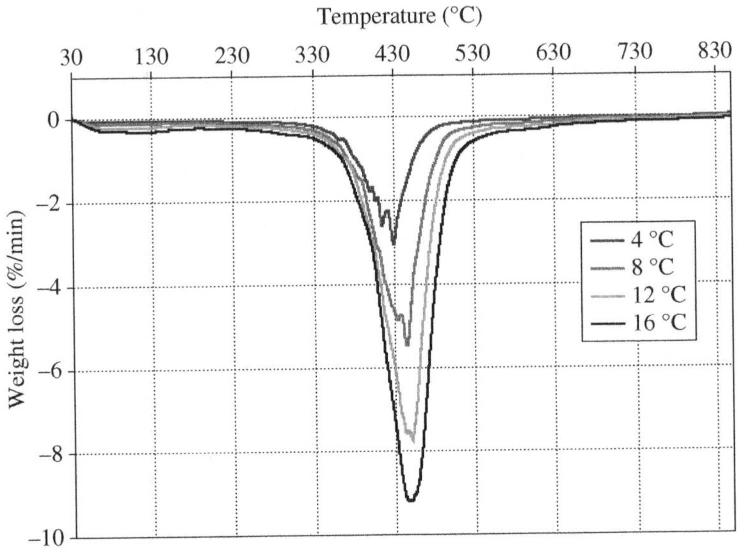

Figure 4.5 TGA of asphaltenes at different heating rates.

bonds are broken and asphaltene molecules are fully destroyed and as a consequence gaseous molecules are obtained until coke prevails as the final product of asphaltenes pyrolysis (Dong et al., 2005). It is also observed that at 4 and 8 °C/min, there are small shoulders that could correspond to resins or aromatics.

Figure 4.6 shows volatilization of asphaltenes at different heating rates, where asymptotic curves are obtained at higher temperatures indicating that coke amount is constant. Values of 90% volatilization are uncertain due to the difficulty in obtaining accurate values at this conversion. Friedman's method gives simultaneously the reaction order, activation energy, and preexponential factor, which are calculated without the knowledge of the rate dependence on conversion. Once that volatilization curves are obtained, linear regression is applied to obtain the activation energy and preexponential factor at different volatilization ratio as observed in Figure 4.7. It is seen that lines are not completely parallel that indicate that activation energy present different values as a function of the degree of volatilization of asphaltenes. The results of activation energy and preexponential factor are summarized in Table 4.1, where the differences are clearly observed. A good correlation coefficient (r) was obtained in all the cases. Schucker (1983) also obtained a similar trend of activation energy values. The differences in activation energy likely reflect changes in the strength of bonds that is intimately related to change in mechanism reaction during coking. This could mean that kinetics of coking is not always well represented by simple kinetic mechanisms having only one activation energy as reported by Miura and Maki (1998) as well. Obtained results are also in agreement with those reported by Shih and Sohn (1980).

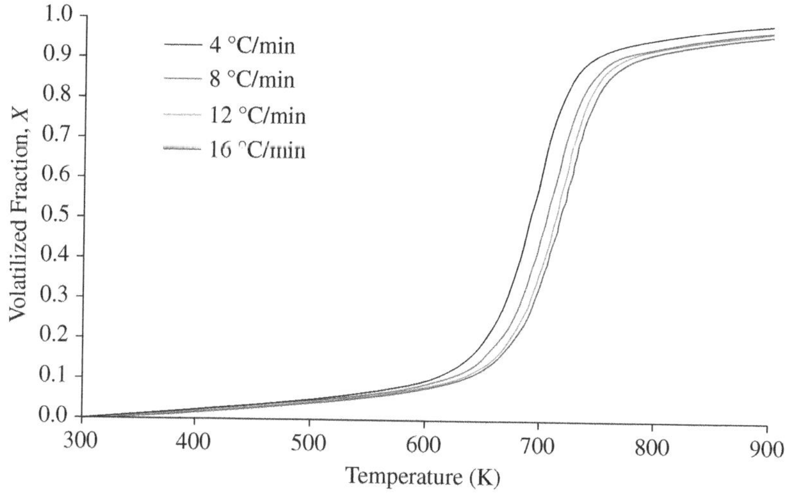

Figure 4.6 Volatilization of asphaltenes from a 13° API crude as a function of temperature.

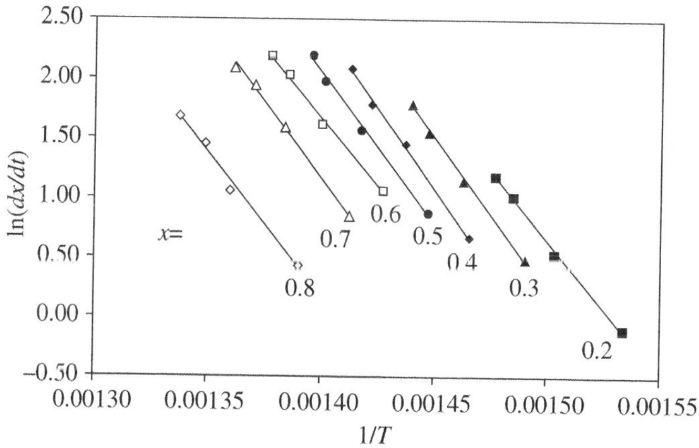

Figure 4.7 Determination of kinetic parameters at different heating rates.

Other attempts to study the pyrolysis reaction are based on the distributed activation energy models (DAEMs), which consider that activation energy is not a unique value; on the contrary, in complex reactions such as pyrolysis, a distribution of activation energy is required. Shih and Sohn (1980) and Braun and Burnham (1987) have stated that activation energy dependent on the volatilization obtained from the Friedman method is consequence of having a distribution of activation energy instead of having only a unique value. However, parameter estimation requires big computational efforts and an approximation of analytical solutions to time–temperature integrals is necessary. To simplify the solution, Lakshmanan and White (1994) have

TABLE 4.1 Activation Energy and Preexponential Factors

x	E_A (kcal/mol)	A (min^{-1})	r
0.1	39.9	6.20×10^{13}	0.996
0.2	46.1	3.23×10^{15}	0.996
0.3	50.9	8.67×10^{16}	0.998
0.4	52.2	1.79×10^{17}	0.999
0.5	49.9	3.03×10^{16}	0.996
0.6	45.5	1.13×10^{15}	0.997
0.7	49.6	1.65×10^{16}	0.998
0.8	48.8	5.04×10^{15}	0.994
0.9	43.3	1.68×10^{13}	0.980

used the Weibull distribution to represent the kinetics of complex thermal transformations. Miura (1995), Maki et al. (1997), and Miura and Maki (1998) have reported an approach and simple method for estimating the distribution of activation energies $f(E_A)$ and preexponential factor $k_0(E_A)$ from three sets of experiments carried out at different heating rates without assuming any functional form for $f(E_A)$ and $k_0(E_A)$. The following equation is based on Miura's method:

$$\ln\left(\frac{\mathrm{HR}}{T^2}\right) = \ln\left(\frac{k_0 R}{E_A}\right) + 0.6075 - \frac{E_A}{R}\frac{1}{T} \tag{4.4}$$

where HR is the heating rate, T the temperature in Kelvin, and R the ideal gases constant. Dong et al. (2005) have used the DAEM based on Miura's procedure to analyze the pyrolysis kinetics and observed that the loss of weight curves were completely predicted by this method.

4.3 EARLY DEACTIVATION OF DIFFERENT SUPPORTED CoMo CATALYSTS

4.3.1 Experimental Procedure

The catalysts were prepared by the incipient wetness technique and coimpregnation method. All catalysts have 10 wt% of MoO_3 and 3 wt% of CoO. Four different supports (alumina, alumina–titania, alumina–silica, and carbon) were used to prepare CoMo catalysts. Both titania and silica were 10 wt% on the alumina support.

The appropriate amount of ammonium heptamolybdate and cobalt nitrate salts were dissolved into distilled water to make a clear solution. The amount of water taken for making salt solution was adequate enough to fill up the pore of support. This clear solution was impregnated on dry support, and impregnated samples were allowed overnight at room temperature. Samples were then dried for 7 h at 120 °C and calcined at 450 °C for 5 h in the presence of air. For the carbon-supported catalyst, calcination was performed at 450 °C in the presence of nitrogen.

BET specific surface area, pore volume, and pore size distribution (PSD) of fresh, spent, and regenerated catalysts were measured by nitrogen adsorption at −196 °C (Automatic Micromeritics ASAP 2100). The percentage of carbon was also measured on the spent catalysts. The spent catalysts were washed with hot toluene by Soxhlet process and dried at 110 °C before carbon and metal analysis. Coke is defined as being carbon content on a spent catalyst. The metals distribution of catalysts was measured by a scanning electron microscope, model XL30ESEM, Philips. The spent catalysts were also characterized by TGA. Weight loss by combustion was measured by Perkin-Elmer model TGA 7-HT. In this experiment, around 50–100 mg of catalyst was heated from room temperature to 1000 °C at a rate of 10 °C/min in the presence of air (flow 50 mL/min).

The physical properties of the catalysts are given in Table 4.2. The carbon-supported CoMo catalyst has a very high surface area of 1288 m^2/g as well as high total pore volume of 1.15 mL/g. Alumina-supported catalyst has the highest average pore diameter. The specific surface area, pore volume, and pore diameter of alumina–titania- and alumina–silica-supported catalysts are in intermediate between alumina- and carbon-supported catalysts.

Heavy oil hydrotreating activity tests were performed in a high-pressure fixed-bed microreactor in upflow mode (Figure 4.8). The oxide catalysts were sulfided *in situ* before actual run was started. Ten milliliters of oxide catalyst was loaded with equal volume of diluent, carborundum (0.2 mm size). Both catalyst and carborundum were mixed and divided into five parts. Each part of the mixture was loaded into the reactor at a time and tapped little bit. The catalyst was then dried for 2 h at atmospheric pressure and 120 °C. After drying, the catalyst was allowed for soaking for 2 h at 150 °C. LGO was used for soaking. This LGO contains 1.7 wt% of sulfur. Actual sulfiding agent was introduced after soaking. The sulfiding agent was LGO with dimethyl disulfide (1 wt%). Sulfidation was done at 28 kg/cm^2 pressure at two different temperatures. The first sulfidation was done at 260 °C for 3 h and finally the catalyst was sulfided at 320 °C for 5 h. The experimental conditions were as follows: total pressure of 54 kg/cm^2, reaction temperature of 380 °C, LHSV of 1.0 h^{-1}, and hydrogen-to-hydrocarbon ratio of 356 m^3/m^3. A mixture of Maya heavy oil with hydrodesulfurized naphtha (50/50 wt/wt) was used for catalyst activity tests. Sulfur, nitrogen, asphaltene, nickel, and vanadium contents of this feed were 1.86, 0.253, 7.96 wt%, 26.17, and 123.35 wppm, respectively. The first balance was taken after stabilization of 9 h. The products were collected at the duration of 12 h each.

Thiophene hydrodesulfurization activity of fresh, spent, and regenerated catalysts was tested in microplant (MP; Figure 4.9). In this plant, 0.5 g of fresh catalyst was sulfided *in situ* into a glass-tubular reactor at atmospheric pressure and 400 °C temperature. For sulfidation, hydrogen was passed through a container having carbon disulfide (CS_2). The saturated mixture of CS_2 and hydrogen was passed through the reactor. Hydrogen flow was 50 mL/h and duration of sulfidation was 2 h. After sulfidation, the catalyst was flushed at this temperature by H_2 until no CS_2 could be detected in the effluent gas. Thiophene feed was introduced through the gas bubblers. H_2 flow rate was 50 mL/min. Reaction products were analyzed by an online gas chromatography using a FID (flame ionization detector) detector.

TABLE 4.2 Properties and Coke and Vanadium Depositions on Catalysts

Properties	$CoMo/Al_2O_3$ (A)			$CoMo/Al_2O_3–TiO_2$ (B)			$CoMo/Al_2O_3–SiO_2$ (C)			CoMo/C (D)	
	F	S	R	F	S	R	F	S	R	F	S
SSA (m^2/g)	309	248	308	312	270	300	416	242	404	1288	39
TPV (mL/g)	0.71	0.37	0.69	0.46	0.37	0.48	0.72	0.26	0.67	1.15	0.12
APD (Å)	92	0.60	91	60	55	64	69	43	66	36	135
PSD (vol%)											
<50 Å	28.47	72.49	29.5	49.66	71.17	53.12	38.8	70.81	58.44	59.04	12.08
50–100 Å	31.99	14.73	44.99	38.82	17.51	34.86	53.32	14.84	34.24	12.68	9.85
100–200 Å	30.31	8.69	20.66	6.78	6.75	7.73	4.62	7.03	4.95	13.55	37.5
200–500 Å	5.38	4.09	3.77	3.17	3.04	3.14	2.23	4.37	1.84	13.86	35.39
500–1000 Å	0.85	0	1.09	1.57	1.53	1.15	1.03	2.95	0.53	0.87	5.18
>1000 Å	3.00	0	0	0	0	0	0	0	0		
Coke (wt%)		13.2			6.1			10.4			–
Vanadium (wt%)		1.11			0.15			0.12			0.09

F, fresh; S, spent; R, regenerated.

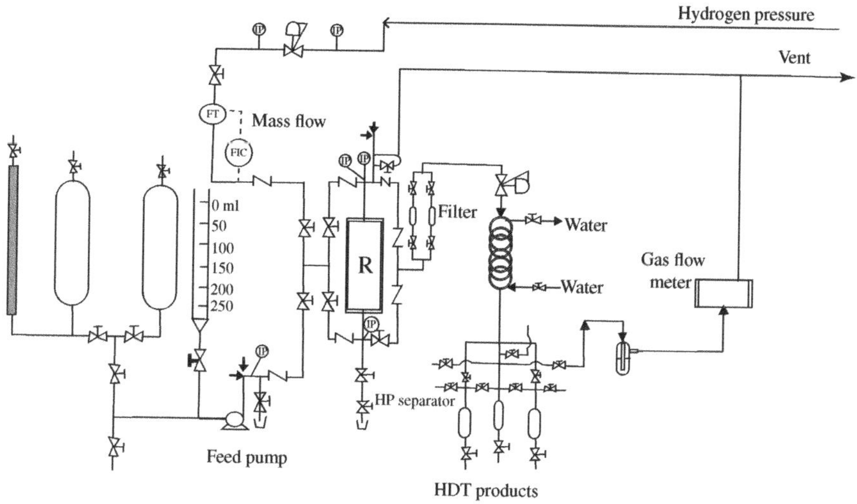

Figure 4.8 High-pressure microplant for hydrotreating experiments.

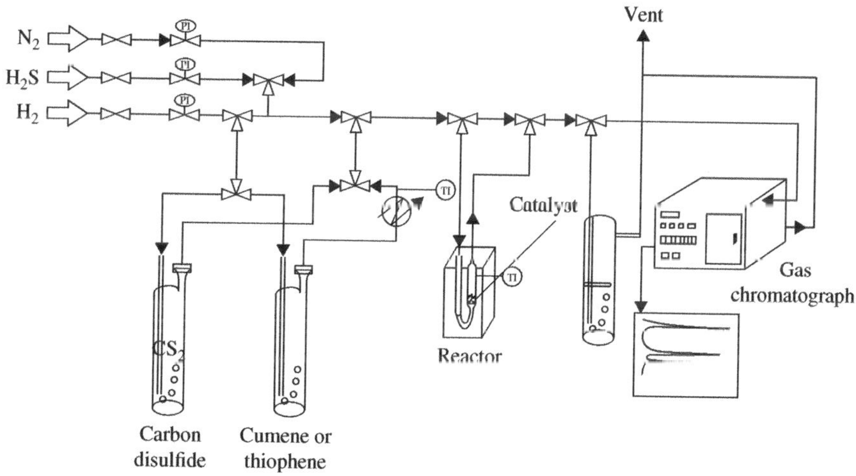

Figure 4.9 Microreactor for hydrodesulfurization of thiophene and hydrocracking of cumene.

The spent catalyst after obtaining from high-pressure microplant was washed and it was tested for HDS of thiophene. Before actual run, the spent catalyst was also sulfided as describe earlier. HDS activity of the spent catalyst was tested at period of 4 h. This same spent catalyst was regenerated by control flow of air at 500 °C and sulfided before testing its activity.

The total amount of metals in the feed and products were measured by Atomic Absorption (Thermoelectron model Solar AA) while sulfur content was analyzed by X-ray fluorescence (HORIBA model SLFA-2100).

4.3.2 Results and Discussion

4.3.2.1 HDM and HDS Activities with Real Feed The HDM activities of different supported catalysts are presented with time-on stream in Figure 4.10. All HDM activities decrease with TOS; however, the trend varies for different catalysts. Alumina–titania-supported CoMo catalyst shows the highest HDM activity. It is also observed that although the initial activity of alumina–silica catalyst is high, the activity decreases rapidly with TOS. Alumina-supported catalyst has also high HDM activity and it is stable with TOS. Carbon-supported CoMo catalyst shows the lowest activity. The HDM activity of four supported CoMo catalysts is in the order of $CoMo/Al_2O_3–TiO_2 > CoMo/Al_2O_3 > CoMo/Al_2O_3–SiO_2 > CoMo/C$.

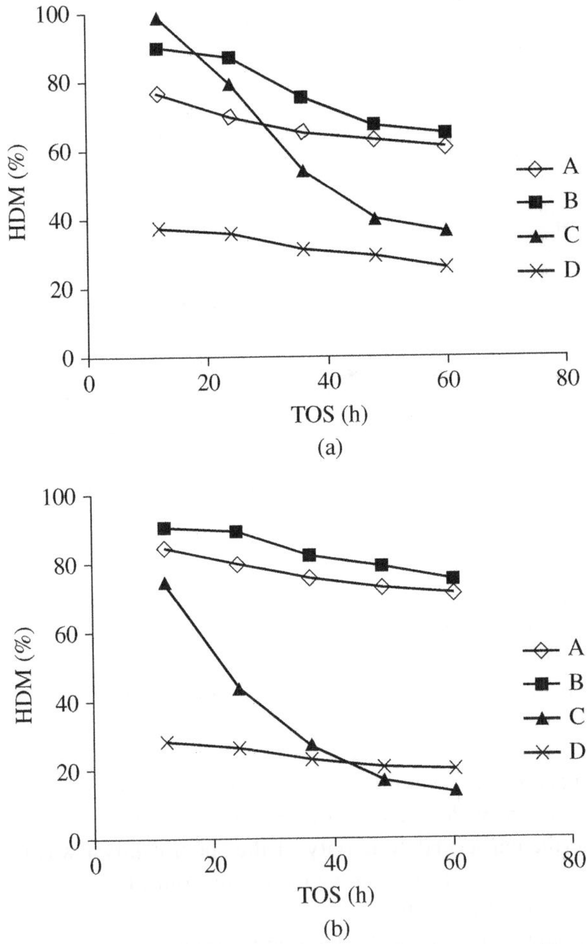

Figure 4.10 (a) HDM and (b) HDS activities of $CoMo/Al_2O_3$ (A), $CoMo/Al_2O_3–TiO_2$ (B), $CoMo/Al_2O_3–SiO_2$ (C), and CoMo/C (D) catalysts.

The HDS activity has also been studied for heavy crude and the results are given in Figure 4.10. Similar to HDM, $CoMo/Al_2O_3$–TiO_2 catalyst has the highest HDS activity and CoMo/C shows the lowest. In this case also a rapid decline in HDS activity is noticed for the alumina–silica-supported catalyst.

The rate of deactivation for all catalysts is also calculated by using following equation (Chen and Hsu, 1997):

$$X_t = X_0 e^{-bt^n} \tag{4.5}$$

where X_t is the conversion at time t, X_0 is the initial conversion, b is the deactivation rate constant, and $n = 1$ (Chen and Hsu, 1997).

$-\ln(X_t/X_0)$ is plotted against TOS for HDM and HDS reactions in Figure 4.11. Although the total reaction time is short (60 h), even within this time different deactivation rates are noticed. The deactivation rates (b value) have also been calculated from the straight lines. The HDM deactivation rates are 0.0047, 0.0076, 0.0222, and 0.0078 for catalysts A, B, C, and D, respectively, while these rates are 0.0036, 0.0041, 0.0357, and 0.0076 for HDS reaction. It is found from the above values that the deactivation rate for HDM reaction is marginally faster than that of HDS reaction. The metal compounds in the heavy crude oils are very large complex molecules. On the other hand, sulfur-containing molecules in heavy fractions are in general of two types: one is bigger in size, attached with asphaltene structure, and the other one is smaller, nonasphaltenic. During hydroprocessing reaction, the solubility of asphaltenes is reduced by removing their aliphatic chains, particularly in the initial period when the catalyst activity is high, causing asphaltenes to precipitate from the product and hence they are deposited on the catalysts. As a result, catalyst is deactivated by this deposition. Therefore, the deactivation rate for HDM reaction is faster than that for HDS because metals are associated with asphaltene moieties. The smaller size of nonasphaltenic sulfur compounds is still having path to enter the pore cavity and hence the deactivation rate is slower. It is also noted that the rate of deactivation of the $CoMo/Al_2O_3$–SiO_2 catalyst is very high for both HDM and HDS reactions. It suggests that the alumina–silica support may have some acidic sites and these sites enhance the formation of coke on the bare support surface. The deposited coke on the spent catalyst given in Table 4.2 is high, around 10 wt%. The lowest deactivation rates for both HDM and HDS reactions are found for carbon-supported catalysts. The carbon is a neutral support and hence it shows low activity and, therefore, the coke deposition is also low. This may be the reason for showing stable performance with TOS.

4.3.2.2 Thiophene HDS Activities on Fresh, Spent, and Regenerated Catalysts

Hydrodesulfurization activities of fresh, spent, and regenerated catalysts were studied and the results are given in Figure 4.12. It is worth to mention that the spent catalysts are obtained from high-pressure microplant, where a mixture of Maya crude with naphtha was treated for 60 h. The HDS activity of regenerated catalyst D is not presented here because carbon-supported catalyst cannot be regenerated by oxidation. It is seen that catalyst A has very high HDS activity. Although HDM

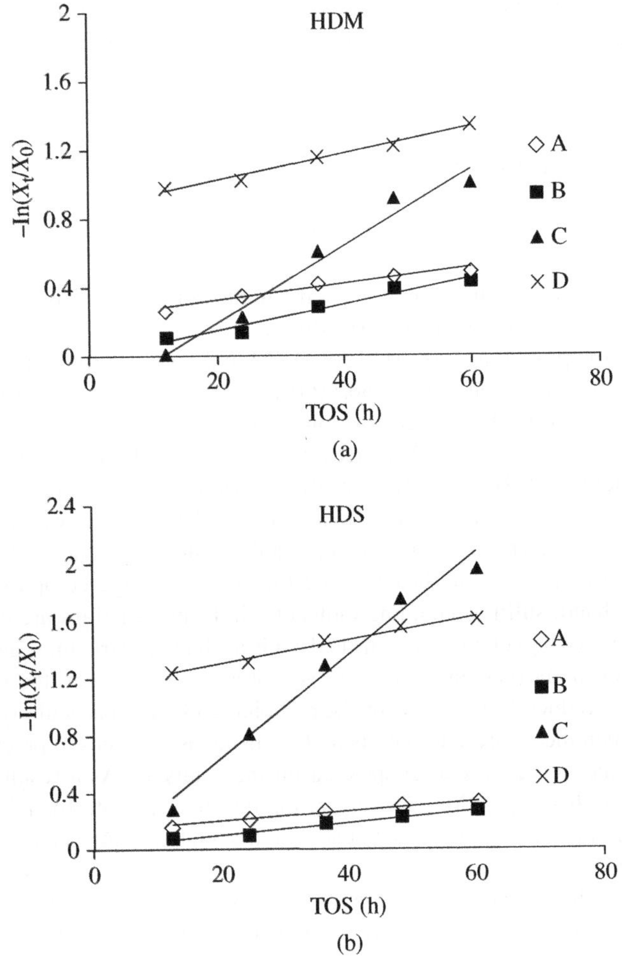

Figure 4.11 Rates of (a) HDM and (b) HDS of $CoMo/Al_2O_3$ (A), $CoMo/Al_2O_3–TiO_2$ (B), $CoMo/Al_2O_3–SiO_2$ (C), and CoMo/C (D) catalysts.

and HDS activities of heavy oil are low for carbon-supported catalyst D, thiophene HDS activity is moderately high. At least it is higher than those of catalysts B and C. HDS activity of catalyst A is drastically reduced due to deactivation. However, when coke in the spent catalyst is burned off, the catalyst literally gains its original activity. The other two catalysts B and C show almost a similar trend, that is, the fresh catalyst has higher HDS activity than the spent and regenerated catalysts. Even though catalyst D has moderate activity, its spent catalyst shows very low activity. Thiophene HDS activity of the fresh sulfided CoMo catalysts is in the order of $CoMo/Al_2O_3 > CoMo/C > CoMo/Al_2O_3–TiO_2 \sim CoMo/Al_2O_3–SiO_2$. The overall results indicate that the different supported catalysts have different HDS activities.

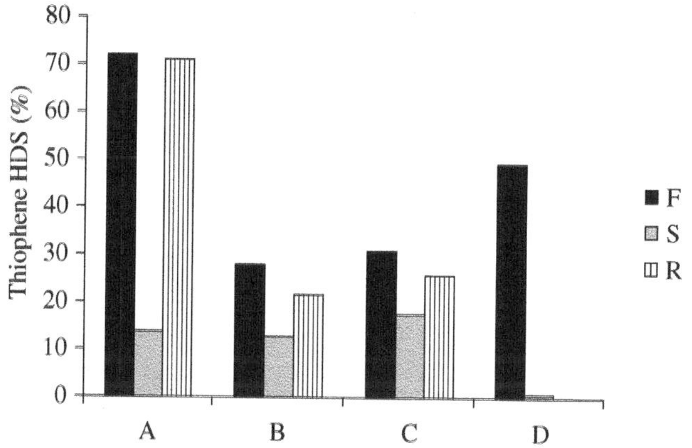

Figure 4.12 Activity for thiophene hydrodesulfurization of fresh (F), spent (S), and regenerated (R) catalysts. $CoMo/Al_2O_3$ (A), $CoMo/Al_2O_3–TiO_2$ (B), $CoMo/Al_2O_3–SiO_2$ (C), and CoMo/C (D).

The deactivation trend is also different for different catalysts. The coke deposition may have adverse effects on the alumina- and carbon-supported catalysts. However, the metal sulfides may also take part at the early stage of deactivation for the other two supported catalysts B and C.

4.3.2.3 Scanning Electron Microscopy The profiles of vanadium and coke deposits are also studied by SEM-EDX and their radial distributions on the spent catalysts are presented in Figure 4.13. It is observed that vanadium is evenly deposited throughout the catalyst particle for catalysts A, B, and C. These catalysts are supported by Al_2O_3, $Al_2O_3–TiO_2$, and $Al_2O_3–SiO_2$, respectively. However, more vanadium deposition occurs at the outer surface of the carbon-supported catalyst. The coke deposition profiles for catalysts A, B, and C are more or less homogeneous throughout the catalyst particle. Since catalyst D is supported by carbon, the carbon deposited during the hydroprocessing reaction cannot be distinguished from the carbon support and hence the coke deposition profile is not presented for catalyst D in this figure.

4.3.2.4 Themogravimetric Analysis Catalyst weight loss due to combustion of the spent catalysts has been measured by TGA. The weight loss and its derivative with temperature are presented in Figure 4.14. All derivative curves clearly show two principal weight losses at around 100 °C and at 400 °C. The second peak at around 400 °C is very sharp and prominent. The weight loss at the first peak is for the loss of water and the second peak is identified as the weight loss due to coke burning. However, the second peak position is not the same for all catalysts. The peak position for catalysts A and B is very near 400 °C, whereas this peak has been shifted to higher temperatures for catalysts C and D , 460 and 440 °C , respectively. Catalyst D is a carbon-supported catalyst and carbon (or coke) has also been deposited during

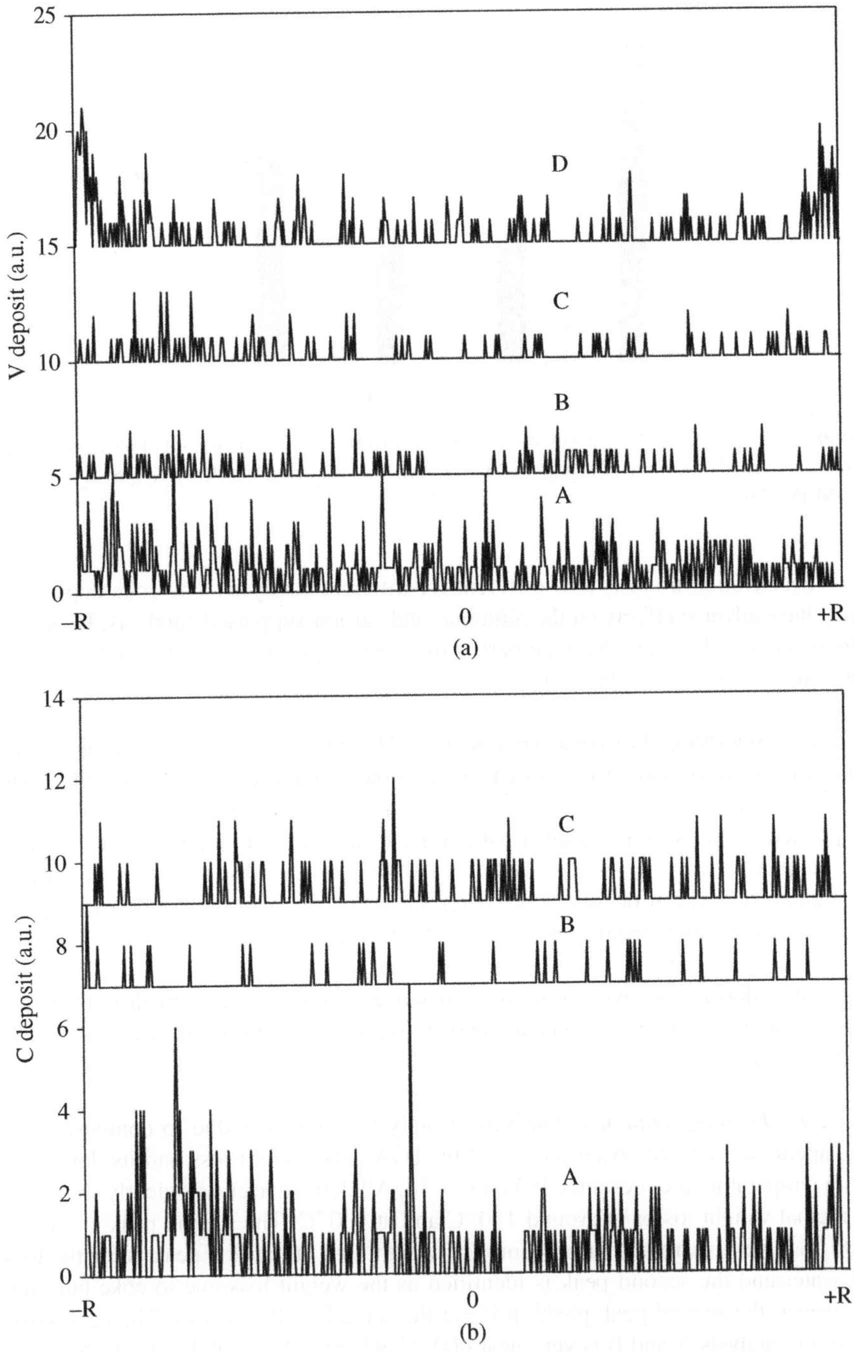

Figure 4.13 Radial distribution of (a) vanadium and (b) coke of $CoMo/Al_2O_3$ (A), $CoMo/Al_2O_3$–TiO_2 (B), $CoMo/Al_2O_3$–SiO_2 (C), and CoMo/C (D) spent catalysts.

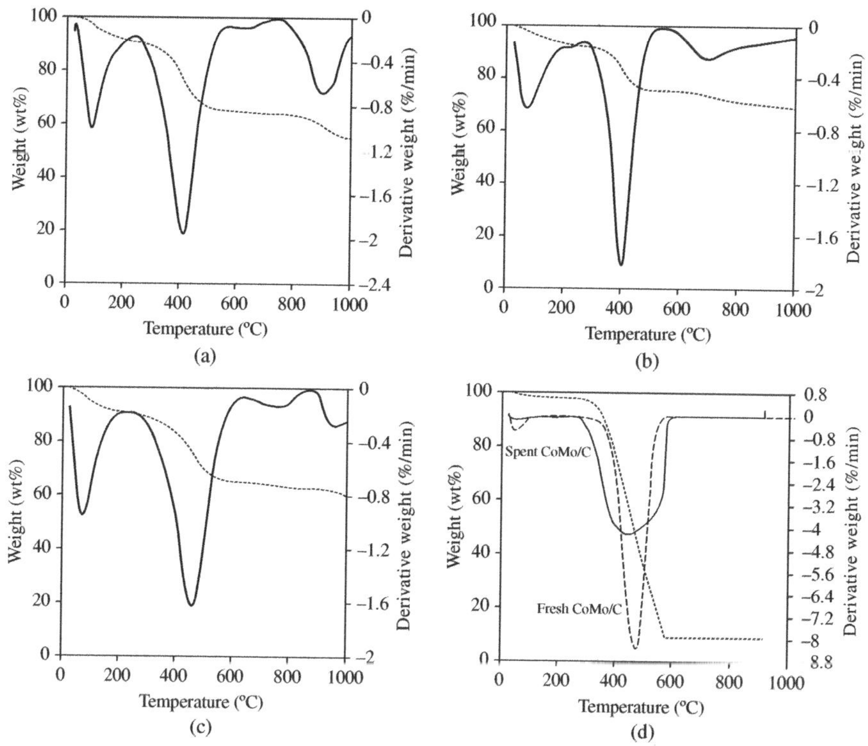

Figure 4.14 Thermogravimetric curves of $CoMo/Al_2O_3$ (a), $CoMo/Al_2O_3–TiO_2$ (b), $CoMo/Al_2O_3–SiO_2$ (c), and CoMo/C (d) spent catalysts.

hydroprocessing reaction. So the net weight loss of this catalyst during TGA analysis is due to the carbon from the support and from coke. To distinguish these two sources of carbon, TGA analysis is also performed on fresh CoMo/C catalyst and the results are compared with the spent catalyst in Figure 4.14d. The nature of the derivative curves for spent catalyst D is very wide compared with the other catalysts and with its fresh catalyst. It indicates that both peaks of carbon support and coke from reaction are superimposed. The combustion of fresh CoMo/C catalyst starts at 290 °C and ends at 550 °C, whereas the combustion of spent CoMo/C catalyst starts at 260 °C and ends at 575 °C. A very wide peak is also observed at higher temperatures for catalysts A, B, and C. These wide and broad peaks appear around 900, 650, and 750 °C, respectively. This broad peak is absent for carbon-supported catalyst D. The characterization of coke by TGA experiment has also been studied by various researchers. Sahoo et al. (2004) observed three different regions during combustion of spent hydrotreating catalyst. Region I was <180 °C, region II was in between 180 and 330 °C and region III was in between 330 and 750 °C. The first region was obviously due to loss of water, whereas region II was due to the weight loss of mobile carbonaceous residues, which are defined as soft coke. The bulky hard coke was combusted at higher temperatures and appeared in region III. TGA experiment on the spent catalysts was studied in detail by Begon et al. (1999). They also used mass spectroscopy to analyze the gaseous products coming out due to burning of spent catalysts. The authors found that the peak at 350–360 °C was due to the weight loss of presulfiding material present in the catalyst and bulk carbonaceous materials was burned off at 480 °C. They also observed other peaks due to leaving out of SO_2 from the samples. In this study, a sharp peak was observed at 400–460 °C, which was the principal peak for combustion of carbonaceous material. The nature of this coke was not similar for all the catalysts. For catalysts A and B, the coke was combusted at a lower temperature (400 °C), whereas it is high (460 °C) for catalyst C. So it can be stated that for alumina–silica supported catalysts, the deposited carbon is relatively hard. On the other hand, for carbon-supported catalysts, both soft and hard coke are present.

4.3.2.5 Physical Properties of Spent Catalysts The specific surface area, total pore volume, average pore diameter, and PSD of fresh, spent, and regenerated catalysts are compared in Table 4.2. The PSD of each supported catalyst has also been compared in Figure 4.15. Table 4.2 shows that the specific surface area and total pore volume of the spent catalysts are reduced with respect to the corresponding fresh catalysts. The loss of these physical properties is not the same for all catalysts. The surface area of spent catalyst D is dramatically reduced from 1288 to 39 m^2/g within a period of 60 h

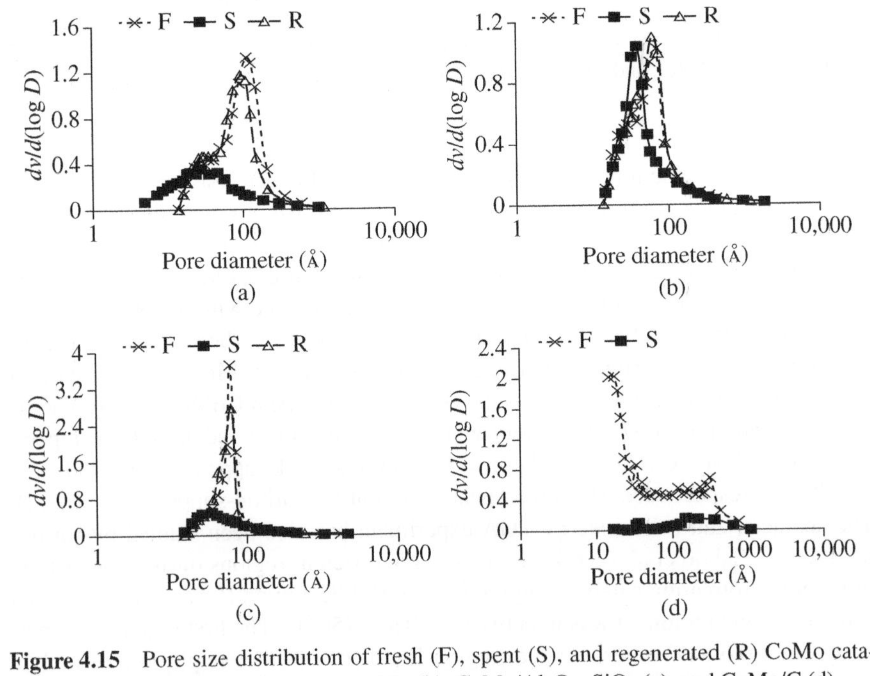

Figure 4.15 Pore size distribution of fresh (F), spent (S), and regenerated (R) CoMo catalysts. CoMo/Al_2O_3 (a), CoMo/Al_2O_3–TiO_2 (b), CoMo/Al_2O_3–SiO_2 (c), and CoMo/C (d).

of reaction with heavy crude oil. A least change of the surface area is observed for catalyst B. Catalyst A regains its surface area and total pore volume by regeneration of the spent catalyst. When HDS activities of fresh and regenerated catalysts A are compared, it is also noticed that both catalysts show almost similar activities.

The PSD of the spent catalyst has been changed compared with the fresh catalyst. The pore volume of the spent catalysts below 50 Å increases, whereas the pore volume having higher pore diameters decreases. When the changes in the total pore volume of the spent catalysts are compared, total pore volume of catalyst D is drastically reduced from 1.15 to 0.12 mL/g, whereas it is the least for catalyst B. Figure 4.15b shows that alumina–titania-supported catalyst has strong ability to resist the changes of its textural properties. The surface area, total pore volume, and its pore distribution of this catalyst do not change significantly during hydroprocessing reaction of heavy crude, particularly at the early stage of reaction.

Alumina–silica-supported catalyst has more coke deposition compared with alumina–titania-supported catalyst. The former may have more acidic sites that enhance cracking of asphaltene molecules and hence coke formation. The microplant activity results also support that though the initial activities of alumina–silica-supported CoMo catalyst are high, they rapidly decrease with TOS. Figure 4.15c also shows that the PSD of catalyst C is also changed considerably upon deactivation. Alumina-supported catalyst A shows very high HDS and HDM activities. The coke and vanadium depositions on this catalyst are also high. The result of thiophene HDS activity indicate that the spent catalyst reduce their activity. However, when this catalyst is regenerated, it recuperates almost all HDS activities like a fresh catalyst. So it suggests that the coke deposit mainly causes deactivation of this catalyst. The PSDs of the fresh, spent, and regenerated catalysts show that the pore structure of catalyst A changes dramatically with deactivation; however, the pore structure becomes almost similar to that of the fresh one when the carbon was burned off from the spent catalyst (Figure 4.15a). Table 4.2 shows that both coke and vanadium depositions are high on spent catalyst A. Thus, it suggests that the change of pore structure is mainly due to the coke formation. Due to the coke deposition, the activity of this catalyst is also drastically reduced and when this coke is eliminated by burning, the catalyst recuperates its original activity. It also hints that though metal deposition is high, it may not take part in deactivation.

The deactivation for the other two supported catalysts (Al_2O_3–TiO_2 and Al_2O_3–SiO_2) is moderate. Thiophene HDS activity of the regenerated catalysts (B and C) is lower than that of the respective fresh catalysts. It reveals that coke and metal deposits take part in deactivation and reduce HDS activity for catalysts B and C. It suggests that the early stage of deactivation is unavoidable and the causes of it may vary from catalyst to catalyst.

The catalyst at the beginning of a reaction is very active and hence conversion is high. At the early stage of the reaction, coke is predominantly deposited on the external surface of the catalyst. In due course of operation, coke gradually penetrates into the interior cavity of the catalyst particle. Therefore, deactivation can occur by both phenomena – pore mouth plugging and core poisoning – depending on the catalyst pore structure. For a small pore, catalyst deactivation occurs via pore mouth plugging.

On the other hand, for a large pore, the catalyst is deactivated by core poisoning. However, the metals from the feed are gradually deposited on the catalyst surface. At the early stage of hydrotreating reaction when the metal deposition is very low, metal sulfides may cover the interior active sites of the particle rather than pore mouth plugging and it causes catalyst deactivation (Maity et al., 2007). Therefore, these results show that both coke and metal depositions are responsible for the early stage of deactivation. However, their influence may not be the same for different supported catalysts. The adverse effect of metal deposition is absent for the alumina-supported catalyst, whereas both coke and metals take part in the deactivation of the alumina–titania- and alumina–silica-supported CoMo catalysts. The reason of catalyst deactivation for the carbon-supported catalyst is not clear; however, a dramatic change in the pore structure is observed for this catalyst and it is mainly due to the coke deposition. It is also noted from SEM analysis that vanadium sulfide may deposit on the surface due to its smaller pore diameter. Gualda and Kasztelan (1996) observed that both metal sulfides and coke deactivated HDM performance of NiMo/Al_2O_3 catalyst but deactivation by metals is more prominent even at the initial stage of operation.

4.3.3 Conclusions

The causes for early-stage deactivation are studied on different supported catalysts by using heavy crude oil. Four different supports (alumina, alumina–titania, alumina–silica, and carbon) are used to prepare CoMo catalysts. Thiophene HDS activity of fresh CoMo/Al_2O_3 catalyst is the highest and it is reduced dramatically for the spent catalyst. However, the regenerated catalyst shows almost equal activity to that of the fresh catalyst. The PSDs of the fresh and regenerated catalysts also show that both catalysts have almost similar pore distributions. It suggests that for CoMo/Al_2O_3 catalyst, the cause of deactivation is mainly coke deposition. On the other hand, both coke and metal sulfide depositions cause the deactivation of CoMo/Al_2O_3–TiO_2 and CoMo/Al_2O_3–SiO_2 catalysts. The HDM and HDS activities of the carbon-supported catalyst are the lowest among all the catalysts. However, its thiophene HDS performance is better than the other two mixed-oxide-supported catalysts and it is only lower than that of the alumina-supported catalyst. Moreover, the spent carbon supported catalyst does not show any thiophene activity.

4.4 CARBON AND METAL DEPOSITION DURING THE HYDROPROCESSING OF MAYA CRUDE OIL

4.4.1 Preparation, Evaluation, and Characterization of Catalyst

Molybdenum–alumina-supported catalysts were prepared by the incipient wetness impregnation method. The promoted Co(Ni)Mo catalysts were also prepared by using coimpregnation procedure. The final catalysts were dried in the presence of air at 120 °C overnight and calcined at 450 °C for 4 h.

The sulfide catalyst activity tests were carried out in a high-pressure fixed-bed upflow reactor (Figure 4.8). The catalyst was sulfided *in situ* with a mixture of dimethyldisulfide, straight-run gas oil, and hydrogen. The reactor was loaded with an oxidic catalyst of 3–5 mm diluted with an equal volume of SiC. The final temperature of sulfidation was 320 °C, which was stabilized for 5 h at 2.8 MPa. After sulfidation, the flow was switched to the feed and the operating conditions were adjusted accordingly. The catalyst activity was measured using a diluted Maya crude with diesel (50/50 wt%) and virgin Maya crude oil feeds at 380 and 400 °C temperature, respectively. The properties of the feedstocks are presented in Table 4.3.

The feed and products were analyzed using flame atomic adsorption spectrometry for metals (Ni, V), ultraviolet fluorescence for sulfur, while nitrogen was measured by oxidative combustion and chemiluminescence method. Asphaltene is defined as the insoluble fraction in *n*-pentane. The catalyst tested in microflow reactor is indicated with the MP, while the catalyst tested in pilot-plant is known as PP. The relative activity (RA) is defined as the ratio of the conversion (HDS, HDM, and HDAs) with time (X_{TOS}) and initial activity (X_i) at a given TOS at 0–200 h:

$$\mathrm{RA} = X_{\mathrm{TOS}}/X_i \tag{4.6}$$

4.4.2 Catalyst Characterization

Surface area, total pore volume, and PSD of the fresh, spent, and regenerated catalysts were measured by nitrogen adsorption–desorption at −196 °C using

TABLE 4.3 Properties of the Feeds

	Maya Crude	HDM (Maya+Diesel, 50/50, wt/wt)
C (wt%)	86.9	84.2
S (wt%)	3.52	2.21
Ni	49.5	26.21
V	273.0	124.78
(Ni+V)	322.5	150.99
Ca	11.26	5.0
Mg	2.04	1.01
Na	44.83	21.2
K	20.25	10.2
Fe	2.16	1.02
Asphaltene, wt% (n-C_7 insol.)	12.7	8.43
Density (20/4 °C)	0.9251	0.88
Pour point (°C)	−30	−15
Ramscarbon (wt%)	10.87	5.45
API gravity	21.31	37.09
Catalyst	$CoMo/Al_2O_3$ (PP)	$CoMo/Al_2O_3$ (MP), $NiMo/Al_2O_3$ (MP)

Quantachrome Nova 2000 equipment. Coke-on-catalyst analysis was carried out with toluene-washed samples in a Leco SC-444 instrument using ASTMC 1408-98 method by direct combustion-infrared detection. The chemical analysis of the fresh and spent catalysts was carried out in order to determine the concentration of metals (V, Ni, Mo, etc.) by using inductively coupled plasma-atomic emission spectrometry (ICP-AES). The surface morphology of catalysts was studied by means of elemental analysis with the SEM-EDS analytical instrument xT Nova NanoLab 200 (FEI Schottky FEG, 30–1 keV). The samples were deposited on a carbon holder and evacuated at high vacuum (10–5 Torr) before images were taken. The analysis was made across the radial line of the extrudate in order to obtain metal, sulfur, and carbon distribution profiles.

The sulfided fresh and spent catalysts were characterized by JEM-2200FS transmission electron microscope with an accelerating voltage of 200 kV. Local chemical analysis and chemical mapping was carried out with an energy-dispersive X-ray spectroscope, which is attached to the microscope, and using the STEM-EDX combination. The samples were grounded, suspended in *n*-heptane at room temperature, and dispersed with ultrasonic agitation; then, an aliquot of the solution was dropped on a 3-mm-diameter lacey carbon copper grid. The nickel and molybdenum contents on the catalysts were analyzed by atomic absorption spectrometry (ASTM D5863).

Solid-state ^{13}C CP/MAS NMR spectra of spent catalyst samples were recorded on a Bruker 400 MHz NMR spectrometer operating at 100.6 MHz for 13C and equipped with a 4 mm broadband CP/MAS probe using a magic angle spinning rate of 10 kHz. Approximately 100 mg of dried and finely powdered sample was packed into the zirconia rotor fitted with Kel-F cap. The effective radio frequency field strengths for high-power proton decoupling and ^{1}H–^{13}C Hartmann–Hahn matched cross-polarization were between 60 and 70 kHz. The typical parameters for ^{13}C CP/MAS NMR measurement were as follows: 20 ms acquisition time, 10 s recycle time, contact time of 2 ms, and sweep width of 40 kHz.

4.4.3 Results and Discussion

4.4.3.1 Properties of the Feedstock The crude oil fractions were separated by using solvent deasphalting, and individual fractions were determined as shown in Figure 4.16. Asphaltenes were extracted from crude oil as a solid, which can be further purified or separated from the carbenes and carboids by using aromatic extraction. Due to the carbonaceous nature and source of heteroatoms, asphaltene is considered the main cause of deactivation during hydroprocessing as shown in Figure 4.17.

Thus, coke formation mechanism is mainly due to the chemistry of asphaltenes. Apart from the coke deactivation, asphaltene fraction typically consists of large percentage of heteroatoms (sulfur, nitrogen, and oxygen) and provides shelter for organometallic constituents in the crude oil and residue (Ancheyta et al., 2009). Hence, catalyst stability is closely related to the asphaltene content and the feed compositions. Coke formation due to the presence of asphaltene may have different

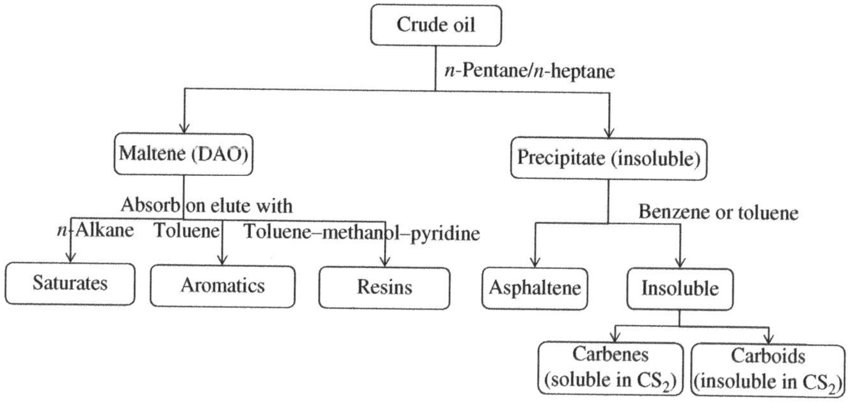

Figure 4.16 Fractionation of crude oil and its separation by chemical treatment.

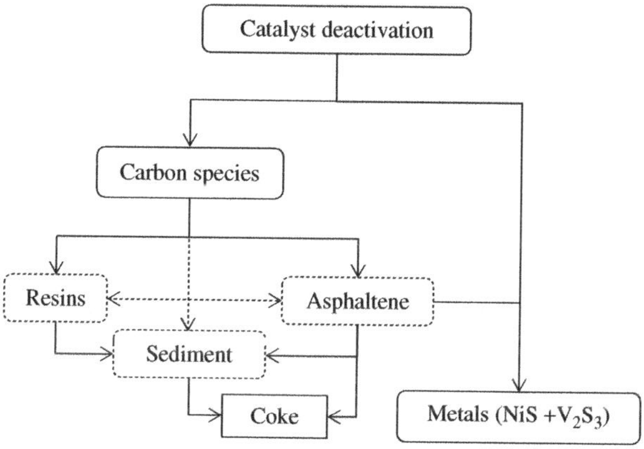

Figure 4.17 Catalyst deactivation species and their instability factor (indicated by dotted lines).

structure than the normal coke generated due to olefin polymerization or aromatization structure. Thus, one of the key indicators for the catalyst deactivation at the start-of-run is the HDM and asphaltene conversion along with the identification of light compound formation. If this conversion is high, the catalyst deactivation is expected to be faster.

4.4.3.2 ***Activity Tests*** To examine the aging effect, the catalysts were tested at 380 and 400 °C; various activities were monitored on each of the two catalysts continuously with TOS. Figure 4.18 shows the relative activities (normalized to the corresponding initial activity) of microflow and bench-scale $CoMo/Al_2O_3$ catalysts as a function of aging time of 125 and 200 h, respectively. Clearly, both temperatures showed aging behavior with TOS. The initial activity was reported after 12 h, the catalyst deactivated relatively quickly in the first 50 h and then maintained an activity level

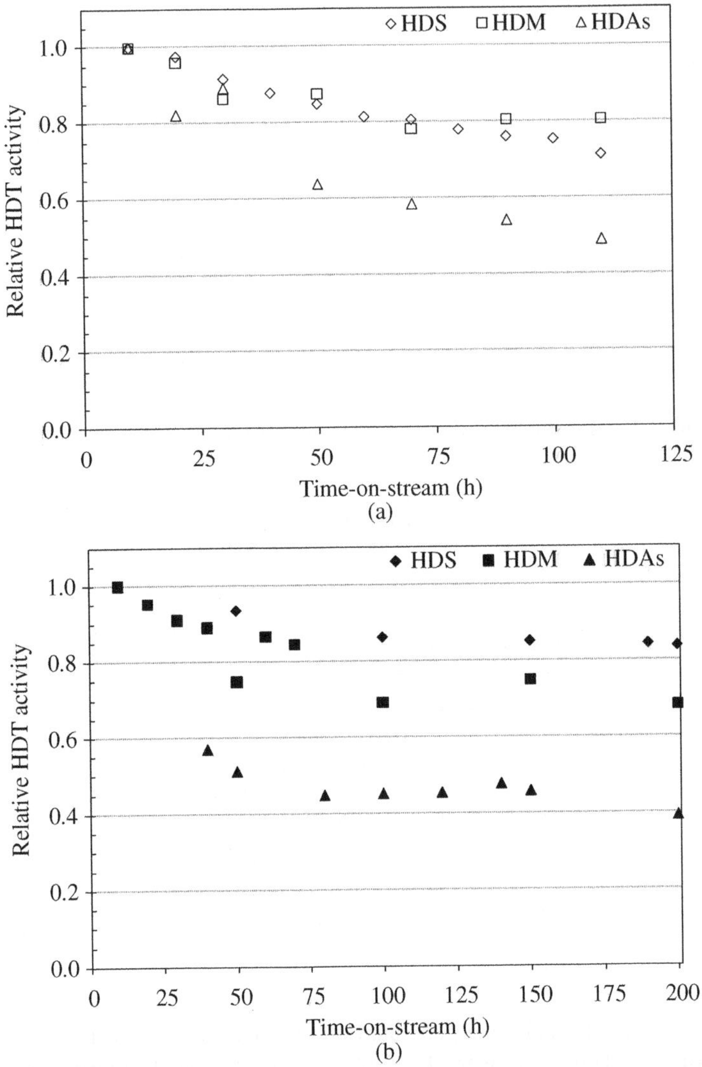

Figure 4.18 Relative catalytic activity with time-on-stream over $CoMo/Al_2O_3$ catalyst at (a) microflow scale and (b) bench scale.

particularly for HDS and HDM. The catalyst activities were investigated for two different processing units, where two different types of feeds were used. The 50% diesel dilution with Maya crude was due to the variation in the asphaltene content in the feed. The use of light hydrocarbon as diluent led to insolubility of asphaltenes while stability effect was not observed when using diesel as diluent (Rayo et al., 2004). The activity results (at 380 and 400 °C) indicated that the catalysts were deactivated with TOS.

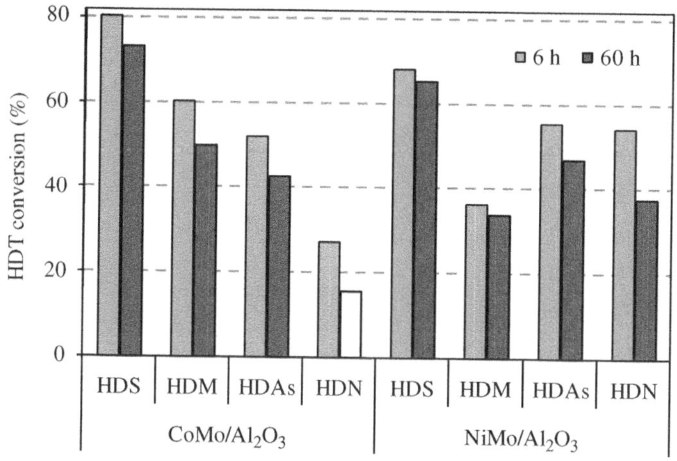

Figure 4.19 Effect of catalyst composition on the hydrotreating of diluted Maya crude activities in microflow reactor (MP).

However, deactivation behaviors are not similar for both feeds, which highlights the effect of the feedstock on the catalyst. The activity results are also compared for the CoMo- and NiMo-supported alumina catalysts with diluted feed (Maya and diesel) at 380 °C, where higher CoMo catalyst showed better conversion for HDS and HDM while NiMo catalyst showed better conversion for HDAs and HDN, where the latter reaction is a function of hydrogenation as shown in Figure 4.19. However, the comparison between the catalyst composition (CoMo/NiMo) on activity variation with 6 and 60 h was not significant, which showed that both catalysts have almost the same type of metal and carbon deposition.

4.4.3.3 Characterization of Spent Catalyst The deactivated Co–Mo/Al_2O_3 catalyst samples were characterized by SEM-EDX, TEM, and NMR, which identify the nature and amount of deactivation species. The details of characterizations are reported in the subsequent sections.

Chemical Analysis and Textural Properties of Catalysts In order to analyze coke and metal profiles as a function of feed, micro-, and pilot plants, unloaded spent catalysts were characterized, which showed changes depending on the unit operating conditions. The physical loss of catalytic activity is due to the coke and metal accumulation, causing partial or full blockage of pores. The deactivated pores can partially be regenerated by burning the coke, but metal accumulated pores will not be regenerated. The total loss in SSA for MP was about 60% while for PP is around 65%, a similar tendency is also shown for the pore volume as reported in Table 4.4. In order to separate the effect of coke and metals on deactivation, spent catalysts were air regenerated and the carbon was removed by burning at 600 °C for 8 h. The deactivation percentage was calculated by using the textural properties of the fresh, spent, and generated

TABLE 4.4 Textural Properties of the Fresh, Spent, and Regenerated Catalyst and the Deactivation on Surface Area and Total Pore Volume

Catalyst	$CoMo/Al_2O_3$ (MP)		$CoMo/Al_2O_3$ (PP)	
	SSA (m^2/g)	PV (mL/g)	SSA (m^2/g)	PV (mL/g)
Fresh catalyst	183.6	0.3173	183.6	0.3173
Spent catalyst	72.25	0.150	65.0	0.1423
Regenerated catalyst	131.0	0.2112	145.0	0.2517
% Deactivation (total)	60.65	52.73	64.60	55.15
(% Loss, coke+metals)	(32.0 + 28.65)	(19.29 + 33.44)	(43.47 + 21.02)	(34.48 + 20.67)
% Recovery	71.35	66.56	78.98	79.33
Metal, % deactivation	10.70	13.84	14.38	24.17
Carbon, % deactivation	89.30	86.16	85.62	75.83

$\%\ \text{deactivation} = 100 - \frac{\text{SSA spent catalyst}}{\text{SSA fresh catalyst}} \times 100 = \%$ loss of SSA by coke + % loss of SSA by metal.

$\%\ \text{recovery} = \frac{\text{SSA regenerated catalyst}}{\text{SSA fresh catalyst}} \times 100.$

% metal deactivation = % recovery − % deactivation (total).

catalysts, indicating that about 14% metals were deposited on the PP unit, while about 11% metals were deposited on the MP unit. On the other hand, the quantitative analyses of metals are reported in Table 4.5. The results of textural properties indicated that metals affect more the pores, while the coke is deposited mostly on the surface of the catalyst, which is eliminated during the regeneration, which metal will not affected the deposited metals. These results are complementary to the quantitative analysis of the metals, indicating that metal deposition is more on the pores.

SEM-EDS Analysis The air-exposed spent catalysts were measured by SEM-EDS and linear analysis of the surface extrudates is shown in Figure 4.20. The deposited species are measured in qualitative approach in which the pilot plant spent catalyst

TABLE 4.5 Chemical Analyses of Spent Catalysts

Catalysts	wt%						% Metals	% Carbon
	Ni	V	Fe	C	Total Metal	Total (Metal + Carbon)		
$CoMo/Al_2O_3$ (PP)	0.30	1.39	0.08	13.3	1.78	15.07	11.78	88.22
$CoMo/Al_2O_3$ (MP)	0.14	0.56	0.02	8.66	0.72	9.378	7.66	92.34
$NiMo–Al_2O_3$	0.15	0.31	0.00	11.0	0.48	11.46	4.01	95.99
$CoMo–Al_2O_3$	0.08	0.43	0.00	10.5	0.51	11.01	4.63	95.37

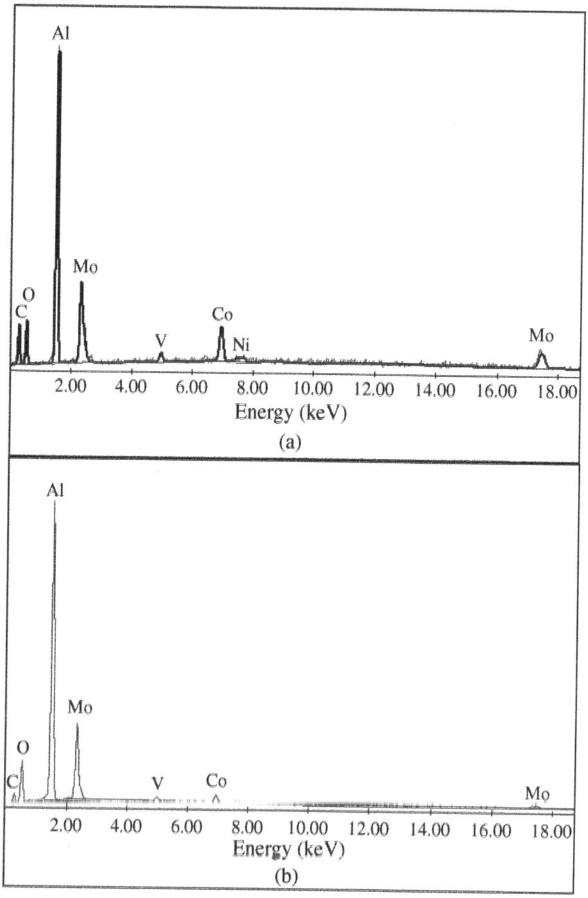

Figure 4.20 SEM-EDX elemental microprobe analysis of the surface of two different spent catalysts extrudates (a) pilot plant (PP) and (b) microplant (MP).

(Figure 4.20a) showed considerable number of higher peaks of elements for metals and carbon deposition than the microplant unit (Figure 4.20b). Based on the EDS-XRD results, it can be stated that the spent catalysts are in the partially oxysulfided phase with a hypothetical formula $MoS_2{-}xO_x$, where the value of x is close to one (Rana et al., 2004). Interestingly, distribution of cross-section line microprobe analysis of the surface indicated that V is mostly deposited on the edges of extrudates, while Ni is relatively homogeneously distributed along the line as shown in Figure 4.21. It appears that deposition of metals is much slower than that of the carbonaceous material. Particularly, V metal is usually deposited on the surface, while Ni is more distributed homogeneously, indicating that the diffusion of Ni is deeper on the catalytic pores or sites. The distribution of metal deposition is more related to the possible existing form of the metals, where Ni and V complexes structural differences are shown in Figure 4.22. The V metal in vanadium porphyrins exist as (VO)+2 in

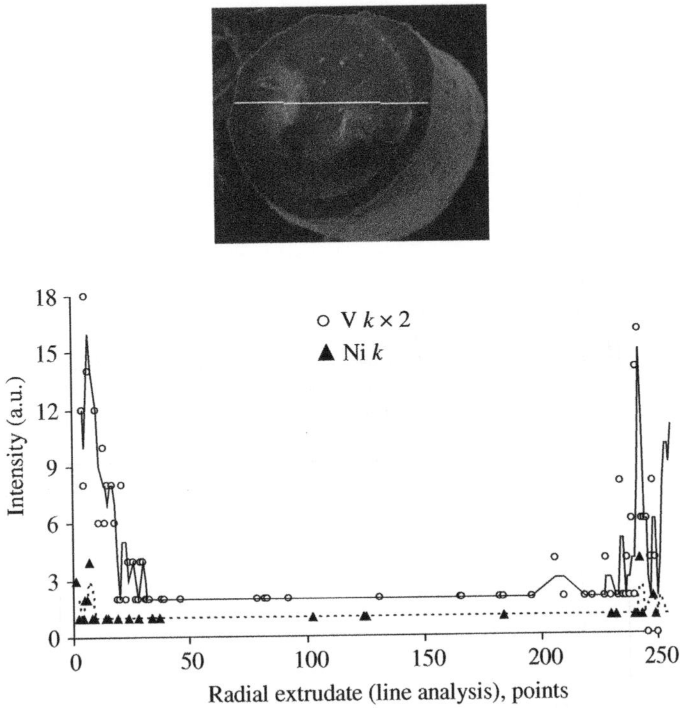

Figure 4.21 Radial analysis of deposited V and Ni in a line analysis by SEM-EDS for the pilot plant (PP) spent catalyst.

which V atom locates upward central location having larger annular plane along with the polar oxygen. In addition, the polar nature of oxygen can interact with V metal complex easily on the surface of catalyst/support and get removed faster, which leads to surface deposition of vanadium sulfide. The previous studies also indicated that V coordination has one O at 0.164 nm, four N at 0.214 nm while one N (or O) at 0.196 nm, and the Ni coordination has six N at 0.209 nm. Hence, variation in structural geometry and the ligands coordination bond distance caused different reactivity behaviors of Ni and V metals removal. Therefore, nickel porphyrins are expected to diffuse into catalyst pore easily and be hydrotreated, which cause the homogeneous distribution of Ni inside as well as outside the catalyst (Rana et al., 2005). On the other hand, Ni complex has low polarity and minor annular plane, indicating that Ni porphyrins can diffuse even in smaller catalyst pores relatively easily and after hydrotreating left Ni sulfides distributed homogeneously. Hence, diffusion of Ni complex molecules is better than that of V, which can be also attributed to its molecular size and inherently slower denickelation kinetic rate. Thus, it is well documented that the larger the fresh catalysts textural pore diameter, the greater the carbon and metal deposition, and hence the greater stability or metal retention capacity of catalyst (Ancheyta et al., 2005).

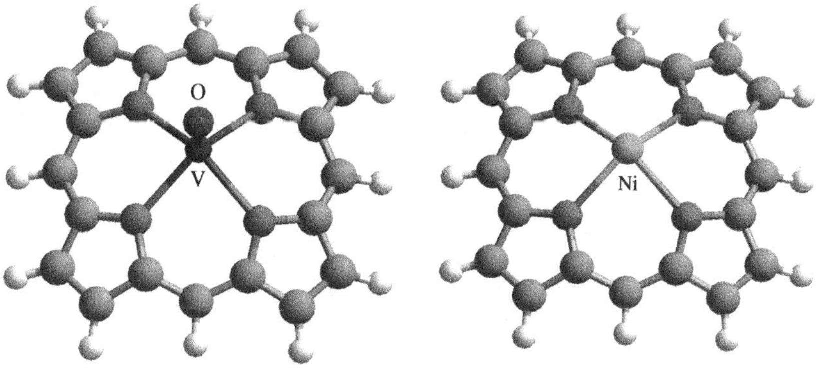

Figure 4.22 Structural differences between (a) V and (b) Ni porphyrin molecules.

It also became evident that metals accumulate more abundantly near the pore mouth granules. Therefore, deactivation due to pore plugging is a result of the decreased accessibility of part of the catalyst pore network, and the reduced rate of reactant molecule transport through the obstacle pores. Usually, metals that cause decline in activity are mainly Ni and V, which are believed to exist in asphaltene as metallic porphyrins (Rana et al., 2007). It is believed that diffusion of metal porphyrins into catalyst pores is an important factor impacting on HDM activity. Moreover, metallic porphyrins (nickel and vanadium) have a reactive structure that is highly reactive at hydroprocessing conditions. The aforementioned two factors play a key role in the HDM reaction and the metal-sulfide-deposited species are selectively seen at the outer surface of the catalyst. The deposition causes the metal accumulation unevenly on the catalyst outer surface, which eventually decreases pores inlet near the pore mouth, which is indicated as pore mouth plugging. Due to pore mouth plugging, catalyst performance drops because reactant molecules can no longer be diffused within the catalyst or deep catalytic sites. V metal concentrates near the edges of the outer surface.

HRTEM Characterization Apart from the Ni and V deposition, other metals such as Fe, Mg, Na, and P are deposited in smaller amount, which are confirmed by the transmission electron microscopy analysis. The distribution of metals is confirmed with the scanning transmission electron microscopy as shown in Figure 4.23. The presence of carbon and heteroelements on the surface of the spent catalyst is clearly observed at nanoscale. Through mapping, it is seen that carbon is the main component of the sample followed by vanadium and nickel. Phosphorous and metals such as sodium, iron, and calcium are dispersed unevenly and present in low percentage compared with carbon. STEM mapping analysis showed a broad distribution of different particle sizes with a mean diameter of about 5 nm. This was also estimated by energy-dispersive spectroscopy on the bright areas and by comparing the result

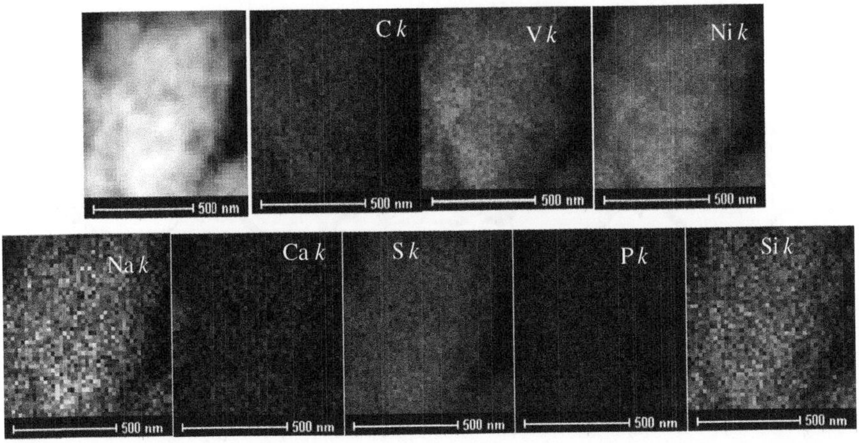

Figure 4.23 STEM chemical maps showing the metal particle distribution within the selected area of spent hydroprocessing catalyst (PP) (carbon (C*k*), vanadium (V*k*), nickel (Ni*k*), calcium (Ca*k*), sulfur (S*k*), potassium (P*k*), and silica (Si*k*)).

with that of the adjacent graphite background. This also indicates that the rate of metal deposition varies from metal to metal, which can be attributed to the differences in Ni- and V-porphyrins type of structure, since in the case of vanadium a perpendicular oxygen atom is linked to it, which forms strong bond with the catalyst surface, and nickel does not have this oxygen link. The deposited metal species (V_xS_x or Ni_xS_y) are strongly expected that the deposited phases provide some autocatalytic activity on the surface of the catalyst. Usually, these metal sulfide species may have some catalytic activity but relatively less active than molybdenum sulfide (MoS_2). The high metal deposition on the catalyst is expected to destroy or bury the existing MoS_2 sites.

The TEM analysis also confirms the existence of the catalytic sites as MoS_2 in the spent catalyst. For comparison, TEM images of the spent and fresh catalysts are shown in Figure 4.24. The fresh catalyst showed a multilayer (3–4) staking of MoS_2 with an average length of 2.2 nm, while multilayer MoS_2 stacking was not seen for the spent catalyst as well as the length of layer was about 3.1 nm, which were either covered with carbon or metals or remain as a single crystal (single slab). The spent catalyst was exposed for 2–3 min in air during unloading, which may exchange some sulfur with the oxygen and can have corresponding changes in the slab size. Nevertheless, the spent catalyst showed relatively stumpy MoS_2 stacking, which indicated that deposited species deform the MoS_2 stacking or catalytic active sites. Hence, these partial slabs with several other metal sulfidic entities are seen in the spent catalyst, which was exposed to heavy crude oil feed.

The solid foreign species (carbon and metal sulfides) were deposited during hydroprocessing, which wrap the existing MoS_2 catalytic sites. The deposited metal sulfides (Ni and V) appeared as microcrystalline sulfides that were expected to deposit close to the MoS_2 site mainly due to HDM catalytic reaction. The metal

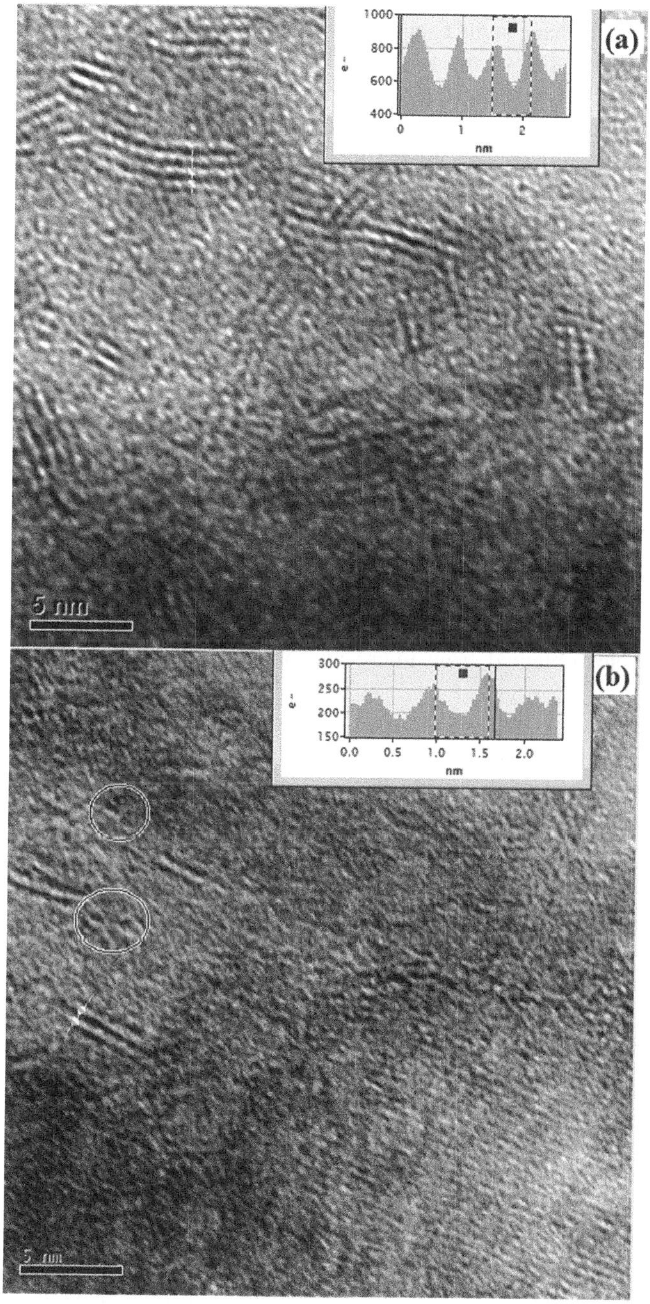

Figure 4.24 Typical HRTEM images of (a) fresh sulfide and (b) spent catalyst (PP).

sulfide entities had not been identified clearly thus, it is just a speculation that these metal sulfides may have some catalytic activities also. However, the HRTEM results do not agree with that of the study by Vogelaar et al. (2010) on thiophene and VGO spent catalysts, where the authors did not find distinct differences between the fresh and spent catalysts. On the other hand, deposition of carbonaceous materials on supported catalyst phases are the main cause for the deactivation and veil the catalytic sites. However, coke deposition mainly depends on the temperature of reactor that is continuously increased in order to compensate for catalyst deactivation and maintain conversion and product selectivity.

NMR Characterization ^{13}C CP/MAS and CP/MAS with dipolar dephasing NMR were used to characterize the chemical nature and composition of coke deposited during micro- and pilot plant operation of hydroprocessing of Maya feedstocks. Figure 4.25 shows ^{13}C CP/MAS NMR spectra of four spent catalyst samples obtained from micro- and pilot plant operations with variable coke loading in CoMo samples. In the ^{13}C NMR spectrum, regions from 0 to 70 ppm and from 90 to 180 ppm are attributed to the aliphatic carbons (saturated and side chain carbons attached to aromatic ring) and aromatic (protonated and nonprotonated) carbons, respectively. In all four spectra, two sets of featureless broad peaks are observed due to aromatic and aliphatic carbons of coke deposits. Comparing spectra from spent catalyst samples obtained from micro and pilot plant operation, two features are distinct.

The intensity of the peak corresponding to aliphatic carbons significantly decreases compared with aromatic carbons in coke deposits obtained from pilot plant operations. Another notable difference in the ^{13}C NMR spectra from spent catalysts samples is the higher signal-to-noise ratio of the spectra in NiMo spent catalyst samples compared with CoMo samples.

Although the amount of coke deposits in the case of CoMo catalyst obtained from pilot plant operation is almost double compared with NiMo catalyst sample, the aromaticity value shows a quite opposite trend. The coke aromaticity for CoMo samples are 0.45 and 0.52 for spent catalysts obtained from micro- and pilot plant operations, respectively. This shows that in pilot plant operations, the coke deposited is more aromatic in nature mostly due to the higher operation temperature (400 °C) compared with micro plant operations (380 °C).

The aromaticity of coke sample, which in general represents the nature of coke molecules, should be significantly higher for CoMo sample than for NiMo sample (Sahoo et al., 2004). As NiMo catalyst samples have better hydrogenation functionality compared with CoMo catalysts, the coke should show less aromatic character for NiMo sample. In contrast, the lower aromaticity value of coke deposits in the case of CoMo sample may be due to the fact that some of the coke molecules formed pericondensed ring structures that are invisible especially in mismatched cross-polarization spectrum. This is supported by that fact that the ^{13}C SPE/MAS NMR spectra of all coke deposited samples show more than 90% aromatic character. The dipolar dephasing ^{13}C CP/MAS NMR spectrum of the coke deposits from two spent catalysts at

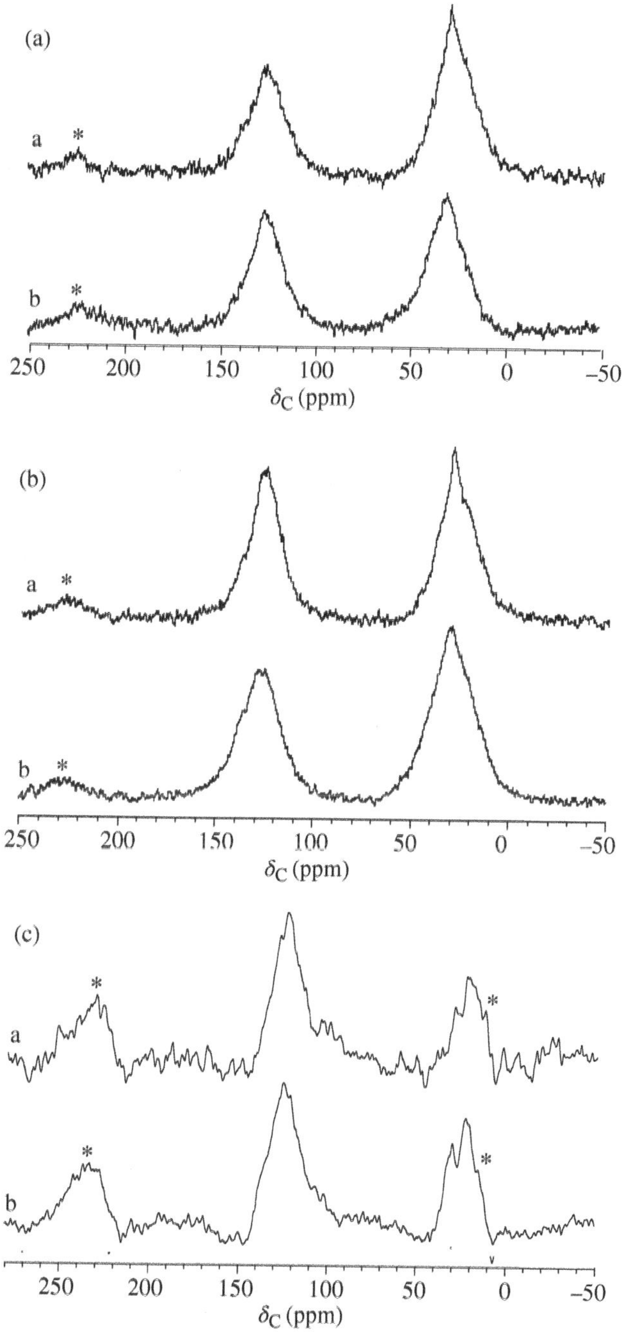

Figure 4.25 ^{13}C CP/MAS NMR spectra of coke deposited on spent alumina-supported catalysts: A: (a) $CoMo/Al_2O_3$ (MP); (b) $CoMo/Al_2O_3$ (PP). B: (a) $CoMo/Al_2O_3$ (MP); (b) $NiMo/Al_2O_3$ (MP). C: With dipolar dephasing time of 40 s of coke deposited on spent catalysts: (a) $CoMo/Al_2O_3$ (MP); (b) $NiMo/Al_2O_3$ (MP).

dipolar dephasing time of 40 s are also shown in Figure 4.25. The protonated aromatic carbons have strong dipolar coupling dephase after 40 s and only observable carbons are nonprotonated aromatic carbons. In these spectra, there is some resolution in the aliphatic region of each spectrum. However, it is hard to obtain any kind of quantitative information from them. Although the coke deposits in pilot plant operations are highly aromatic, it is still much less aromatic compared with the coke deposits in the case of acid catalysts. This could also imply that the low aromatic coke in CoMo catalyst can be due to the asphaltenes behavior in the presence of light hydrocarbons (i.e., paraffin) and/or stronger hydrogenation function carried out by NiMo catalyst that has lower aromatic coke. Apart from this, when the light paraffinic hydrocarbon comes into contact with asphaltene, it gets precipitated as a solid particle, which subsequently acts as coke. Thus, the coke identification is entirely different because asphaltene properties are very similar to that of coke. There is a very thin line difference between the asphaltene, sediment, and coke compositions, which can be confirmed by the H/C ratios $1–1.2 > 0.98–0.9 > 0.7–0.3$, respectively. The identification of these solid species becomes more difficult in the refinery conditions, where the long exposure of temperature and pressure are likely. It has been also reported that H/C ratio of coke on spent catalyst is higher than hydrotreated asphaltene, even some time higher than virgin asphaltene, which could be due to the presence of cracking and strong catalytic hydrogenation functions.

4.4.4 Conclusions

The heavy oil and residue catalyst deactivation is a phenomenon in which the structure and state of the catalyst change permanently, leading to the loss of active sites on the catalyst's surface and thus causing a decrease in the performance. Hence, catalyst deactivation is a result of a number of unwanted chemical and physical changes. Coke formation during the hydroprocessing reaction is a critical point to the catalyst deactivation and reactor plugging, which may turn out as pressure drop or reactor plugging. The detrimental effect of coke is a reduction of support porosity, leading to diffusional limitations, and finally blocked access to active sites. About 20–30% of the initial surface area and pore volume is lost due to coke deposition. Vanadium deposition on spent catalyst is relatively more on the surface, while Ni is distributed deeper into the pores of the catalyst. Ni and V removal showed different behaviors, which were attributed to an oxygen atom in vanadium and its influence on the vanadium porphyrin structure.

4.5 CHARACTERIZATION STUDY OF NiMo/SiO_2–Al_2O_3 SPENT HYDROPROCESSING CATALYSTS FOR HEAVY OILS

4.5.1 Samples of Spent Catalysts

Four supports were synthesized by homogeneous coprecipitation method varying the composition of silica and the four respective nickel–molybdenum supported catalysts were prepared by the incipient wetness coimpregnation method; more details about operating conditions and synthesis of materials can be found elsewhere (Leyva et al., 2012). The NiMo/SiO_2–Al_2O_3 spent catalysts were obtained after the hydroprocessing of Maya crude oil (21.31°API, 12.7 wt% asphaltenes, 322 ppm Ni+V, 3.52 wt% sulfur) carried out in a stainless steel tubular fixed bed upflow reactor (1.3 cm, inner diameter). The operating conditions used during the reaction were as follows: temperature of 380 °C, pressure of 5.4 MPa, LHSV of 1.0 h^{-1}, H_2-to-oil ratio of 356 m^3/m^3, and TOS of 204 h. The spent catalysts were unloaded from the reactor after hydroprocessing and washed by Soxhlet method with a mixture of toluene/heptane (2:1) at 96 °C during 8 h. Hereafter, the spent catalysts are denoted as NiMoSA-5-S, NiMoSA-10-S, NiMoSA-25-S, and NiMoSA-50-S, where the number (5, 10, 25, 50) represents the composition of silica in the support.

4.5.2 Catalyst Characterization

4.5.2.1 Electron Microscopy of the Energy Dispersive X-Ray The dispersion of nickel and vanadium in the spent catalysts as well as the content of carbon were determined by an SEM-FIB analytical instrument xT Nova NanoLab 200, using SEM-EDX analysis.

4.5.2.2 Textural Properties Nitrogen adsorption–desorption analysis at liquid nitrogen temperature (−196 °C) was done with a Quantachrome 4000 unit. Prior to analysis, the pellets were outgassed in vacuum at 350 °C for 3 h. BET specific surface area, total pore volume, and PSD were obtained by the BJH method.

4.5.2.3 Atomic Absorption (AA) Nickel and vanadium contents were measured by atomic absorption using a SOLAAR AA Series Spectrometer analyzer. Solid samples were heated to 550 °C for elimination of possible organic material. After that, the samples were digested in an acid mixture (HCl/HNO_3) with heating until total dissolution. Finally, they were filtered and analyzed.

4.5.2.4 Fourier Transformed Infrared Spectroscopy: CO Adsorption Fresh sulfided and spent catalysts were studied using CO adsorption at a low temperature (−196 °C) with increasing amounts of small CO doses up to an equilibrium pressure of 133 Pa. Spectra were scanned using a Nicolet 710 FTIR spectrometer equipped with MCT detector with a resolution of 4 cm^{-1}. For comparison, all the spectra were normalized to a disk of 10 mg/cm^2. More details about the procedure of catalyst sulfidation are given elsewhere (Leyva et al., 2012).

4.5.2.5 Hydrocracking of Cumene The cumene hydrocracking reaction was conducted in a continuous flow glass reactor operating at atmospheric pressure and 400 °C, in which 200 mg of catalysts were loaded (Figure 4.9). Prior to reaction, the catalysts were sulfided at 400 °C for 3 h in a flow of 50 mL/min of hydrogen saturated with CS_2.

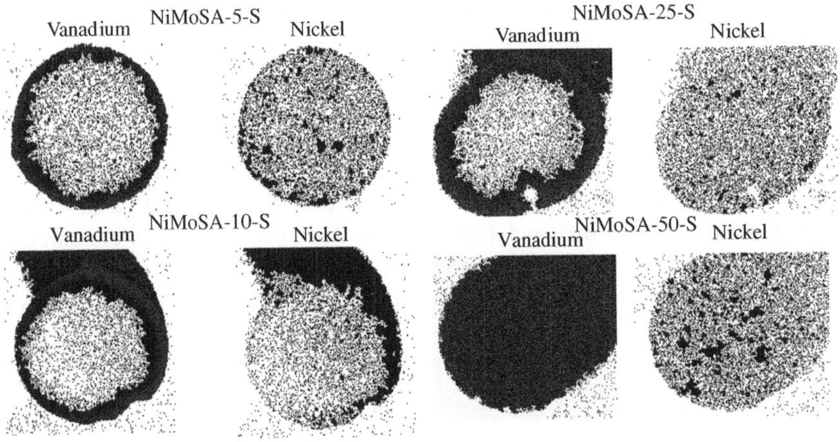

Figure 4.26 SEM-EDX profiles of silica content spent catalysts.

4.5.3 Results and Discussion

4.5.3.1 Scanning Electron Microscopy The SEM-EDX mapping of NiMoSA catalyst extrudates is illustrated in Figure 4.26, where the $Ni_{K\alpha}$ and $V_{K\alpha}$ distribution reveals how nickel and vanadium were deposited on these extrudates. For NiMoSA-5 and NiMoSA-10 catalysts, while nickel was deposited along the extrudate, vanadium was deposited on the surface of the catalyst, which confirms that the pores are plugged by vanadium metal (Maity et al., 2012). This indicates an internal diffusion limitation of these two spent catalysts that are characterized by small pore diameter and low silica content. The large pore diameter NiMoSA-50-S catalyst led to higher amounts of metal deposit. The profiles of both metals are homogeneous along the extrudate, confirming that macropores promote the diffusion of large molecules into the pores. NiMoSA-25-S had an intermediate behavior between the small pores catalysts (NiMoSA-5-S and NiMoSA-10-S) and the large pore catalyst: the distribution of nickel was homogeneous along the extrudate and while vanadium was more concentrated at the outer surface, its internal concentration was significantly higher than that of NiMoSA-5-S or NiMoSA-10-S catalysts. These results indicate that the diffusion of Ni containing molecules was larger than that of vanadium molecules, which could be due to its molecular size and/or intrinsically slower de-nickelation kinetic rate. It is also known that the Ni complex molecule is more refractory than vanadium.

4.5.3.2 Textural Properties A comparison of the textural properties of the fresh and spent catalysts is presented in Table 4.6. The decrease in specific surface area of NiMoSA-5-S and NiMoSA-10-S is large (70% and 80%, respectively); this is mainly due to the deposit of coke and metals on the catalyst surface and the presence of small diameter pores (5.6–5.8 nm). For high-silica-content catalysts (NiMoSA-25-S and NiMoSA-50-S), spent catalyst surface area decreases by 58% and 56%, respectively. These two catalysts were relatively less affected than low-silica-content catalysts, which could be explained by their macroporosity that allows large molecules

TABLE 4.6 Textural Properties of Fresh (F) and Spent (S) NiMoSA Catalysts

Catalyst	NiMoSA-5		NiMoSA-10		NiMoSA-25		NiMoSA-50	
	F	S	F	S	F	S	F	S
SSA (m^2/g)	327	89.6	349	65.5	255	106.0	167	73.1
TPV (cm^3/g)	0.48	0.16	0.49	0.14	0.54	0.27	0.58	0.30
APD (nm)	5.8	7.0	5.6	8.4	8.5	10.1	14.0	16.1
Micropore[a] (%)	1.1	16.2	1.6	13.0	1.2	4.9	1.5	2.3
Mesopore[a] (%)	98.3	79.5	97.7	81.2	93.3	46.8	73.8	65.2
Macropore[a] (%)	0.59	4.4	0.66	5.8	5.5	48.3	24.7	32.5
Pore size distribution (%)								
<5 nm	43.0	27.4	46.8	22.1	19.7	2.4	9.0	8.5
5–10 nm	47.8	48.8	44.4	37.9	31.2	2.9	15.8	11.4
10–25 nm	7.5	13.7	7.0	26.9	33.1	14.3	27.8	24.8
25–50 nm	1.02	5.7	1.1	7.3	10.5	32.1	22.7	22.8
50–100 nm	0.4	2.3	0.38	3.9	4.3	27.5	15.9	20.5
>100 nm	0.19	2.1	0.28	1.9	1.2	20.8	8.8	12.0

F, fresh catalyst; S, spent catalyst.
[a]Micropore < 2 nm, mesopore 2–50 nm, macropore > 50 nm.

to diffuse into the macropores instead of blocking up micro-/mesopore mouths. The porous structure of spent catalysts exhibits an increase in the fraction of micropores, which could be explained by substantial meso- and macropore blocking caused by the carbon and metal deposit during the reaction. The mesopores decrease for all catalysts in the following order: NiMoSA-25-S, NiMoSA-5-S, NiMoSA-10-S, and NiMoSA-50-S, which could be related to the metal deposited on the surface of the catalyst (Leyva et al., 2012).

PSD of spent and fresh catalysts are shown in Figure 4.27. A large decrease in small pore diameter (<5 nm) is shown for the NiMoSA-5-S and NiMoSA-10-S samples. This confirms the deactivation at the pore mouth, which is affecting the maximum utilization of internal surface of the catalyst, and consequently, blocks the active sites. This figure also indicates that the deposition of metal around the pore mouth increases with the decrease in pore diameter.

The PSD for NiMoSA-25-S shows a significant decrease in the small pores in the range of 5–10 nm, around 85% with respect to the fresh catalyst. On the other hand, NiMoSA-50-S was less affected due to its macroporosity structure; larger pore diameter has lower pore mouth plugging and better tolerance for metal retention. The most common deactivation is either due to the blocking of catalytic sites by deposit of species or due to the reactant diffusion limitation, which is more severe for small-pore-diameter catalysts than high-silica-content catalysts.

4.5.3.3 Atomic Absorption The nickel and vanadium contents deposited on the catalysts obtained by atomic absorption are presented in Figure 4.28; coke deposition is also shown in the same figure. All the values correspond to fresh basis, that is, grams of metal/100 g of fresh catalyst, and nickel amount of the fresh

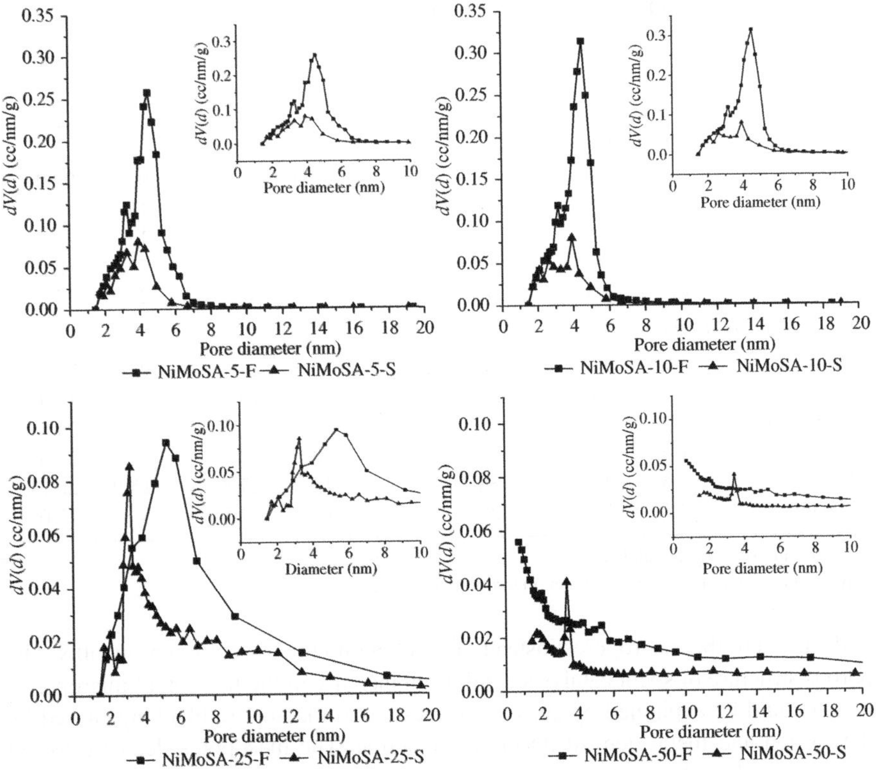

Figure 4.27 Pore size distribution of fresh (■) and spent (▲) catalysts.

catalysts has been discounted. On the one hand, a trend is observed between the amount of deposited metals and the acidity and mesoporosity of the NiMoSA-5-S, NiMoSA-10-S, and NiMoSA-25-S catalysts. This fact is rather normal due to the presence of wide pores as the silica content increases. On the other hand, NiMoSA-50-S has lower amount of deposited metals than NiMoSA-25-S but higher than NiMoSA-5-S and NiMoSA-10-S, which can be attributed to its highest acidity that provokes higher carbon deposits, causing those pores available for metal deposition be first occupied by coke. Also mesoporosity of this catalyst, particularly its pores of less than 10 nm, is about a half of that of NiMoSA-25-S catalyst.

4.5.3.4 FT-IR: CO Adsorption The deactivation of heavy oil hydroprocessing catalysts is not only due to deposition of metals such as nickel and vanadium but also due to coke deposit. To elucidate the species of carbon deposited on the surface, the spent catalysts were characterized by FTIR using CO as molecule probe for studying the sulfided phase after reaction.

Carbon Deposition Figure 4.29 illustrates the carbonaceous species present on the surface of the catalyst (C—C stretching region and C—H stretching region). According to the literature, the 1377 cm^{-1} band is characteristic of the catalysts poisoned by

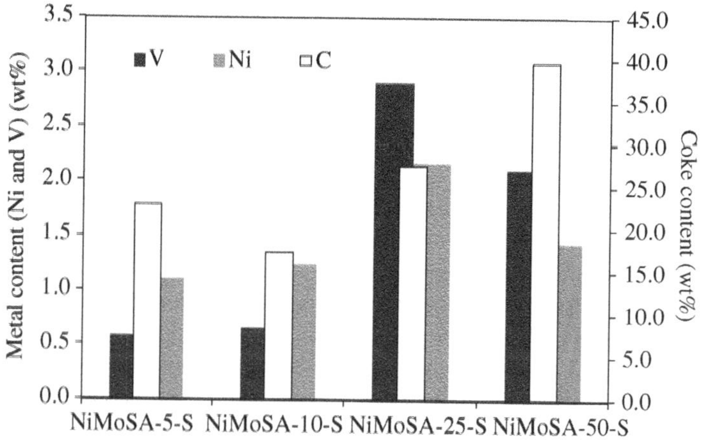

Figure 4.28 Metal and carbon depositions on spent catalysts.

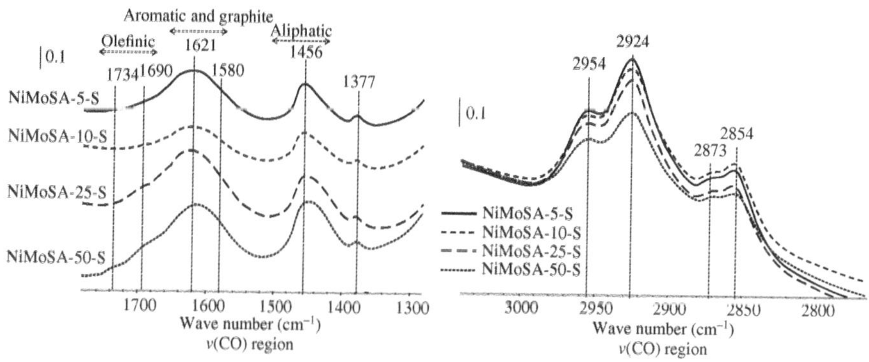

Figure 4.29 FTIR spectra of NiMoSA spent catalysts.

coke and is ascribed to the symmetric bending CH vibration of CH_3 groups and characterized as "soft coke." No specific band at 1580 cm^{-1} could be observed. The latter is usually characteristic of crystalline graphite (hard coke). This shows that no significant amount of hard coke was deposited. Since hard coke is expected to build up at long TOS, this absence might be due to the short TOS (204 h) used in the experiments. The frequency band of CH_2 and CH_3 groups are related to the 1500–1400 cm^{-1} region. The vibration mode of methyl/methylene groups in long aliphatic chains are characterized by the bands located near 1456 cm^{-1}, which corresponds to the asymmetric bending C—H vibration bond and the bands in the region of 1375 cm^{-1} are due to the symmetrical bending mode. The band at 1734 cm^{-1} could be related to the ν (C=O) vibration of the carboxylic acids (Mariey et al., 1996). The appearance of bands between 1500 and 1640 cm^{-1} can be ascribed to aromatics C—C stretching vibrations over polyaromatic compounds and/or to vibration of C—O (at around

1620–1550 cm^{-1}) due to the presence of oxygenated molecules. The similarity of the spectra indicates that the four spent catalysts have the same type of carbon species.

The area of the bands characteristic of the soft coke present on spent catalysts are (a.u.): NiMoSA-5 = 1.13, NiMoSA-10 = 0.98, NiMoSA-25 = 1.43, and NiMoSA-50 = 1.33.

It is seen that the amount of soft coke deposited on the four catalysts is of the same order of magnitude. The deposit on high-silica-content catalysts, however, is significantly higher than that on low-silica-content catalysts, which can be related to acidity and the specific surface area, since high-silica-content catalysts possess higher acidity but a much lower specific surface area. The trend of these results matches well with those plotted in Figure 4.28, except for the highest silica content catalyst. From SEM-EDX analysis, NiMoSA-50-S had the highest coke deposition, while from FTIR results NiMo-SA-25-S exhibited coke deposition a little bit higher than NiMoSA-50-S catalyst. This different behavior is because FTIR results correspond only to soft coke.

In the CH stretching region (3000–2800 cm^{-1}), four bands were detected by infrared spectroscopy. The band at 2924 cm^{-1} corresponds to the asymmetric CH_2 stretching vibrations while the band at 2954 cm^{-1} is attributed to asymmetric CH_3 stretching vibrations. The 2873 and 2854 cm^{-1} bands are related to CH stretching bond in both CH_3 and CH_2 groups.

The broad adsorption bands in the range of 3000–2800 cm^{-1} are assigned to aliphatic moieties containing CH_3 and CH_2 groups; the relative intensity of the CH_2 and CH_3 characteristics bands may give some indications on the amount and chain length or branching of aliphatic moieties.

The computed areas are shown in Figure 4.30. These results indicate that the amount of both CH_3 and CH_2 decreases as the silica content increases. In parallel, the amount of deposited carbon and soft coke increases. This trend could be correlated to the effect of silica acidity, which may favor both aromatic coke deposition and cracking of aliphatic chains.

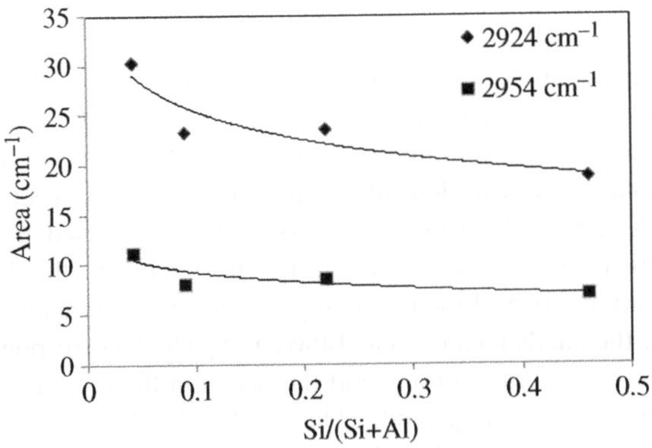

Figure 4.30 Integrated intensities of the bands at 2924 cm^{-1} (CH_2 groups) and 2954 cm^{-1} (CH_3 groups) in the spectra of spent catalysts as a function of Si/(Si+Al) ratio.

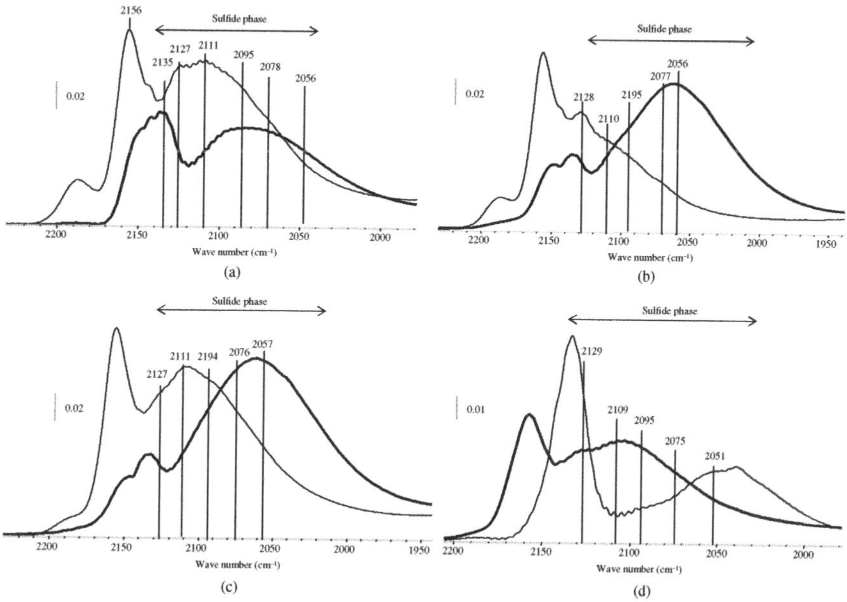

Figure 4.31 FTIR spectra of CO adsorbed on fresh (—) and spent (---) catalysts: (a) NiMoSA-5, (b) NiMoSA-10, (c) NiMoSA-25, and (d) NiMoSA-50.

Sulfide Phase of Spent Catalyst The difference between the fresh and spent catalysts is illustrated in Figure 4.31, which shows the spectra of CO adsorbed at saturating conditions (1 Torr in equilibrium at −196 °C of CO) on both types of catalysts. The sulfide phase of spent catalysts was analyzed after the thermoevacuation of the samples at high temperatures (400 °C) in order to remove some weakly adsorbed carbon species of the surface and to analyze them with the adsorption of CO at low temperatures. The IR spectra obtained on the fresh catalysts display bands at ~2190 cm^{-1} (Al^{3+} Lewis acid sites), ~2155 cm^{-1} (surface silanol and hydroxy groups of the support) and bands at lower frequencies characteristic of nonpromoted molybdenum sites and promoted sites of the NiMoS (2127 cm^{-1}) sulfide phase. The spectra obtained for the spent catalysts show a strong change in the amount of surface sites. In particular, the area of the CO bands characteristic of NiMoS species and unpromoted molybdenum are shifted toward lower frequencies, indicating an increase in the sulfide phase electron density. It is worth noticing that such a frequency shift also leads to an increase in the CO absorption coefficient, indicating that the total amount of sulfide phase sites is lower on spent catalysts compared with fresh catalysts. Similarly, surface OH groups and Lewis acidic sites from the support exhibit a considerable decrease. This shows that the deposit of coke and metals causes the poisoning of the support surface sites. In general, deactivation is due to the strongly adsorbed species, for example, N-compounds, coke, and metals, which occupy active sites, resulting in a loss of catalytic activity. These results also show that the nature of the sulfide phase could be affected, in particular, the promoted NiMoS sites that show enhanced

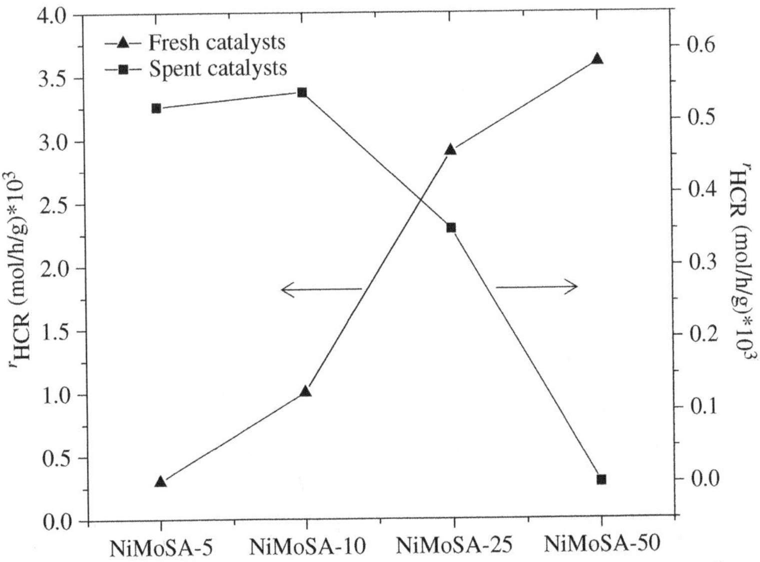

Figure 4.32 Comparison of hydrocracking activity of cumene for fresh and spent catalysts.

electron density. While the origin of this enhancement is not definitely established, it can be tentatively assigned to the deposit of electron-rich deposits (e.g., coke or reduced metals) in contact with the sulfide phase.

4.5.3.5 Cumene Hydrocracking Activity The decrease of catalytic acid sites, which are selectively poisoned by carbon deposition on Brønsted sites of silica–alumina, was confirmed by cumene hydrocracking as illustrated in Figure 4.32. It shows a drastic decrease in catalyst activity of spent catalysts compared with fresh catalysts. Moreover, while in the case of fresh catalysts, activity increases according to the increasing acidity of the catalyst, spent catalysts show the opposite trend, that is, decrease of the catalytic activity with the silica content. This fact is related mainly to the coke and metals deposited on the surface of the most acidic catalysts (NiMoSA-25-S and NiMoSA-50-S). It is observed that the NiMoSA-10-S has a slightly higher catalytic activity among all catalysts; this fact can be attributed to the lower carbon deposition on its surface.

4.5.4 Conclusions

The silica–alumina NiMo catalysts for hydrocracking of heavy crude oils are strongly affected by carbon and metal deposition. Due to the short TOS used in the experiments, only soft coke was detected by FTIR, which was found to be in the same order of magnitude for the four spent catalysts. Hard coke is more common at long TOS, which is why no specific band was detected. The deactivation rate increases with the silica content of the catalyst; the acidity favors the accumulation of aromatic carbon on the surface strongly affecting the high-silica-content catalyst, while lower-silica-content catalysts are relatively less affected by carbon deposition. Low-silica-content catalysts are more sensitive to metal deposition because of their porous structure, which provokes pore mouth plugging. Another important aspect is the decrease of the total amount of NiMoS sites due to the accumulation of electron-rich deposits of carbon and reduced metals, which is more severe for high-silica-content catalysts.

Hence, in order to minimize the deactivation of silica–alumina-supported catalysts, a trade-off is required between acidity and meso-macroporosity, both of which increase with the silica content.

4.6 CHARACTERIZATION OF SPENT CATALYSTS ALONG A BENCH-SCALE REACTOR

4.6.1 Experimental Procedure

4.6.1.1 Samples of Spent Catalysts An atmospheric residue (312 °C+) obtained from a 13°API crude oil distillation was hydrotreated in a bench-scale reactor at T = 360–410 °C, P = 100 kg/cm^2, LHSV = 0.25 h^{-1}, and H_2/HC ratio = 5000 ft^3/bbl (826.7 NL/L). The main properties of the residue are API gravity of 5.7, sulfur content of 5.93 wt%, nitrogen content of 0.582 wt%, nC_7 insolubles of 22.12 wt%, Conradson carbon content of 21.48 wt%, nickel content of 144, and vanadium content of 144 pm.

A NiMo/γ-Al_2O_3 tetralobular extrudate catalyst was packed as front-end catalyst in a fixed-bed reactor operated continuously in concurrent downflow (Figure 4.33). The reactor was shut down after 1100 h of operation. A flow of naphtha was passed through the spent catalysts inside the reactor. Afterward the spent catalysts at different heights of the reactor were sampled by suction; the samples were kept in toluene to prevent oxidation. The spent catalyst samples were placed in cellulose extraction thimbles and then they were washed with hot toluene in a Soxhlet extractor during 20 h. After cooling to room temperature, the washed samples were recovered and then dried at 110 °C and atmospheric pressure during 12 h. The so treated catalysts were halved; a portion is labeled as spent catalysts, while the other half was regenerated. The regeneration was performed by heating at 350 °C for 2 h and then at 550 °C for 8 h in presence of air using 5 °C/min as heating rate.

4.6.1.2 Characterization of Catalysts BET specific surface area, total pore volume, and PSD were determined in Quantachrome Nova 4000 equipment by N_2 physisorption at temperature of liquid nitrogen. Prior to adsorption, the catalysts were pretreated in vacuum and temperature for 3 h, spent at 150 °C and regenerated at 300 °C. The composition and surface morphology of the spent catalysts were analyzed by SEM-EDX using a Nova NanoLab 200 microscope.

Nickel and vanadium contents in the spent catalysts were measured by flame atomic absorption spectrometry (ASTM D5863). C and S contents in the catalysts were determined by direct combustion-infrared detection methods (ASTM C1408 and ASTM D1552).

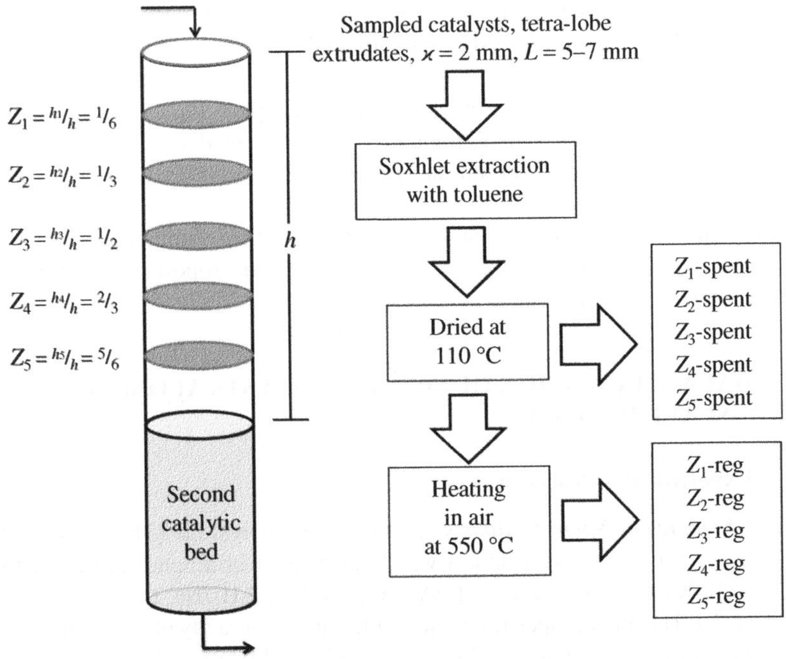

Figure 4.33 Catalysts sampling.

The spent samples were heated in flowing dry air from room temperature to 500 °C with a heating rate of 5 °C/min and ramps at room temperature, 200, 300, 400, and 500 °C. This regeneration process was monitored by DRIFT by means of a Harrick diffuse reflectance accessory model Praying Mantis equipped with KBr windows. The spectra were collected in an 8700 Nicolet FTIR spectrophotometer with 300 scans and a resolution of 4 cm^{-1}.

Analysis for both spent and regenerated catalysts were obtained using a Siemens 5005 diffractometer, using Cu kα radiation, in the 2θ scale from 4 to 70, at 2θ/min scan rate.

The types of carbon present in the coke deposited on spent catalysts were determined by ^{13}C NMR spectroscopy using the Cross-polarization with MAS technique. The spectra were recorded in a Bruker Avance 400 NMR spectrometer operating at a frequency of 100 MHz using a 4 mm probe at a rotor spinning rate of 10 kHz and 1 s relaxation delay.

The adsorption of pyridine analyzed by FTIR spectroscopy was used to quantify the amount of Lewis and Brønsted acid sites in the regenerated catalysts. For these experiments, the regenerated powdered catalysts were pressed into self-supporting wafers. The wafers were introduced into a special IR cell and sulfided at atmospheric pressure and 380 °C and the samples were then outgassed at 380 °C. Afterward, a pulse of pyridine was introduced in the cell, and then the process of desorption under outgassing at increasing temperatures was monitored. Spectra were taken at room temperature, 100, 200, 300, and 380 °C in a Nicolet 8700 FTIR spectrophotometer using a resolution of 4 cm^{-1} and 50 scans.

Thermal gravimetric analysis was performed in NETZSCH model STA409PC. TPO were performed with Zeton Altamira AMI-200 equipment coupled to an Ametek mass spectrometer. Both TGA and TPO analyses were performed from room temperature to 800 °C, under flowing dry air and using 5 °C /min as heating rate.

4.6.1.3 Catalytic Activity To explore the remaining sulfur removal capacity and acidity, in both spent and regenerated catalysts, thiophene HDS and cumene HDC (Hydrocracking) were used. Both reactions were performed in a fixed-bed glass reactor operated at atmospheric pressure and temperature of 350 °C (Figure 4.9). A feed consisting of 60 mL/min of a mixture of 5 mol% thiophene in hydrogen was used for thiophene HDS, while for cumene HDC the feed was a mixture of 0.23 mol% cumene in hydrogen with a flow of 50 mL/min. Prior to activity test, the catalysts (spent and regenerated) were sulfided at 400 °C for 2 h using a flow of 40 mL/min of CS_2/H_2 gas mixture. Catalyst particle was sized to 20–40 mesh. The progress of the reaction was monitored by online gas chromatography using a Varian 4700 gas chromatograph equipped with an FID.

4.6.2 Results

A high amount of coke deposited on the catalyst is expected in view of the contents of asphaltenes and Conradson carbon of the feed (22.12 and 21.48 wt%, respectively). An important contribution to catalyst deactivation by metals is also expected since vanadium and nickel contents are also high.

Spent catalysts corresponding to the first bed of the hydroprocessing reactor were sampled according to Figure 4.33. The bed height was normalized as $Z_i = h_i/h$, where h stands for the total bed length and h_i the height of the ith sample. Hereafter, catalyst samples were identified as their corresponding normalized height plus either "spent" or "reg" for spent and regenerated samples, respectively.

4.6.2.1 Deposition of C, S, and Metals The profiles of the amount of V, Ni, C, and S retained in the spent catalysts are plotted in Figure 4.34. Clearly from top to bottom of the catalytic bed, the retained amount of V, Ni, and S decreases whereas the C content augments. The values are expressed per gram of fresh catalyst. This fresh basis has the advantage of rendering directly the retained amount capacity of the catalyst. Since for instance, a value of 0.4 of V implies that 0.4 g of V were retained in 1 g of fresh catalyst (Ramírez et al., 2009). Henceforth, the deposition is discussed in terms of the amount retained per gram of fresh catalyst. Near the entrance of the reactor (Z_1) the V retained in the catalyst almost reaches 0.5 g and then this amount decreases to 0.2 g at Z_5. The Ni retained amount is lower compared with that of V, consistently with the corresponding contents in the feed (144 wppm Ni and 549 wppm V). Nonetheless, as the catalyst contains Ni in its original formulation, a correction was performed in order to avoid counting Ni original present in the fresh catalyst

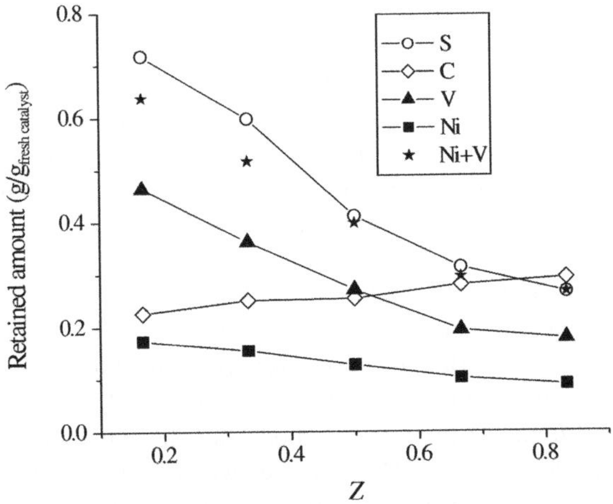

Figure 4.34 Profiles of the deposition of Ni, V, C, and S in fresh basis.

(Ramírez et al., 2009). The Ni retained ranks from 0.17 g in Z_1 to 0.07 g in Z_5. Sulfur is the element predominantly retained, the sulfur amount required for the whole sulfidation of the Mo and Ni in the fresh catalysts was not accounting as S retained. In Z_1, the retained S amount was nearly 0.7 g. In the last S content measured Z_5, the S retained falls to 0.3 g. Interestingly, the amounts of sulfur and the sum Ni+V show the same trend suggesting that sulfur could be trapped by sulfidation of the metals retained in the catalyst. In this research, the amount of coke is defined as the C content. The coke retained is 0.23 g at Z_1 and then it continuously grows up to 0.29 g at Z_5.

Figure 4.35a shows the subtraction of the Z_i-spent and the fresh sample spectra taken at room temperature. Three main peaks attributed to the contribution of C—H stretching of aliphatic hydrocarbons are detected (2957, 2925, and 2855 cm^{-1}). These peaks are related to nonrefractory hydrocarbons (soft coke). Clearly the relative amount of soft coke increases toward the reactor outlet. Figure 4.35b shows the subtraction of the spectra of the Z_i-spent and the fresh sample spectra taken at 300 °C. In this region, the absorption due to stretching of C═O (1750–1650 cm^{-1}), aromatic C═C (1630–1500 cm^{-1}) and the "coke peak" (1585 cm^{-1}) (Marafi et al., 2007) is expected. As the catalyst is under an oxidation process, the three peaks detected at 1782, 1622, and 1554 cm^{-1} can be attributed to graphite oxide (Zsabó et al., 2005; Acik et al., 2011). These peaks are related to the amount of the refractory carbonaceous compounds (hard coke). So, the relative amount of hard coke also increases toward the reactor outlet. Hard and soft coke profiles through the bed length match the carbon profile determined by ASTM D5291, and it can also be inferred that the ratio of soft and hard coke is nearly constant given that both display an increasing tendency through the reactor bed.

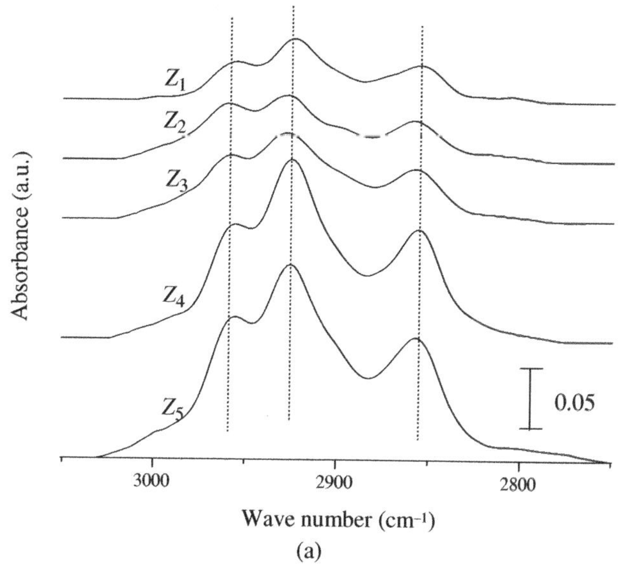

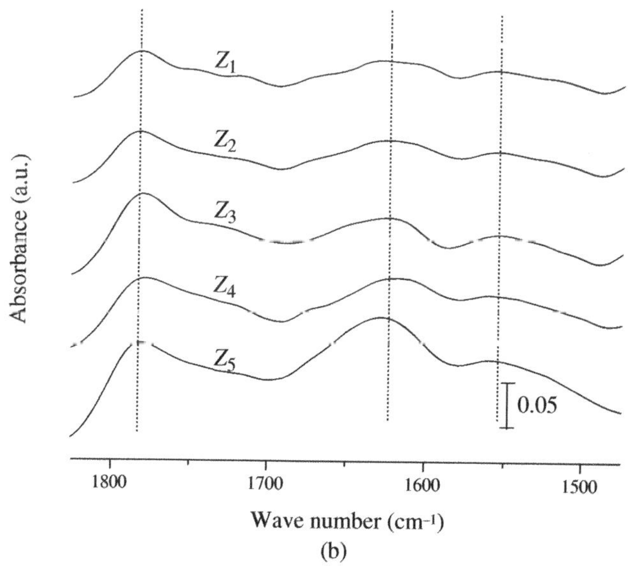

Figure 4.35 DRIFT spectra of spent catalysts during regeneration: (a) spectra in the zone of aliphatic carbon taken at room temperature and (b) spectra in the zone of aromatic carbon taken at 300 °C.

TABLE 4.7 Textural Properties Referred to the Fresh Catalyst

Properties	Z_1-Spent	Z_2-Spent	Z_3-Spent	Z_4-Spent	Z_5-Spent	Z_1-Reg	Z_2-Reg	Z_3-Reg	Z_4-Reg	Z_5-Reg
S_{BET}[a] (%)	37	41	43	49	51	19	21	27	33	44
PV[b] (%)	21	24	26	29	32	18	23	30	35	45
APD (Å)	101	102	102	101	108	172	195	196	185	181
PSD[c]										
<50 Å	414	544	531	448	381	106	4	0	0	0
50–100 Å	40	52	55	67	70	9	10	8	16	16
>100 Å	15	15	16	18	20	21	27	37	41	53

[a]BET surface area percentage of the fresh catalyst = BSA $Z_i \times 100$/BSA fresh catalyst.
[b]Pore volume percentage of the fresh catalyst = PV $Z_i \times 100$/PV fresh catalyst.
[c]Pore size distribution expressed as the percentage of the corresponding value in the fresh catalyst.

4.6.2.2 Textural Properties of Catalysts In order to visualize the changes in the specific surface area as a function of the reactor length, S_{BET} profile is expressed as the percentage of the remaining BET surface area from the fresh catalyst (Table 4.7). Assuming that initially S_{BET} must be 100%, a drastic decline in the specific surface area is detected at the beginning of the catalytic bed (Z_1); beyond this point, the specific surface area slightly increases with the reactor length. The same trend is observed for both spent and regenerated catalysts. Interestingly, spent catalysts show higher S_{BET} values compared with the corresponding regenerated samples. This fact obeys to the contribution to S_{BET} supplied by small pores in spent catalysts as it will be explained later.

For the discussion of porosity, three diameter ranges will be used; pores with a diameter bigger than 100 Å called big pores, pores with a diameter between 50 and 100 Å named midsize pores, and small pores corresponding to diameters lower than 50 Å. PSDs are plotted in Figure 4.36, while Table 4.7 shows the percentages of the three pore ranges related to the corresponding values in the fresh catalyst. Clearly, in spent catalysts big pores are primary affected by metal and coke deposition; the remaining percentage of big pores is 15% for the spent samples Z_1 to Z_3, then it grows to 20% for Z_5-spent. The percentage of remaining midsize pores increases from 40% in Z_1-spent to 70% in Z_5-spent. On the contrary, the amount of small pores is considerably higher than the corresponding in the fresh catalyst. This enhancement shows a maximum value in Z_2-spent, which corresponds to more than five times that of the fresh catalyst. By comparing PSDs of regenerated and spent samples, the next fact can be stated: the excess the amount of small pores detected in spent catalysts disappears in regenerated samples. In fact, only Z_1-reg owns small pores with a volume equivalent to that in the fresh catalyst. While for Z_2 to Z_5 the contribution of small pores is negligible. The pore volume corresponding to midsize pores is lower for regenerated than for spent samples, the keeping of midsize porosity ranks from 9% in Z_1-reg to 16% in Z_5-reg. In addition, while the diminishing in the contribution of big pores in spent samples is important, a clear recovery of big pores in regenerated samples is

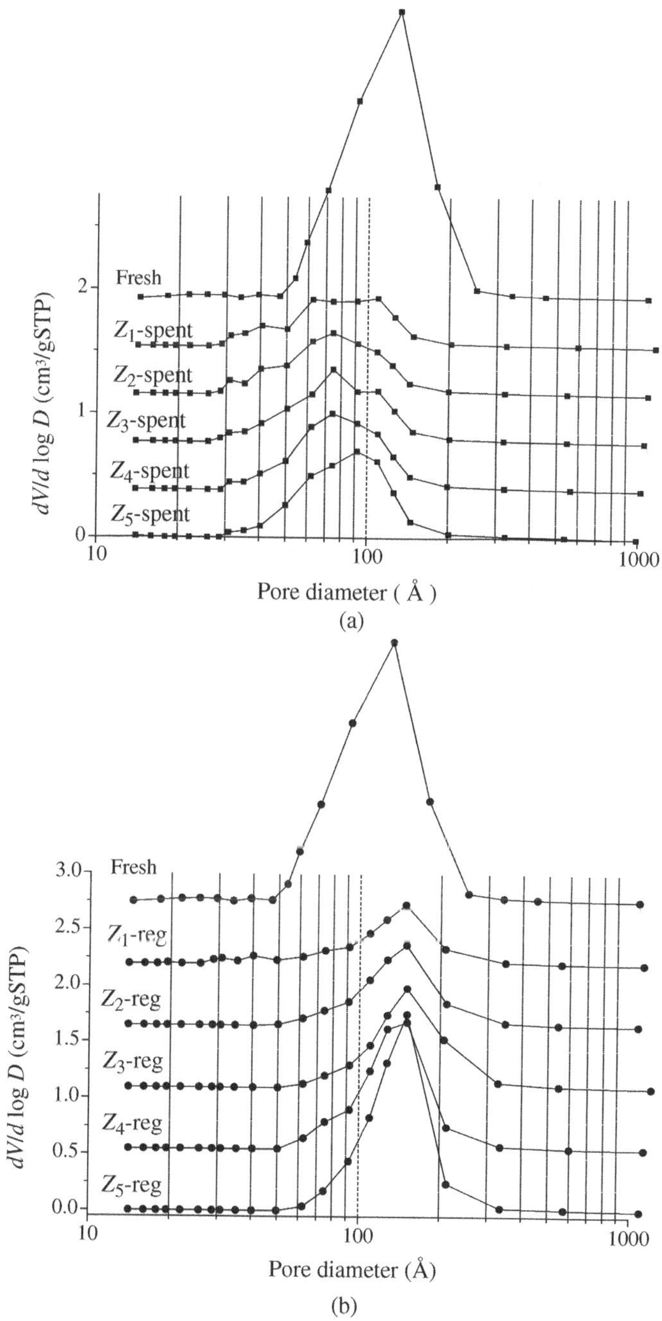

Figure 4.36 Pore size distributions: spent (a) and regenerated (b) catalysts.

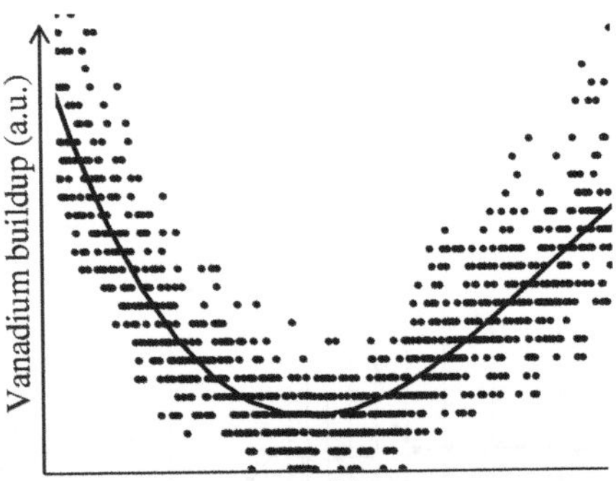

(a)

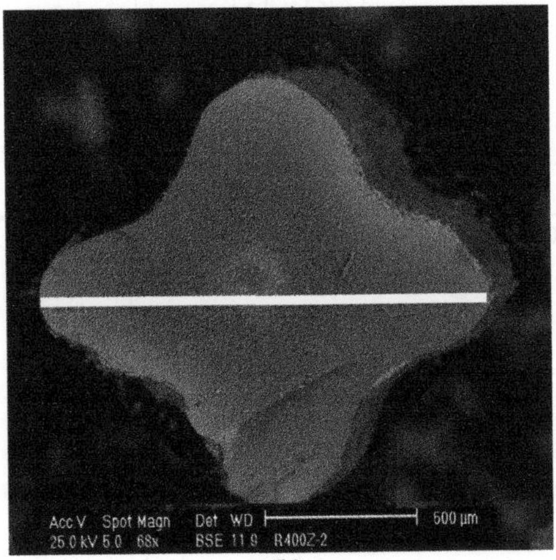

(b)

Figure 4.37 Intraparticle V distribution (a) and Z_2-spent micrograph (b).

detected. Furthermore, the recuperation level of these pores increases with the length of the catalytic bed.

The intraparticle distributions of V, Ni, C, and S of spent catalysts were obtained by elemental analysis performed by SEM-EDX. Figure 4.37 shows a transversal sight of a tetralobular pellet of a spent sample taken at Z_2; the intraparticle distribution of the relative amount of vanadium deposited inside the pellet is also included; clearly, vanadium deposition decreases from the edge to the center of the particle. The intraparticle concentration of sulfur is very similar to that of vanadium, conversely a uniform distribution is obtained with nickel and carbon.

4.6.2.3 Deeper Characterization Figure 4.38 shows the diffraction patterns for spent and regenerated samples. For fresh catalysts, only diffraction peaks corresponding to γ-Al_2O_3 were detected with peaks around 19.5, 31, 38 39.5, 46, 61, and 67 in the 2θ scale. In the case of spent catalysts, apart from γ-Al_2O_3, V_3S_4 was clearly

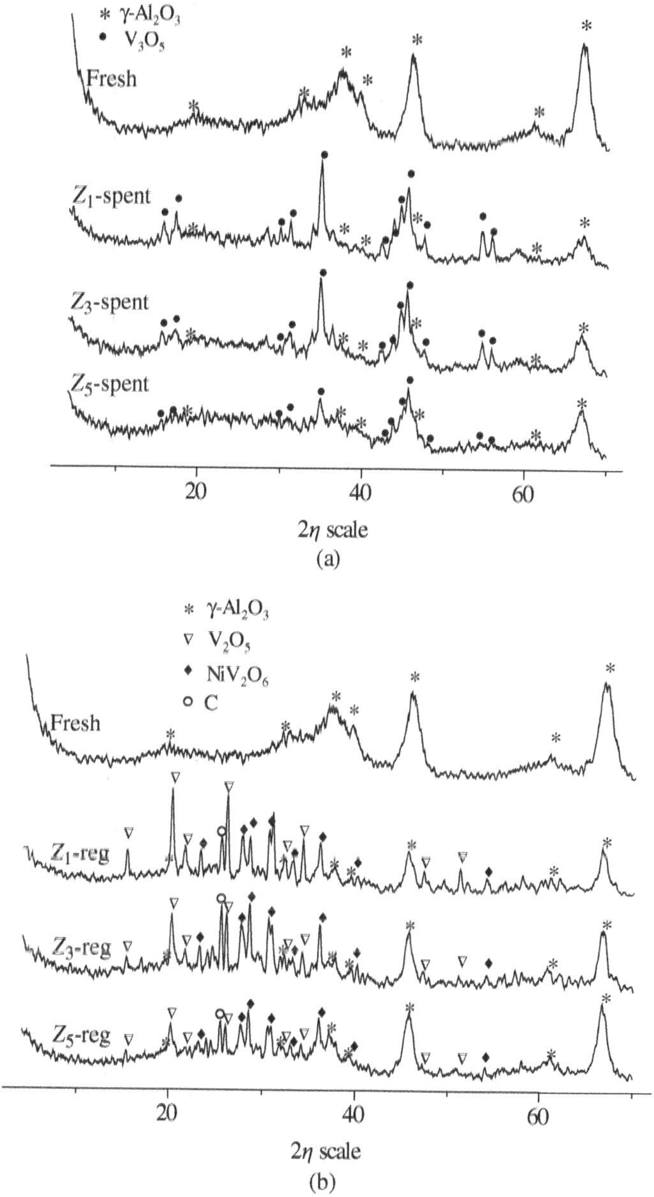

Figure 4.38 XRD results for spent (a) and regenerated (b) catalysts.

detected matching the main peaks marked in Figure 4.38a; 15.7, 18.3, 31.2, 34.8, 34.9, 43.6, 44.6, 45.3, 45.4, 47.5, 54.7, and 56. Clearly, the peaks assigned to V_3S_4 declined through the bed length in agreement with the profile of the deposited metal previously determined. In the case of regenerated catalysts, besides γ-Al_2O_3, two

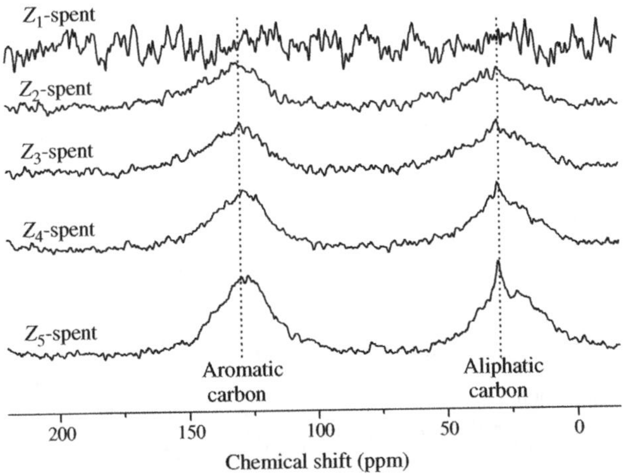

Figure 4.39 Solid-state ^{13}C CP-MAS NMR spectra of spent catalysts.

oxide phases were identified; V_2O_5 with main peaks at 15.5, 20.5, 22, 26.5, 32.3, 34.3, 47.3, and 51.2 in the 2θ scale, and NiV_2O_6 with peaks at 23.6, 27.6, 28.2, 30.6, 31, 32.9, 36.1, 36.2, 40.2, and 54.1. Again the signals related to V_2O_5 and NiV_2O_6 diminished toward the bed outlet coherently with the profile of metal sulfides for spent catalysts. In addition, a sharp peak at 25.6 in the 2θ scale was detected; this peak can be attributed to coke in graphite-like structures.

The solid-state ^{13}C NMR spectra for spent samples are shown in Figure 4.39. No peaks are detected in Z_1-spent spectrum, whereas for Z_2-spent to Z_5-spent two peaks are detected at 30 and 130 ppm corresponding to aliphatic and aromatic carbon, respectively. Both peaks increase through the bed length, which suggests the coke buildup increases from the top to bottom of the catalytic bed in accordance with the profile previously determined. Both aliphatic and aromatic contribution increase toward the bed outlet in the same proportion; therefore, it can be inferred the aliphatic/aromatic ratio is nearly constant through the length of the catalytic bed.

The results for TGA for fresh, Z_1-spent, Z_3-spent, and Z_5-spent catalysts are shown in Figure 4.40a, while Figure 4.40b shows a qualitative analysis of TPO for Z_2-spent sample monitored by mass spectrometry. To relate the weight loss observed in TGA with a specific oxidation process, the results are analyzed with the help of the gases evolved detected by TPD-MS results. The analysis is based on six temperature ranges defined by the weight loss behavior observed in the TGA results. Table 4.8 shows the weight losses and the process associated to each zone, the weight losses for spent samples are expressed in fresh basis to perform an adequate comparison, since the amount of deposited foulants changes through the bed length.

The first zone defined from room temperature to 225 °C is related mainly to release of moisture from the catalysts. The second zone between 225 and 350 °C can be associated to hydrocarbons decomposition leading to CO_2, H_2O, and SO_2; the third zone between 350 and 425 °C in which the oxidation carbonaceous compounds yields

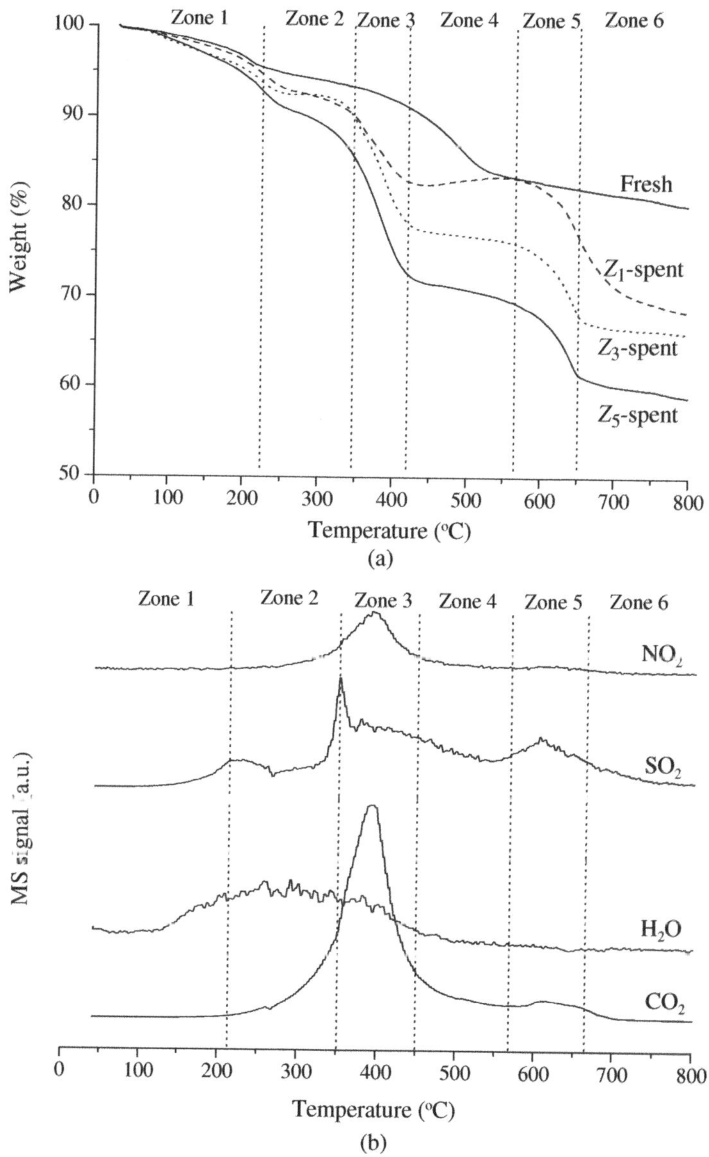

Figure 4.40 Progress of spent catalysts oxidation followed by TGA (a) and TPO-MS (b).

NO_2, SO_2, H_2O, and a high proportion of CO_2. This temperature range can be related mainly to the combustion of the carbonaceous compounds that can be transformed at relatively low temperature (soft coke), the maximum combustion rate of this type of coke is centered near 400 °C according to TPD-MS results. In this zone, the weight loss in the fresh catalyst is notably lower compared with the spent samples. The

TABLE 4.8 Weight Loss in TGA by Zones and Evolved Gases by TPO-MS

Temperature Range		Fresh Catalyst	Z_1-Spent	Z_3-Spent	Z_5-Spent	Desorbed Gases[a]
Zone 1 RT-225 °C	Weight loss[b] (%)	4.6	5.3	6.3	7.2	H_2O
	Fresh basis weight loss[c] (%)	4.6	18.6	16.6	15.3	
Zone 2 225–350 °C	Weight loss[b] (%)	2.1	4.7	4.1	7.5	H_2O, SO_2
	Fresh basis weight loss[c] (%)	2.1	16.6	10.8	16.0	
Zone 3 350–425 °C	Weight loss[b] (%)	2.4	7.1	11.8	13.1	Mainly CO_2, SO_2, NO_2, H_2O
	Fresh basis weight loss[c] (%)	2.4	25.0	31.2	28.0	
Zone 4 425–575 °C	Weight loss[b] (%)	7.7	0	2.0	3.0	CO_2, SO_2, NO_2, H_2O
	Fresh basis weight loss[c] (%)	7.7	0.1	5.4	6.5	
Zone 5 575–650 °C	Weight loss[b] (%)	1.1	12.8[d]	8.6	8.4	CO_2, SO_2
	Fresh basis weight loss[c] (%)	1.1	44.9[d]	22.8	17.9	
Zone 6 650–800 °C	Weight loss[b] (%)	1.9	1.6[e]	1.3	1.9	None
	Fresh basis weight loss[c] (%)	1.9	5.5[e]	3.3	4.2	
Total	Weight loss[b] (%)	19.8	31.6	34.0	41.1	H_2O, CO_2, SO_2, NO_2
	Fresh basis weight loss[c] (%)	19.8	110.7	90.1	87.9	
Removable foulants	Fresh basis weight loss[b] (%)	16.5	58.3	62.9	65.1	H_2O, CO_2, SO_2, NO_2
Irremovable foulants	Fresh basis weight loss[c] (%)	3.3	52.4	27.2	22.7	CO_2, SO_2

[a]Desorbed gases detected by TPO in Z_2-spent sample.
[b]Weight loss measured directly from TGA in spent samples.
[c]Weight loss expressed in fresh basis as weight loss (g) * 100/mass of fresh catalyst (g).
[d]For Z_1-spent temperature range zone 5 = 575–730 °C and zone 6 = 730–800 °C.
[e]For Z_3-spent temperature range zone 5 = 575–665 °C and zone 6 = 665–800 °C.

weight loss observed in zone 3 in spent samples increases toward the bed outlet (see zone 3 row in Table 4.8). Zone 4 from 425 to 570 °C is connected to a slow oxidation of hydrocarbons producing NO_2, SO_2, H_2O, and CO_2. In this region, the weight loss in the fresh sample is important, but spent catalysts practically do not show weight loss.

The oxidation of refractory carbon species in zone 5, located at 570 and 665 °C, contains the oxidation of hard coke and also the final transformation of the trapped metal sulfides to metal oxides. Based on TPD-MS, the burning of carbon species of this zone produces CO_2 but not water, which implies very low hydrogen to carbon ratio. In this zone, the weight loss for the fresh catalyst is negligible while for spent samples the values decrease from the top to bottom (see zone 5 row in Table 4.8). Finally, in the last zone, located above 650 °C, a minimal weight loss for all the samples is detected.

The weight losses corresponding to room temperature to 550 °C and from 550 °C to 800 °C have been included in Table 4.8 since the regeneration of catalysts was performed at 550 °C. The values represent a measure of the amounts of removable and irremovable foulants by regeneration. From top to bottom of the bed, the amount of removable foulants increases, whereas the amount of irremovable foulants decreases.

The acidity in regenerated catalysts was examined by means of pyridine adsorbed. Figure 4.41 shows FTIR spectra of samples desorbed at 100 °C; six bands due to the pyridine interaction with the catalysts were detected at 1640, 1610, 1578, 1540, 1490, and 1450 cm^{-1}. The bands at 1540 and 1450 cm^{-1} were assigned exclusively to Brønsted acid sites and Lewis acid sites, respectively; the integrated absorption areas of these bands were used to estimate the amount of acid sites. Table 4.9 contains a summary of the amounts of Lewis and Brønsted sites determined; for the fresh catalyst the amount of Lewis sites is high, but the amount of Brønsted sites is low. The amount of Lewis acid sites in regenerated catalysts is lower than half the value in the fresh catalysts for almost all the temperatures studied. But interestingly, a higher amount of Brønsted sites compared with the fresh catalyst were observed in the regenerated catalysts at all the temperatures analyzed. In fact, it was noticed that along the bed length the tendency in the amount of Brønsted acid sites measured at 380 °C in regenerated catalysts matches the trend in hydrocracking activity (Figure 4.42).

4.6.2.4 Remaining Catalyst Activity Figure 4.43 shows the results of thiophene HDS expressed as the reaction rate of the spent or regenerated catalyst per gram of the fresh catalyst divided by the reaction rate of the fresh catalyst. This way to express the activity obeys that only Mo is considered responsible for HDS activity. Indeed, Mo is provided only by the fresh catalyst since it is not contained in the feedstock and thus cannot be deposited on spent catalysts. The remaining HDS activity of the regenerated catalyst increases from the top to bottom of the catalytic bed, whereas activity of spent catalysts diminishes at the beginning and then it appears to remain in a steady value around a quarter of the activity of the fresh catalyst.

The remaining acidity in the catalyst was evaluated by HDC of cumene and is also presented in Figure 4.43. In this case, the active acid sites cannot be absolutely assigned only to a particular element in the catalyst. For this reaction, the activity of

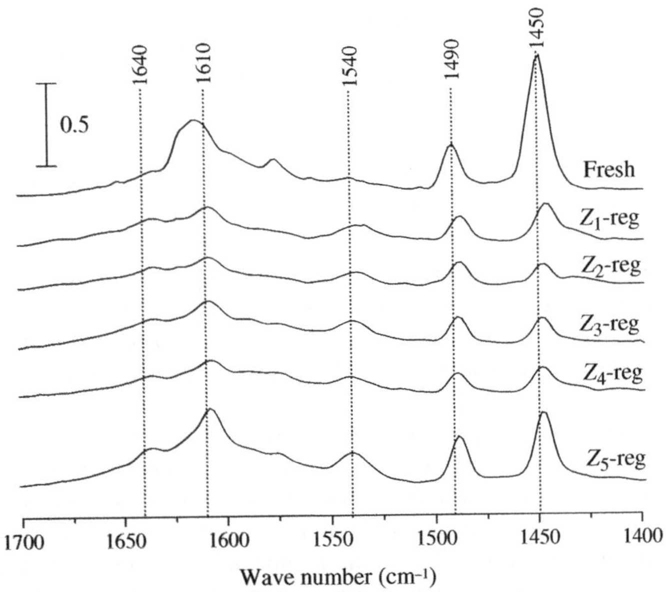

Figure 4.41 FTIR spectra of pyridine desorbed at 100 °C from regenerated catalysts.

TABLE 4.9 Acid Sites Determined by Pyridine Adsorption ($\mu mol/g_{cat}$)

		Desorption Temperature (°C)				
		50	100	200	300	380
Fresh	Brønsted	12	10	8	6	5
	Lewis	850	358	175	152	138
Z_1-reg	Brønsted	58	54	52	47	43
	Lewis	133	119	109	89	87
Z_2-reg	Brønsted	58	57	48	46	39
	Lewis	156	48	24	17	13
Z_3-reg	Brønsted	56	55	46	9	6
	Lewis	229	98	56	19	15
Z_4-reg	Brønsted	49	41	34	20	11
	Lewis	262	151	67	49	21
Z_5-reg	Brønsted	52	48	44	24	5
	Lewis	406	182	99	83	67

spent and regenerated catalysts is expressed as the quotient of the reaction rate per gram of either spent or regenerated catalyst and the reaction rate of the fresh catalyst. The HDC activity of all the spent catalysts is near a quarter of that in the fresh catalyst. Regenerated catalysts, on the contrary, show higher activity than the fresh catalyst; Z_1-reg and Z_2-reg is around 75% upper, while the regenerated Z_3 to Z_5 catalysts show nearly the same activity of the fresh catalyst.

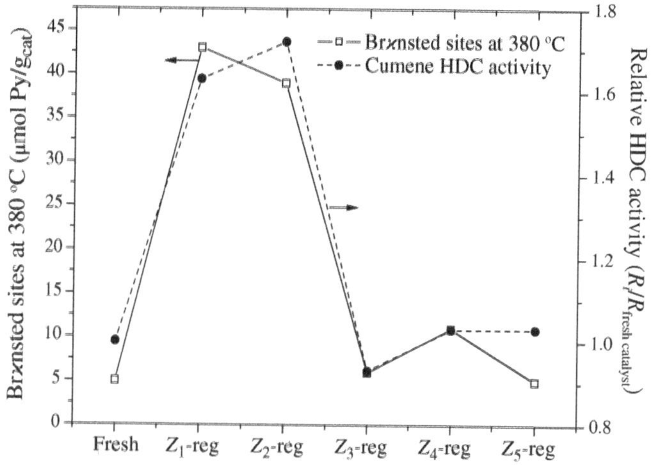

Figure 4.42 Correlation of hydrocracking activity and Brønsted acid sites.

4.6.3 Discussion

Metal and coke deposition shows contrary trends; while the former is high at the beginning and then it decays, the latter starts low and then rises, consistently with profiles previously reported (Marafi et al., 2010; Maity et al., 2013; Nuñez-Isaza et al., 2000; Zsabó et al., 2005). Both hard and soft coke amounts also increase toward the bed outlet, although the soft/hard ratio seems constant through the entire bed. Several factors can contribute to form a higher amount of coke through the bed length; the declining hydrogen concentration toward the bed outlet, a sequential mechanism of coke formation via intermediates, and also the poisoning of active acid sites by intermediate basic N-compounds (Furimsky and Massoth, 1999). The impact of metals and carbon depositions on textural properties obeys to the lessening in the amount of midsize and big pores, resulting in profiles for S_{BET} and pore volume with drastic falling at the bed inlet and then a smooth recovery toward the outlet of the catalytic bed. Bearing in mind that regeneration process removes coke deposits, by comparing the PSDs between spent and regenerated samples, it seems that coke deposition occupies or blocks mainly the big pores and that metals are lodged mostly at midsize pores. Nonetheless, even after regeneration only about 50% of the original big pores are recovered, which implies half of them are filled or blocked by metals. In addition, in spent catalysts, coke deposition creates small pores, which may be the consequence of narrowing of midsize and big pores by the deposition of metals and coke in the pore walls. This behavior has been previously reported (Ancheyta et al., 2002). Besides, a contribution of micropores inherent to the coke is possible.

While nickel and carbon are deposited homogenously inside the pellet, vanadium buildup radially decays to the pellet center. The vanadium profile is well known and it may indicate slow diffusion of V-metalloporphyrins inside the pores, which can

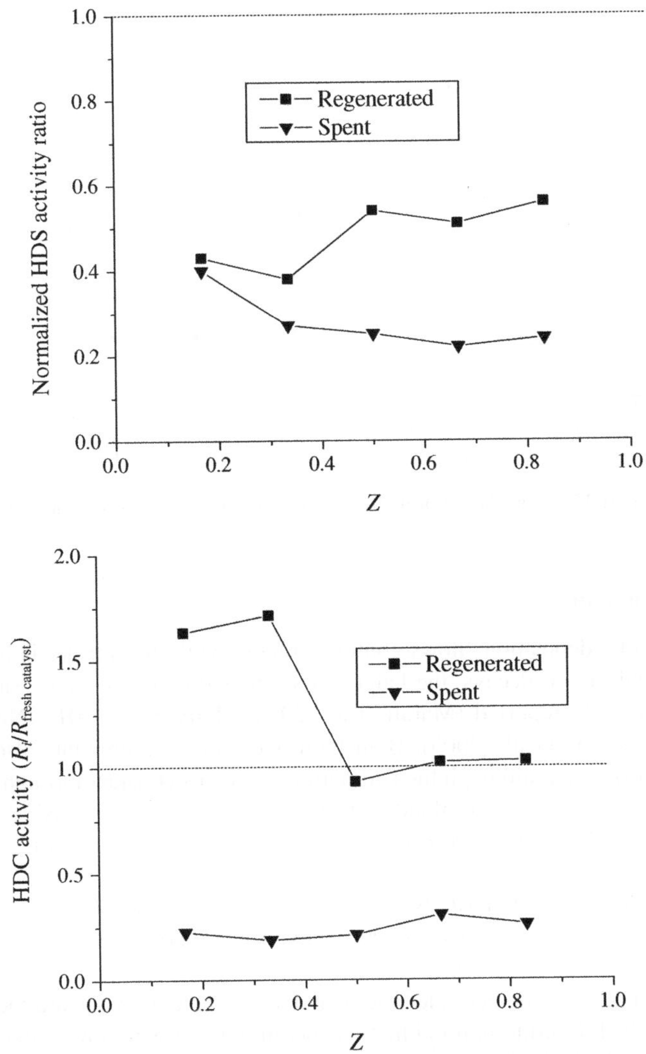

Figure 4.43 Profiles of remnant catalyst activity in (a) HDS of thiophene and (b) HDC of cumene.

result in a certain level of mouth pore plugging, especially of midsize pores. The tendency in the sulfur content in spent catalysts is explained as a reaction between sulfur, perhaps in H_2S form, and the Ni and V retained in the pores. In other words, once the metals are transformed by HDM they are sulfided.

Based on the catalytic activity in both thiophene HDS and cumene HDC, the deactivation level is around 75% for the spent samples taken from Z_2 to Z_5. This value is in line with the decrease in pore volume. By the regeneration process, these samples recover HDS capacity up to 50% of that in the fresh catalyst. Coherently, the retrieval in HDS activity through the reactor bed follows the same trend as the deposited carbon. The Z_1-spent sample simultaneously owns the lowest values of S_{BET} and pore volume and the highest contents of caught Ni and V, which suggests the highest deactivation level. Nonetheless, as Ni and V sulfides may contribute to the overall HDS activity, the Z_1-spent sample shows the highest HDS activity of all the spent catalysts.

Hydrocracking of cumene is a well-known reaction used to test acidity of hydrotreating catalysts, and it has been observed a relation between hydrocracking activity and Brønsted acidity, so the results of HDC activity are related directly to the surface acidity. The regeneration of spent catalysts yields higher HDC activity than that measured for the fresh sample. Therefore, a generation of acid sites instead of a recovery of the original acid sites is detected. After regeneration and resulfidation of the samples, a vanadium sulfide or oxysulfide could be formed. In view of the related profiles of HDC activity and V-deposition, a V-compound could be responsible for the extra acidity observed. Accordingly, Marafi et al. (2007) found enhanced hydrotreating activity when spent catalysts were regenerated and they suggested the formation of new active sites involving V, Mo, and Ni.

XRD results confirm the decreasing trend in retained metals through the bed length previously observed. Besides, it was observed that the metals trapped in catalysts remain as metal sulfides in spent catalysts. Several classifications of coke deposits have been employed in the scientific literature (Furimsky and Massoth, 1999). Two types of coke are defined when reactivity toward oxidation is used as criteria for classification; coke is referred to as reactive, reversible or soft and refractory, irreversible or hard. In the present study, two types of coke were detected: a coke that can be oxidized below 550 °C and whose maximum oxidation speed is located near 400 °C according to TPO results and a refractory form of coke that can be oxidized above 550 °C.

The chemical nature of the deposited carbon is commonly analyzed by ^{13}C NMR. Nonetheless, it has been reported that carbon in the graphite form is not detectable by NMR (Fonseca et al., 1996). This invisibility is attributed to the paramagnetic character of graphite, and it is also pointed out that deposited metals hinder the analysis, resulting in broadening of ^{13}C signal (Marafi et al., 2007). Hence, quantitative analysis of carbonaceous compounds in spent hydrotreating catalysts requires special cautions such as the use of model compounds as standards or by combining NMR with other analytical techniques. Solid-state ^{13}C CP-MAS NMR was employed as a qualitative tool to examine the possible changes in the chemical nature of carbon deposits through the bed length. Nonetheless, the aromaticity could be underestimated due to lack of the contribution to the signal from the nondetectable graphitic carbon (signal at 130 ppm).

The Z_1-spent NMR spectrum in Figure 4.39 is too noisy due to the interference caused by the high amount of metals deposited on this sample. Aliphatic and aromatic carbons were clearly detected in Z_2-spent to Z_5-spent samples. The aromaticity of carbonaceous deposits is commonly referred to as the aromatic to total carbon ratio. Hence, the aromatic/aliphatic ratio renders a measure of the coke aromaticity. It was found that both types of carbon species increase toward the bed outlet but keeping the aromatic to aliphatic proportion along the reactor bed. Hence, it is possible to state that the deposited coke owns nearly the same aromaticity throughout the catalytic bed.

Soft coke constituted mainly by nonrefractory hydrocarbons was removed by the regeneration process. Therefore, the remaining carbonaceous compounds correspond to refractory coke whose crystallinity could be enhanced by the regeneration process. In addition, in regenerated catalysts hard coke is no more diluted with soft coke. For these reasons, carbon as crystalline graphite was only detected by XRD in regenerated samples.

In TGA analysis, the amount of soft coke is assumed proportional to the weight loss in zone 3, the amount of removable foulants is taken as the weight loss up to 550 °C and the amount of irremovable foulants as the weight loss between 550 and 800 °C. It was determined that the amounts of both soft coke and removable foulants increase from the top to bottom of the catalytic bed. Irremovable foulants are ascribed to the oxidation to refractory coke and also to a final step in the oxidation of the sulfide metals trapped in the catalysts. The trend through the bed length is decreasing for the amount of deposited metals but rising for refractory coke; the net result is irremovable foulants decline toward the bed outlet. Accordingly, the deactivation degree for regenerated samples must follow the same trend. Consistently, the remnant HDS activities in regenerated catalysts tend to grow through the bed length.

^{13}C NMR results were not determining about the presence of graphitic carbon. However, XRD patterns revealed the presence of graphitic carbon in regenerated samples. Furthermore, thermal analysis disclosed the presence of carbon species with very low hydrogen to carbon ratio that can be only oxidized at temperatures higher than 550 °C. Such refractory carbonaceous species are assumed as hard coke formed by graphite-like structures.

In the present research, the catalysts regeneration procedure established an oxidation up to 550 °C.

According to the results, after regeneration a fraction of carbonaceous compounds remains lodged in the catalysts and the transformation of metal sulfides to metal oxides are not completed at the regeneration temperature employed. It is possible that a fraction of the metal sulfides stays embedded into the refractory coke delaying its transformation to oxides. The broadening of the oxidation process detected for Z_1-spent and Z_3-spent samples in zone 5 toward higher temperatures is a confirmation of the metal oxidation delays. In fact, the higher the metal content the higher the temperature shift. With respect to the spent sample with the lowest amount of deposited metal (Z_5-spent), the temperature shifts are 80 °C for Z_1-spent and 15 °C for Z_3-spent.

The remaining activity of spent catalysts in the hydrodesulfurization of thiophene and hydrocracking of cumene has been related to the remaining textural properties. However, the activity of regenerated catalysts was not adequately explained. As mentioned, the remaining HDS activity of regenerated catalysts follows the opposite trend to metal deposition profile. This trend is explained by assuming that active sites for HDS reactions are only provided by the fresh catalyst surface, and, consequently, the irreversible deactivation is maximum at the bed entry and then it declines leading to an increasing tendency in the remaining HDS active sites through the bed length.

The modified surface of regenerated catalysts contains, apart from its original components, refractory carbon and also vanadium and nickel in the form of oxides and sulfides. By measuring the acidity of regenerated samples by FTIR of pyridine adsorbed, it was found that the deposited compounds transformed by regeneration produce Brønsted acid sites. It is well-known that there is a relationship between hydrocracking activity and Brønsted acidity. The results show that the tendency in the amount of Brønsted acid sites of regenerated catalysts measured at 380 °C matches the trend in hydrocracking of cumene along the bed length.

4.6.4 Conclusions

The deactivation of the catalytic bed studied comprises two deactivation causes: metal and coke depositions. Metal accumulation is high at the top of the first catalytic bed and then it declines toward the bed outlet. Meanwhile, coke buildup, of both hard and soft coke, increases through the bed length. In this deactivation by simultaneous metal and coke depositions, there is a tendency of metals to deposit in pores of midsize, whereas coke deposition seems to fill big pores. Inside the catalytic pellets, vanadium and sulfur are accumulated from the edge to the center of the particle, while nickel and carbon are homogeneously distributed. The deactivation extent is closely related to textural properties, especially to the pore volume, whose tendency correlates with the activity of spent catalysts in thiophene HDS and cumene HDC.

Different levels of retrieval in HDS or HDC activities after regeneration process are obtained; the activity in thiophene HDS is recovered around half the fresh catalyst value, whereas the activity in cumene HDC equals, and in some cases surpasses, the activity of the fresh catalyst. So, in regenerated catalysts the contribution from the deposited Ni and V species to the overall HDS activity is minimal but in the case of HDC activity it is noteworthy. It is hypothesized that the vanadium retained in the catalysts can generate acid sites by the regeneration procedure.

Metal and coke depositions are analyzed by characterization of catalysts recovered along the reactor bed. Metals trapped in the catalysts remain as metal sulfides in the spent catalysts and their amount decreases from the top to bottom of the bed length. Deposited species that can be removed by regeneration at 550 °C are called removable foulants and their amount increases through the bed length. On the contrary, the irremovable foulants by regeneration are high at the front end and then decline; the irremovable foulants correspond to a fraction of metal sulfides embedded into the refractory coke; for this reason, irremovable foulants and metal deposition profile coincide. Carbonaceous deposits are also classified according to removability: deposits of carbon that can be oxidized below 550 °C (soft coke) and deposits of refractory carbon that can only be oxidized above 550 °C (hard coke). By solid-state ^{13}C CP-MAS NMR, it was determined that coke contains aliphatic and aromatic carbon. It was found that the amount of coke increases from the inlet toward outlet of the catalytic bed keeping the same aromatic/aliphatic proportion, which implies coke has the same aromaticity throughout the catalytic bed.

The profile of remaining hydrodesulfurization activity of regenerated catalysts agrees with the deactivation degree expected by the irreversible deactivation due to irremovable foulants, that is, the higher the deactivation degree, the lower the remaining HDS activity. On the other hand, regenerated catalysts display equal or higher activity in hydrocracking of cumene. This extra hydrocracking activity is the result of the creation of Brønsted acid sites after regeneration. The activity of regenerated catalysts in the hydrocracking of cumene correlates with the trend in amount of Brønsted acid sites measured at 380 °C.

4.7 HYDRODESULFURIZATION ACTIVITY OF USED HYDROTREATING CATALYSTS

4.7.1 Experimental Procedure

4.7.1.1 Used Catalysts Several used catalysts (NiMo) were tested to study the hydrodesulfurization reaction. The used catalysts were obtained from a pilot plant, where different types of feeds (21, 13, and 7°API) were used (Table 4.10). In this reactor, oxide catalyst was sulfided *in situ* before actual run was started. Hundred milliliters of oxide catalyst were loaded by dividing into three zones. The catalyst was then dried for 2 h at atmospheric pressure and at 120 °C. After drying, the catalyst was allowed for soaking for 2 h at 150 °C. LGO was used for soaking. This LGO contains 1.7 wt% sulfur. The actual sulfiding agent was introduced after soaking. The sulfiding agent was LGO with dimethyl disulfide (1 wt%). Sulfidation was done at a pressure of 28 kg/cm^2 at two different temperatures. First sulfidation was performed at 260 °C for 3 h and finally the catalyst was sulfided at 320 °C for 5 h. The experimental conditions were as follows: total pressure of 100 kg/cm^2; reaction temperature of 380, 400, and 420 °C; liquid hourly space velocity of 1.0 h^{-1}; and hydrogen-to-hydrocarbon ratio of 356 m^3/m^3. Details about the pilot plant reactor are given elsewhere (Ancheyta et al., 2001). After 280 h of hydroprocessing reaction, the catalyst was downloaded from the reactor, then washed with toluene by the Soxhlet process, and dried at 110 °C. The catalyst obtained after drying were used for characterization and HDS reactions.

4.7.1.2 HDS of Thiophene Hydrodesulfurization of thiophene and SRGO has been studied on NiMo/alumina used catalyst. A fresh catalyst contains 10.66 wt%

TABLE 4.10 Properties of the Feeds Used to Obtain the Deactivated Catalysts

	21°API	13°API	7°API
Specific gravity (20/4 °C)	0.9234	0.9787	1.0189
API gravity	21.27	12.67	6.98
Sulfur (wt%)	3.44	5.25	4.73
Ni (wppm)	54.7	87.11	87.60
V (wppm)	300	448	411
Asphaltenes (Ins. C_7 wt%)	10.91	21.83	17.12

MoO_3, 2.88 wt% of NiO, and 3.73 wt% of TiO_2. Thiophene hydrodesulfurization was carried at microplant scale (Figure 4.9). In this plant, 0.5 g of catalyst was sulfided *in situ* in a glass tubular reactor at atmospheric pressure and at 400 °C. For sulfidation, hydrogen was passed through a container having CS_2. The saturated mixture of CS_2 and hydrogen passed through the reactor. Hydrogen flow was 40 mL/h and duration of sulfidation was 2 h. After sulfidation, the catalyst was flushed with H_2 until no CS_2 was detected in effluent gas. Thiophene feed was introduced through the gas bubblers. Hydrogen flow rate was 50 mL/min. The reaction products were analyzed by an online gas chromatograph using an FID detector.

4.7.1.3 HDS of SRGO Five grams of fresh catalyst were sulfided *ex situ* for each experiment to study hydrotreating activity of straight-run gas oil. The catalyst was sulfided at the same conditions as those used with thiophene experiments. Hydrotreating of SRGO was carried out in a batch reactor (autoclave). Three hundred grams of SRGO were put into the reactor vessel (1 L capacity). The sulfided catalyst was then transferred into the reactor in nitrogen atmosphere quickly to avoid the contact of the catalyst with air for long time. The reactor was tightened properly and checked for leakage. It was then purged two/three times with hydrogen gas to ensure that there was no air left inside the reactor. The heating was started from room temperature to the required temperature at the rate of 3 °C/min. Stirring was started when the temperature reached the set point (reaction temperature) and the time was noted as the beginning of the reaction at this point. Hydrotreated products of SRGO were collected every hour and its sulfur content was analyzed. The experimental conditions for microplant and batch reactor are given in Table 4.11. The properties of SRGO used in this work are as follows: API gravity of 29.99, sulfur of 1.46 wt%, and nitrogen of 585 wppm.

The HDS activity of thiophene and SRGO has also been compared with the regenerated catalysts. An appropriate amount of used catalyst was regenerated in a flow of oxygen at 550 °C for 5 h in atmospheric glass unit. After regeneration, the catalyst was again sulfided and activity was measured.

TABLE 4.11 Reaction Conditions for Hydrodesulfurization of Thiophene and SRGO

	Microplant	Batch Reactor
Feed	Thiophene	SRGO
Catalyst weight (g)	0.5	5
Feed weight (g)	–	300
Temperature (°C)	400	360
Pressure (MPa)	Atmospheric	5.9
H_2 flow with saturated feed (mL/min)	50	–
Duration (h)	4	6

The metal content in the feed and in the products was determined by Atomic Absorption (Thermoelectron model SOLAAR AA). Sulfur was analyzed by X-ray fluorescence (HORIBA model SLFA-2100). PSD of fresh and used catalysts was measured by nitrogen adsorption at −196 °C (Quantachrome Nova 2000).

4.7.2 Results and Discussion

4.7.2.1 Characterization of Fresh and Used Catalysts The fresh catalyst used in this study has a specific surface area of 137 m^2/g, total pore volume of 0.53 mL/g, and average pore diameter of 155 Å (15 nm). It has been used for hydrotreating reaction of different heavy oils and residue. Textural properties of the fresh, used, and regenerated catalysts are compared in Table 4.12. It is shown that the physical properties of the used catalyst change due to the hydrotreating of heavy oils and residue. It is also observed that the total surface area decreased more when the catalyst processed the crude of 13° API. This feed has the highest amounts of metals and asphaltenes. PSDs of fresh, used (21° API feed), and regenerated catalysts are presented in Figure 4.44. It has been observed that when used catalyst was regenerated, some part of pore that was blocked by coke was cleaned and again contributed to the total pore volume. PSD of fresh catalyst has been compared with that of regenerated and used catalysts deactivated with 13° API feed obtained from the three different reactor zones. The results also suggest that pore structure is modified due to the deactivation of catalyst during hydrotreating reaction of heavy crude oil even as low as 280 h of operation. Changes of pore structure among the used catalysts from the three different reactor zones are not very prominent. It is also noted in all cases that microporosity increases in the used catalysts. It suggests that some of the large pores are blocked by deposits, which results in increased porosity in the micropore range. The PSDs of regenerated catalysts of the three different zones are similar. Figure 4.44 shows PSD of the

TABLE 4.12 Textural Properties of Fresh (FC), Used (UC), and Regenerated (RC) Catalysts

Feed		21° API		13° API		7° API	
Catalyst	FC	UC	RC	UC	RC	UC	RC
SSA (m^2/g)	137	97	132	78	124	88	119
TPV (mL/g)	0.53	0.19	0.39	0.11	0.28	0.16	0.32
APD (Å)	155	86	118	66	90	78	108
PSD (vol%)							
<50 Å	17.6	53.6	39.5	33.4	32.6	32.9	28.8
50–100 Å	47.2	32.1	36.1	45.7	46.	35.4	42.1
100–200 Å	27.6	8.5	16.5	12.5	16.2	23.9	21.5
200–500 Å	6.2	4.4	5.4	5.9	4.7	6.1	4.8
500–1000 Å	1.1	1.4	2.3	1.6	0.5	1.7	2.1
>1000 Å	0.3	0	0.2	0.9	0	0	0.7

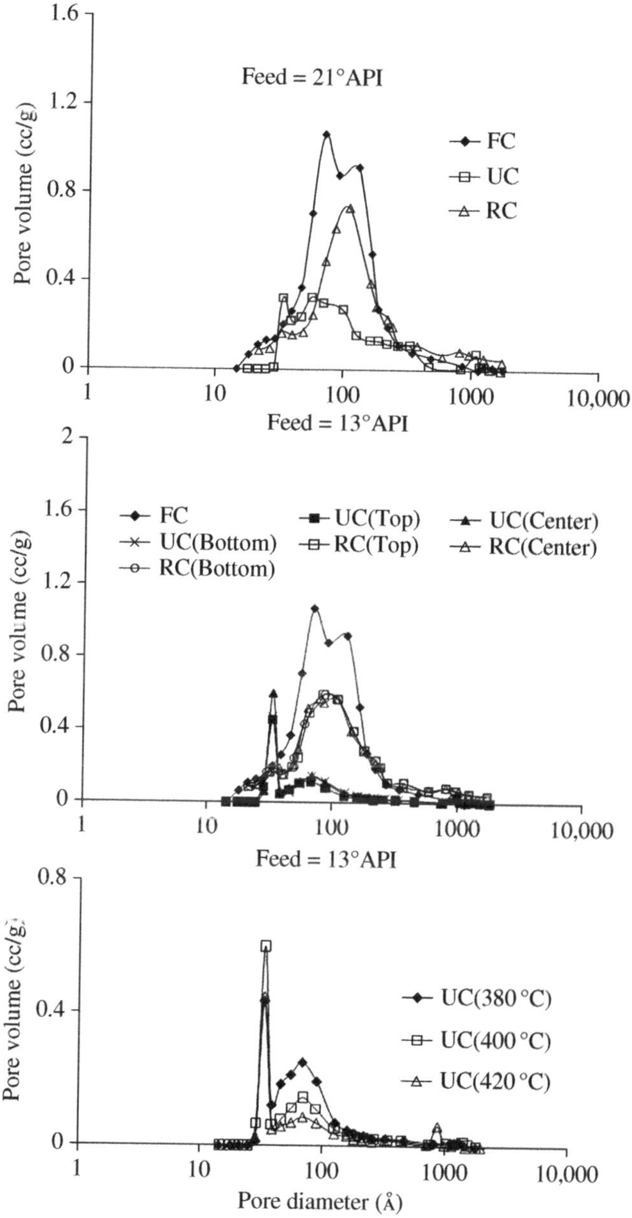

Figure 4.44 Pore size distribution of fresh, used (processed with 21°API feed), and regenerated catalyst (FC = fresh catalyst, UC = used catalyst, and RC = regenerated catalyst).

used catalysts obtained after reaction at three different temperatures (380, 400, and 420 °C). The results reveal that the catalyst is indeed deactivated at a faster rate when the reaction temperature is high. It implies that the pore blockage by coke or metal deposition is more severe at 420 °C compared with that at 380 °C of reaction temperature.

Carbon and vanadium depositions were measured on the used catalysts and the results are presented in Figure 4.45. The vanadium deposition on the catalyst processed with 21° API heavy crude is around 2 wt%. The percentage of metal deposition increases with an increased amount of contaminants in the feed. However, the increasing trend is very minimal. A similar tendency is observed for the carbon deposition. Maximum amount of carbon is deposited on a catalyst treated with 13° API crude and it is around 19 wt% on the fresh catalyst basis. At this point, it is worth to mention the work done by Wei (1991). He reported that if 20 wt% of coke and 20 wt% of vanadium are deposited uniformly on a catalyst having 200 m^2/g surface area, the deposits are approximately equivalent to 5–6 monolayer. At this condition, the catalyst would be completely covered by deposits and total loss of activity is expected. However, in practice, the catalyst remains with some activity. Therefore, it suggests that the deposition does not occur uniformly, rather metal sulfides (or carbon) deposit at some specific sites and the crystallites size of these sulfides grow with time rather than spread uniformly throughout the catalyst. The formation of coke during hydrotreating of heavy crude oils having different asphaltene content has been studied by Maity et al. (2007). They found that the deposition of coke increases rapidly with an increased asphaltene content in a feed, while that of vanadium increases slightly. Regarding this, Table 4.10 shows that heavy crude oil of 13° API contains high percentage of asphaltene. Therefore, carbon deposition on the used catalyst treated with this feed is the highest among all catalysts studied in this work.

The percentage of vanadium and carbon on used catalysts from three different zones (top, center, and bottom) is also given in Figure 4.45. Used catalysts were obtained by hydrotreating of crude of 13° API at 400 °C. Results show that the vanadium deposit is higher at the top of the reactor where fresh feed enters and it is low at the bottom of the reactor. However, a reverse trend is observed for carbon deposition, that is, the percentage of carbon deposition is lowest at the top and highest at the bottom of the reactor. Simpson (1991) and Núñez et al. (1997) have also reported a similar tendency of metal and coke depositions along the reactor bed. The catalyst activity at the top of the reactor is high. When the reactants (feed and hydrogen) enter at the top of the reactor, the catalyst effectively removes metals from the feed. However, at the bottom, due to the presence of lower amount of hydrogen and higher amount of H_2S, the activity of the catalyst is low. Therefore, the metal deposition is always high at the top of the reactor. On the other hand, due to the lower hydrogen partial pressure, the coke formation at the bottom is higher. It is known that coke formation occurs by free radical mechanism. With increasing temperature, free radical mechanism becomes favorable. However, with the presence of hydrogen, the free radicals are hydrogenated rather than taking part in polymerization (Absi-Halabi et al., 1991).

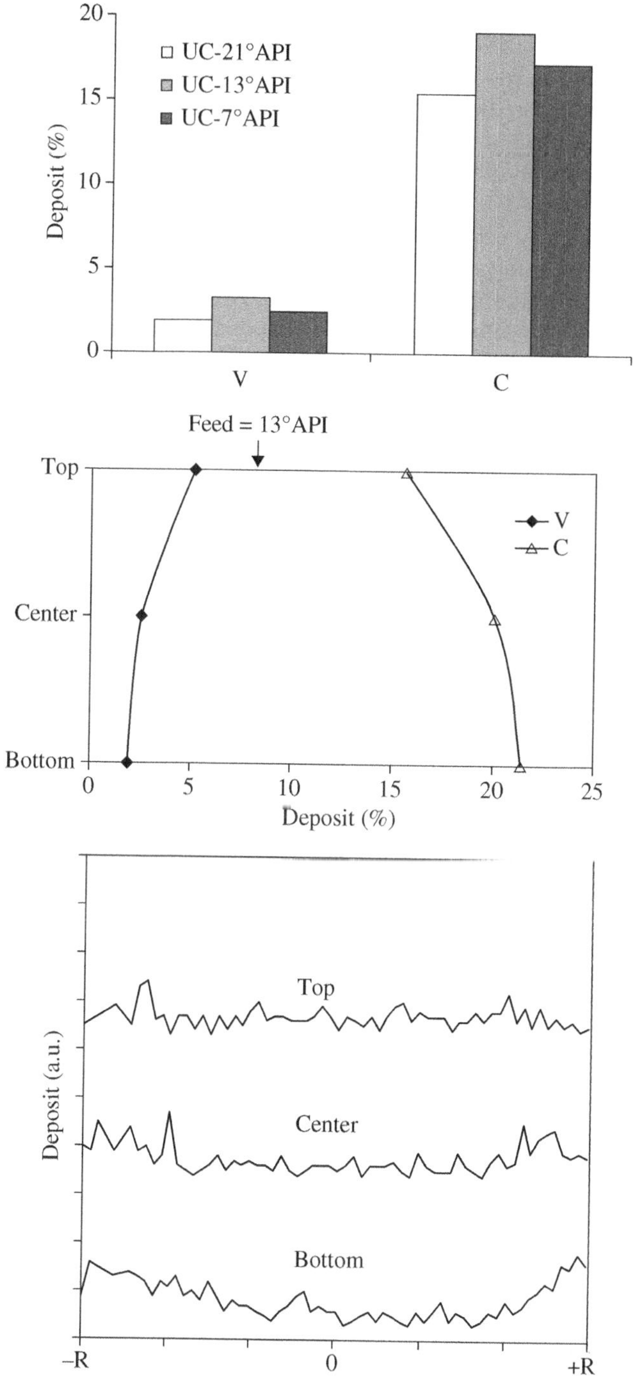

Figure 4.45 Coke and vanadium deposition for three different feeds, from three different reactor zones and radial distribution of vanadium deposition.

Profiles of vanadium deposits at the top, center, and bottom of a catalyst bed were also studied by SEM, and its radial distribution of the deposits is presented in Figure 4.45. The uniform profiles of deposits indicate that the catalyst pore mouth is sufficiently large for vanadium to enter, and hence the pore mouth plugging is not observed (Núñez et al., 1997). If the metals are only deposited on the edge of the extrudate (generally it occurs for small pore diameter catalysts), the vanadium deposition profile would be having an "M" shape. The radial distribution profile, particularly at the bottom of the catalyst, shows slightly an M shape. It is also observed that at the end of the catalyst bed coke deposit is higher than at the top of the bed. This coke gradually penetrates into the interior portion of the catalysts and hence hinders the vanadium atoms to enter the pore cavity deeply. It causes deposition of vanadium atoms outside the catalyst particle.

4.7.2.2 HDS of Thiophene Hydrodesulfurization of thiophene model molecule has been studied on the used catalysts and the results are given in Figure 4.46. In this figure, the HDS of fresh and regenerated catalysts are also presented. Thiophene HDS with a fresh catalyst is 28.33% and it decreases to 8.84% due to deactivation with heavy crude of 21°API. The activity decreases more when the nature of a feed becomes heavier. The lowest activity is observed on the used catalyst obtained after hydrotreating of 13°API crude. Thiophene HDS regains substantially due to the regeneration of the used catalyst, particularly for the sample obtained by hydrotreating of 21°API crude. Activity differences between the used and the regenerated catalysts for 21°API crude and 7°API are noticeable. This suggests that the principal catalyst-deactivation-mechanism is mainly due to the coke formation at the initial stage of reaction. However, deactivation due to metals is observed when a treated feedstock has high percentages of vanadium and nickel (e.g., 13°API crude). It is also worth to mention that this feed contains around 535 wppm metals and the percentage of vanadium deposits on the catalyst is approximately 3 wt% after 280 h of operation.

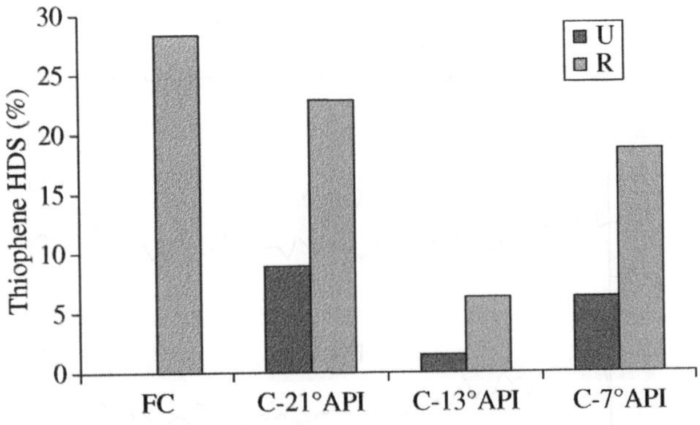

Figure 4.46 Thiophene HDS of fresh, used, and regenerated catalysts ($T = 400$ °C, $P =$ atmospheric pressure, FC = fresh catalyst, C-21°API = catalyst processed with 21°API feed, U = used, and R = regenerate).

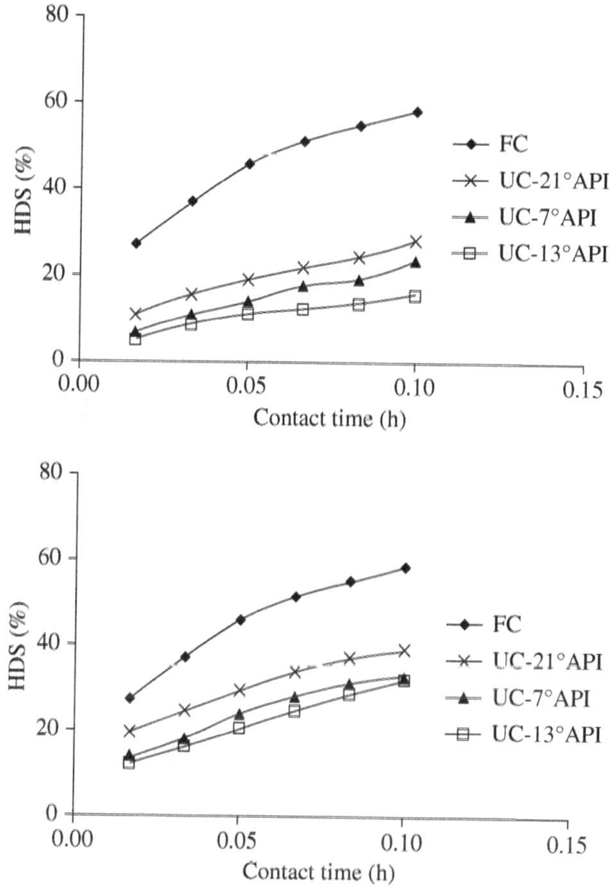

Figure 4.47 Gas oil HDS of fresh, used, and regenerated catalysts ($T = 360\,°C$, $P = 60\,kg/cm^2$).

The catalyst processed with the 13°API feed is deactivated faster compared with the other two feeds due to the presence of high content of metals and asphaltenes.

4.7.2.3 HDS of SRGO The hydrodesulfurization of SRGO oil with the fresh and the used catalysts is presented in Figure 4.47. In this figure, conversions are plotted against contact time, which is calculated for a batch reactor by the following equation (Gualda and Kasztelan, 1996):

$$\text{Contact time} = \frac{\text{(Catalyst weight)(Total reaction time)}}{\text{Feed weight}} \tag{4.7}$$

The activity increases with increasing contact time as expected. The HDS with the fresh catalyst at 0.1 h contact time is 58.5% and it is reduced to 28.62% for the used catalyst processed with the 21°API crude. The activity decreases for the catalysts processed with 7°API and 13°API feeds. Hence, it leads that the catalyst used for hydrotreating of heavier crude oil is deactivated more. Stephens and Stohl (1985) also reported that deactivation greatly depends on the nature of the feed used for the reaction. They observed that processing of a feed below 290 °C cut using a $NiMo/Al_2O_3$ catalyst caused a 23% loss of hydrogenation activity compared with the fresh catalyst. However, the HDS remains unchanged. If the same catalyst is used to process a feed below 454 °C cut, the losses of HYD and HDS activities are 82% and 70%, respectively. It is also observed that the rate of coke formation of similar type of feeds is higher for a feed containing high percentage of aromatics. It is reported that the rate of deactivation depends on the thickness of coke deposition during hydrotreating reaction. The crystalline thickness of Ni-promoted MoS_2 is about 0.615 nm (Furimsky and Massoth, 1999). If coke deposition on the catalyst is less than 0.615 nm, it has no effect on deactivation. On the contrary, if it is thicker than 0.7 nm, deactivation is very fast. The HDS activity of the used $CoMo/Al_2O_3$ catalysts with different percentage of metal deposits has also been studied by Ammus et al. (1986). They reported that with increasing metal deposits, the HDS catalyst activity decreased linearly. The metal sulfides gradually narrow the pores and thus slow down the diffusion of the reactant molecules into the undeactivated interior sites. At the end-of-run, the catalyst may still have some activity. In our case, the 13°API crude contains high amount of metals and that may be the reason why the catalyst processed with this feed is deactivated more. Moreover, the aromatic content of the feeds is also another factor that affects the degree of deactivation.

The HDS of SRGO was also studied on the regenerated catalyst and the results are also presented in Figure 4.47. A comparison of HDS of used and regenerated catalysts at 0.1 h of contact time is made in Figure 4.48. A substantial gain of HDS is observed due to the regeneration of the used catalysts.

The evaluation of the fresh and used catalysts has also been carried out at three different temperatures: 380, 400, and 420 °C for the gas oil HDS, and the results are

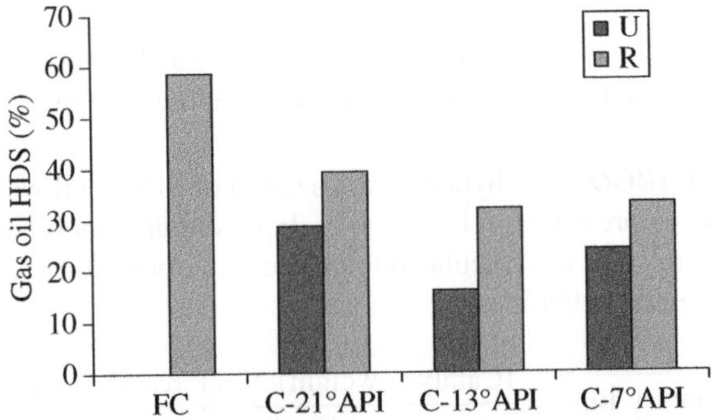

Figure 4.48 Comparison of gas oil HDS between used and regenerated catalysts at 0.1 h contact time (FC = fresh catalyst, C-21°API = catalyst processed with 21°API, U = used, and R = regenerated).

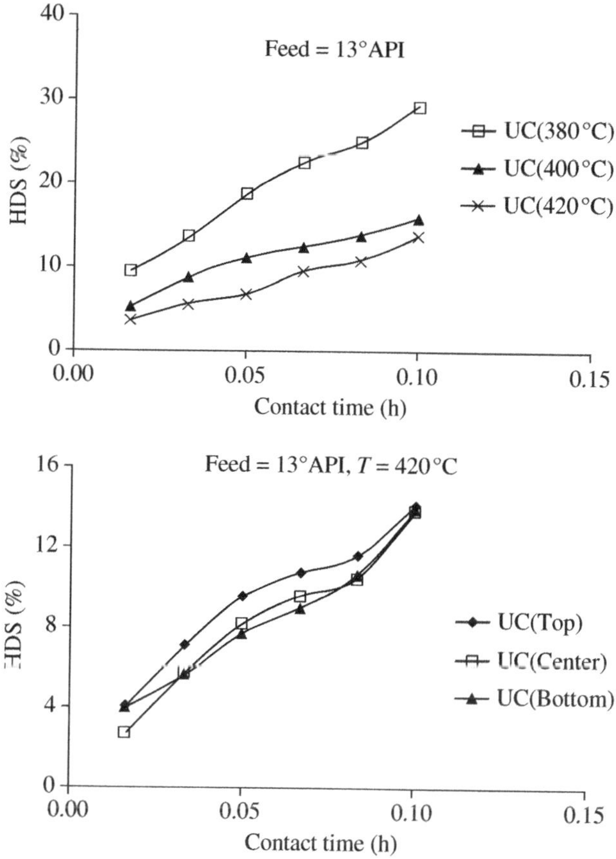

Figure 4.49 Gas oil HDS of used catalysts obtained after reaction at three different temperatures and from three different reaction zones.

given in Figure 4.49. It is noted that the catalyst was deactivated more extensively when the hydrotreating reaction was performed at high temperatures, which is in good agreement with the findings reported.

4.7.2.4 Effect of Catalyst Position within the Reactor It has been mentioned in the Experimental section that the used catalysts for this study were obtained from a pilot plant reactor, where the fresh catalyst was loaded by dividing three separate reaction zones (top, center, and bottom). The gas oil HDS experiments were also performed using the used catalysts obtained from these zones and the results are given in Figure 4.49. It is noted that there is not a wide difference of activity among the used catalysts coming from different reactor zones. However, slightly less HDS activity is observed for the bottom-bed used catalyst and the highest activity is obtained for the top-bed catalyst. The pure hydrogen and feed enter the reactor at the top of the reactor in the pilot plant. When the feed passes through the catalyst bed from the top to bottom, hydrogen is consumed and H_2S is produced. Thus, at the bottom, the hydrogen partial pressure is low compared with that at the top of the reactor. On the contrary, the partial pressure of H_2S is higher at the bottom. Due to higher hydrogen partial pressure and lower H_2S at the reactor inlet, the catalytic activity at the top of the reactor is always higher than that at the bottom. The catalyst activity at the bottom is less due to lower hydrogen and higher H_2S partial pressure, which provokes the formation of carbon. That is why the carbon deposition at the bottom is higher.

Both HDS activities of thiophene and gas oil show that the used catalysts have lost activity during hydrotreating of heavy oils and residue. The catalyst processed with the heaviest feed (with respect to metals and asphaltenes) is deactivated more and hence it loses more activity. However, when the catalyst is regenerated, some activity regains. It indicates that when deposited coke is burnt off the loss of activity due to the coke deposition is recovered. Nevertheless, if the catalyst is processed with the feed having high metal content, it does not gain activity in the same extent. HDS activity of fresh $NiMo/Al_2O_3$ catalyst is 28% and it is reduced to around 1% for the catalyst processed with 13°API feed. This activity increases to only around 6% upon regeneration of this deactivated catalyst. It is significant to notice that the catalyst activity is adversely affected by metal deposit. However, the degree of deactivation is not so adverse for the catalyst treated with 21°API feed. It is worth to mention that metal concentrations of 21 and 13°API feeds are 355 and 535 wppm, respectively. So the above result indicates that metals either block the pore mouth or can cover the active sites of the used catalysts. The SEM results show that vanadium deposit is distributed along with catalyst radius uniformly, indicating that pore mouth blockage does not occur in this case. This suggests that metal deposit may cover some of the active sites particularly when the deposits are high (Maity et al., 2007). It is a well known fact that the rapid coke deposit is the principal cause for initial deactivation of the catalyst. While Gualda and Kasztelan (1996) observed that both metal sulfides and coke deactivated HDM performance of $NiMo/Al_2O_3$ catalyst, but deactivation by metals was more prominent even at the initial stage of operation. It is worth to explain little more about the nature of the feeds and deactivation. Though the feed having 7°API gravity is the heaviest crude used in this study, the deactivation is observed the highest for the crude having 13°API gravity. It is important to mention that the used catalyst studied in this case was obtained after 280 h TOS with different heavy and residues. Within this period, metals coming from the feed are deposited on the support. In the due course, these metals enter the pore cavity and block some of the active sites. Hence, the catalyst is deactivated and the degree of deactivation depends on the concentration of metals in the feed. The results also show that coke or metal deposition is higher at high reaction temperatures. Coke is generally formed by polymerization of free radicals. Formation of free radicals is favorable at higher temperatures. It is also interesting to notice that the vanadium deposition is highest at the entrance of the reactor bed and it is least at the end. The reverse trend is observed for coke deposition, that is, it is least at the top and highest at the end of reactor.

4.7.3 Conclusions

The hydrodesulfurization activities of the used catalysts were studied by using thiophene and gas oil as feeds. The used catalysts for this study are obtained from pilot plant, where the catalyst was processed with heavy crude oils and residue at different temperatures. Both results of thiophene and gas oil HDS show the high degree of catalyst deactivation with a feed having high content of asphaltenes and metals. The regenerated catalyst regains its original HDS activity indicating that the principal reason for deactivation of this catalyst is coke deposition. The vanadium deposition is high at the inlet of the reactor, whereas the coke deposition was high at the outlet of the reactor. Radial distribution of vanadium measured by SEM-EDX shows that vanadium is uniformly distributed throughout the catalyst particle. An average pore diameter of the fresh catalyst is 15 nm and it is large enough for the reactants to enter the pore cavity. Therefore, vanadium distribution is even on the catalyst particle.

REFERENCES

Absi-Halabi, M., Stanislaus, A., and Trimm, D.L. 1991. Coke formation on catalyst during the hydroprocessing of heavy oils. Appl. Catal. 72:193–215.

Acik, M., Lee, G., Mattevi, C., Pirkle, A., Wallace, R.M., Chhowala, M., Cho, K., and Chabal, Y. 2011. The role of oxygen during thermal reduction of graphene oxide studied by infrared absorption spectroscopy. J. Phys. Chem. C 115:19761–19781.

Al-Dalama, K., and Stanislaus, A. 2006. Comparison between deactivation pattern of catalysts in fixed-bed and ebullating-bed residue hydroprocessing units. Chem. Eng. J. 120(1–2) 33–42.

Amemiya, M., Korai, Y., and Mochida, I. 2003. Catalyst deactivation in distillate hydrotreating (Part 2): Raman analysis of carbon deposited on hydrotreating catalyst for vacuum gas oil. J. Jpn. Pet. Inst. 46(2):99–104.

Ammus, J.M., Androutsopoulos, G.P., and Tsetseku, A.H. 1986. An investigation of the deactivation phenomena associated with the use of commercial HDS catalysts. Ind. Eng. Chem. Res. 26:1312–1323.

Ancheyta, J., Betancourt, G., Marroquín, G., Pérez, A., Maity, S.K., Cortez, M.T., and Río, R. 2001. An exploratory study for obtaining synthetic crudes from heavy crude oils via hydrotreating. Energy Fuels 15:120–127.

Ancheyta, J., Betancourt, G., Centeno, G., Marroquín, G., Alonso, F., and Garciafigueroa, E. 2002. Catalyst deactivation during hydroprocessing of Maya heavy crude oil. 1. Evaluation at constant operating conditions. Energy Fuels 16(6):1438–1443.

Ancheyta, J., Betancourt, G., Centeno, G., and Marroquín, G. 2003. Catalyst deactivation during hydroprocessing of Maya heavy crude oil. (II) Effect of temperature during time-on-stream. Energy Fuels 17(2):462–467.

Ancheyta, J., Rana, M.S., and Furimsky, E. 2005. Hydroprocessing of heavy petroleum feeds: tutorial. Catal. Today 109:3–15.

Ancheyta, J., Trejo, F., and Rana, M.S. 2009. Asphaltenes: chemical transformation during hydroprocessing of heavy oils, Taylor & Francis Group LLC, New York.

Begon, V., Warrington, S.B., Megaritis, A., Charsley, E.L., and Kandiyoti, R. 1999. Composition of carbonaceous deposits and catalyst deactivation in the early stages of the hydrocracking of a coal extract. Fuel 78:681–688.

Braun, R.L., and Burnham, R.L. 1987. Analysis of chemical reaction kinetics using a distribution of activation energies and simpler models. Energy Fuels 1(2):153–161.

Brown, F.R., Makovsky, L.E., and Rhee, K.H. 1977. Raman spectra of supported molybdena catalysts: II. Sulfided, used, and regenerated catalysts. J. Catal. 50(3):385–389.

Chen, Y.W., and Hsu, W.C. 1997. Hydrodemetalation of residue oil over CoMo/alumina-aluminum phosphate catalysts in a trickle bed reactor. Ind. Eng. Chem. Res. 36:2526–2532.

Chua, Y.T., and Stair, P.C. 2003. An ultraviolet Raman spectroscopic study of coke formation in methanol to hydrocarbons conversion over zeolite H-MFI. J. Catal. 213(1):39–46.

Clark, F.T., Springman, M.C., Willcox, D., and Wachs, I.E. 1993. Interactions in alumina-based iron oxide–vanadium oxide catalysts under high temperature calcination and SO_2 oxidation conditions. J. Catal. 139(1):1–18.

Cotte, E.A., and Calderon, J.L. 1981. Pyrolysis of Boscan asphaltenes: process description and nature of the products. Prepr. – Am. Chem. Soc., Div. Pet. Chem. 26(2):538–547.

Digne, M., Marchand, K., and Bourges, P. 2007. Monitoring hydrotreating catalysts synthesis and deactivation using Raman spectrometry. Oil Gas Sci. Technol. 62(1):91–99.

Dong, X.G., Lei, Q.F., Fang, W.-J., and Yu, Q.-S. 2005. Thermogravimetric analysis of petroleum asphaltenes along with estimation of average chemical structure by nuclear magnetic resonance spectroscopy. Thermochim. Acta 427(1–2):149–153.

Egiebor, N.O., Gray, M.R., and Cyr, N. 1989. ^{13}C-NMR of solid organic deposits on spent hydroprocessing catalysts. Chem. Eng. Commun. 77(1):125–133.

Espinat, D., Dexpert, D., Freund, E., Martino, G., Couzi, M., Lespade, P., and Cruege, F. 1985. Characterization of the coke formed on reforming catalysts by laser Raman spectrometry. Appl. Catal., A 16(3):343–354.

Flynn, J.H., and Wall, L.A. 1966. General treatment of the thermogravimetry of polymers. J. Res. Natl. Bur. Stand., Sect. A 70(6):487–523.

Fonseca, A., Zeuthen, P., and Nagy, J.B. 1996. ^{13}C NMR quantitative analysis of catalyst carbon deposits. Fuel 75:1363–1376

Friedman, H.L. 1964. Kinetics of thermal degradation of char-forming plastics from thermogravimetry. J. Polym. Sci. C6:183–195.

Furimsky, E., and Massoth, F.E. 1999. Deactivation of hydroprocessing catalysts. Catal. Today 52:381–495.

Gonçalves, M.L.A., Ribeiro, D.A., Teixeira, A.M.R.F., and Teixeira, M.A.G. 2007. Influence of asphaltenes on coke formation during the thermal cracking of different Brazilian distillation residues. Fuel 86(4):619–623.

Gualda, G., and Kasztelan, S. 1996. Initial deactivation of residue hydrodemetallization catalysts. J. Catal. 161:319–337.

Jiménez-Mateos, J.M., Quintero, L.C., and Rial, C. 1996. Characterization of petroleum bitumens and their fractions by thermogravimetric analysis and differential scanning calorimetry. Fuel 75(15):1691–1700.

Lakshmanan, C.C., and White, N. 1994. A new distributed activation energy model using Weibull distribution for the representation of complex kinetics. Energy Fuels 8(6):1158–1167.

Leyva, C., Ancheyta, J., Travert, A., Maugé, F., Mariey, L., Ramírez, J., and Rana, M.S. 2012. Activity and surface properties of NiMo/SiO_2–Al_2O_3 catalysts for hydroprocessing of heavy oils. Appl. Catal., A 145–146:1–12.

Li, S., and Yue, C. 2003. Study of different kinetic models for oil shale pyrolysis. Fuel Process. Technol. 85(1):51–61.

Maity, S.K., Pérez, V.H., Ancheyta, J., and Rana, M.S. 2007. Catalyst deactivation during hydrotreating of Maya crude in a batch reactor. Energy Fuels 21:636–639.

Maity, S.K., Blanco, E., Ancheyta, J., Alonso, F., and Fukuyama, H. 2012. Early stage deactivation of heavy crude oil hydroprocessing catalysts. Fuel 100:17–23.

Maity, S.K., Ancheyta, J., Alonso, F., and Rayo, P. 2013. Hydrodesulfurization activity of used hydrotreating catalysts. Fuel Process. Technol. 106:453–459.

Maki, T., Takatsuno, A., and Miura, K. 1997. Analysis of pyrolysis reactions of various coals including Argonne premium coals using a new distributed activation energy model. Energy Fuels 11(5):972–977.

Marafi, M., Al-Omani, S., Al-Sheeha, H., Al-Barood, A., and Stanislaus, A. 2007. Utilization of metal-fouled spent residue hydroprocessing catalysts in the preparation of an active hydrodemetallization catalyst. Ind. Eng. Chem. Res. 46:1968–1974.

Marafi, M., Stanislaus, A., and Furimsky, E. 2010. Handbook of spent hydroprocessing catalysts, Elsevier, Amsterdam

Mariey, L., Lamotte, J., Chevreau, T., and Lavalley, L.C. 1996. Reaction kinetics. Catal. Lett. 59:241–446.

Massoth, F.E. 1981. Characterization of coke on coal catalysts by an oxidation technique. Fuel Process. Technol. 4(1):63–71.

Miura, K. 1995. A new simple method to estimate $f(E)$ and $k_0(E)$ in the distributed activation energy model from three sets of experimental data. Energy Fuels 9(2):302–307.

Miura, K., and Maki, T. 1998. A simple method for estimating $f(E)$ and $k_0(E)$ in the distributed activation energy model. Energy Fuels 12(5):864–869.

Núñez, M., Gómez, E., Suarez, C., and Carmona, E. 1997. Catalyst deactivation by metals and coke during hydrodemetallation, in: C.H. Bartholmew, G.A. Fuentes (Eds) Catalyst deactivation, Elsevier, Amsterdam 351–357.

Nuñez-Isaza, M., Pachon, Z., Kafarov, V., and Resaco, D.E. 2000. Deactivation of Ni–Mo/Al_2O_3 catalysts aged in a commercial reactor during the hydrotreating of deasphalted vacuum residuum. Appl. Catal., A 199:263–273.

Ramírez, S., Leyva, C., Ancheyta, J., and Centeno, G. 2009. On the importance of calculating fresh basic catalyst composition from spent catalyst analysis. Fuel 88:2311–2314.

Rana, M.S., Ancheyta, J., Rayo, P., and Maity, S.K. 2004. Effect of alumina preparation on hydrodemetallization and hydrodesulfurization of Maya crude. Catal. Today 98:151–160.

Rana, M.S., Ancheyta, J., Maity, S.K., and Rayo, P. 2005. Characteristics of Maya crude hydrodemetallization and hydrodesulfurization catalyst. Catal. Today 104:86–93.

Rana, M.S., Samano, V., Ancheyta, J., and Diaz, J.A.I., 2007. A review of recent advances on process technologies for upgrading of heavy oils and residua. Fuel 86:1216–1231.

Rayo, P., Ancheyta, J., Ramírez, J., and G.-Alejandre, A. 2004. Hydrotreating of diluted Maya crude with NiMo/Al_2O_3–TiO_2 catalysts: effect of diluent composition. Catal. Today 98:171–179.

Sahoo, S.K., Ray, S.S., and Singh, I.D. 2004. Structural characterization of coke on spent hydroprocessing catalysts used for processing of vacuum gas oils. Appl. Catal., A 278(1):83–91

Saito, T., and Kumata, F. 1998. Bench scale plant tests of three kinds of different pore size HDS catalyst and characteristics of coked catalysts: effect of extraction using LGO. Prepr. – Am. Chem. Soc., Div. Pet. Chem. 43(4):628–633.

Sato, M., Takayama, N., Kurita, S., and Kwan, T. 1971. Vanadium and nickel deposits distribution in desulfurization catalysts. Nippon Kagaku Zasshi 92(10):834.

Schucker, R.C. 1983. Thermogravimetric determination of the coking kinetics of Arab heavy vacuum residuum. Ind. Eng. Chem. Process Des. Dev. 22(4):615–619.

Shih, S.M., and Sohn, H.Y. 1980. Nonisothermal determination of the intrinsic kinetics of oils generation from oil shale. Ind. Eng. Chem. Process Des. Dev. 19(3):420–426.

Simpson, H.D. 1991 Implications of analysis of used resid demetallation catalyst, in: C.H. Bartholomew, J.B. Butt (Eds), Catalyst deactivation, Elsevier, Amsterdam, 265–272.
Snape, C.E., Tyagi, Y.R., Castro-Díaz, M., Hall, P.J., Murray, I.P., and Martin, S.C. 1999. Characterisation of coke on deactivated hydrodesulfurisation catalysts and a novel approach to catalyst regeneration. Prepr. Symp. – Am. Chem. Soc., Div. Fuel Chem. 44(4):832–836.
Speight, J.G. 1987. Initial reactions in the coking of residua. Prepr. – Am. Chem. Soc., Div. Pet. Chem. 32(2):413–418.
Stephens, H.P., and Stohl, F.V. 1985. Catalyst deactivation in direct coal liquefaction process: a quantitative approach to understanding the rate of activity loss. Prepr. Pap. – Am. Chem. Soc., Div. Fuel Chem. 30 (3):465–472.
Trejo, F., Rana, M.S., and Ancheyta, J. 2008. CoMo/MgO–Al_2O_3 supported catalysts: an alternative approach to prepare HDS catalysts. Catal. Today 130(2–4):327–336.
Vogelaar, B.M., Eijsbouts, S., Bergwerff, J.A., and Heiszwolf, J.J. 2010. Hydroprocessing catalyst deactivation in commercial practice. Catal. Today 154:256–263.
Wei, J. 1991. Catalyst deactivation, C.H. Bartholomew, J.B. Butt (Eds), Elsevier, Amsterdam.
Yoshimura, Y., Matsubayashi, N., Yokokawa, H., Sato, T., Shimada, H., and Nishijima, A. 1991. Temperature-programmed oxidation of sulfided cobalt-molybdate/alumina catalysts. Ind. Eng. Chem. Res. 30(6):1092–1099.
Zsabó, T., Berkesi, O., and Dékány, I. 2005. DRIFT study of deuterium-exchanged graphite oxide. Carbon 43:3186–3189.

5

MODELING CATALYST DEACTIVATION

5.1 INTRODUCTION

Catalyst deactivation is undoubtedly the most important concern during hydroprocessing of heavy oils, since the life of the catalyst and the entire economy of the process strongly depend on it. Deactivation rate is a function of feed properties, particularly metal and asphaltene contents. It has been previously reported that not only the amount of asphaltenes affects catalyst activity decay but also the properties and structure of asphaltenes. For instance, feeds with low relative solubility index, that is, the H/C atomic ratio of asphaltenes divided by that of the corresponding maltenes, tend to form more coke. Also continental type of asphaltenes having core with more than seven aromatic rings are able to stack-forming columns, while archipelago type of asphaltenes, those with less than seven aromatic rings, form planar aggregates in asphaltene solutions. It is also known that the higher the amount of metals in the feed, the faster the deactivation rate. Figure 5.1 shows typical catalyst life trends for different feeds. Light feeds, for example, naphtha and gas oil, do not deactivate the catalyst by metal deposition but by coke deposition; however, the deactivation rate is slow and the unit can operate for several years without catalyst replacement. When hydrotreating (HDT) light feeds, the plant is commonly shut down due to high ΔP of the reactor instead of by catalyst deactivation. While for heavy feeds, for example, crude oil and residua, deactivation of catalyst is the main cause of plant shutdown. From these well-documented facts reported in the literature, it is anticipated that different feeds will cause different deactivation patterns.

Kressman et al. (1998) found that asphaltenes content, asphaltene-to-resin ratio, and metal content in the feed play important roles in the catalyst performance. They

Deactivation of Heavy Oil Hydroprocessing Catalysts: Fundamentals and Modeling, First Edition. Jorge Ancheyta.

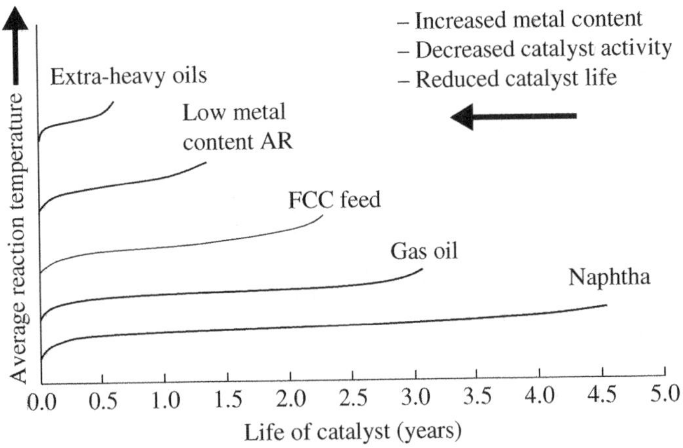

Figure 5.1 Catalyst life of for different petroleum feeds.

reported that a large decrease in hydrodesulfurization and hydrodemetallization reaction rates occur when the asphaltene content of the feed increases. Nuñez et al. (2000) studied the deactivation of NiMo catalysts for HDT of deasphalted vacuum residuum and concluded that catalyst deactivation depends on chemical properties of the feed, chemical and physical properties of the catalyst, as well as the catalyst position along the reactor length. Studies on catalyst deactivation with Maya heavy crude oil (HCO) as feed confirmed by continuously monitoring the product quality during time-on-stream and spent catalyst characterizations that deactivation by coke occurs only at short time-on-stream while metal deposition commands catalyst deactivation at long time-on-stream (Ancheyta et al., 2002, 2003).

There are other documents reported in the literature, although not many, which deal with studies of catalyst deactivation during hydroprocessing of heavy oils. Some of the most recent contributions are summarized in Table 5.1. The main objectives of these studies have been in general to determine the metal tolerance capacity of catalysts, modeling catalyst deactivation, and study the effect of reaction conditions.

It is well known that the main disadvantage of using fixed-bed reactors for hydroprocessing of heavy feeds is the catalyst aging during time-on-stream, caused by metals (Ni+V) and coke deposits. Coking generally occurs during the first few hours of operation and then apparently reaches a pseudo-steady state, in contrast, metal buildup continues through the whole cycle. In industrial practice, the gradual loss in catalytic activity is offset by increasing reaction temperature. The process is shut down upon the attainment, at any position of the reactor, of a maximum allowable temperature (MAT), which is established by the metallurgical characteristics of the reactor vessel. However, at end-of-run (EOR) conditions, the catalyst is not necessarily completely deactivated because metals tend to accumulate in the front end of the catalyst system; this is reflected in a descending-type axial profile of metals-on-catalyst (MOC). In this sense, front-end hydrodemetallization catalysts with high-metal uptake capacity (layered catalyst systems) were introduced to make better use of the catalyst inventory.

TABLE 5.1 Studies on Catalyst Deactivation During Hydroprocessing of Heavy Oils

Author	Type of Feed	Type of Reactor	Reaction Conditions	Type of Catalyst	Main Findings
Nuñez et al. (2000)	Deasphalted vacuum bottoms	Two fixed-bed industrial reactors	T_{max} = 400 °C, P = 103 4 bar, LHSV = 1.0 h^{-1}, T-O-S = 241 days	Six Ni–Mo/Al_2O_3 catalysts with different content of Mo and Ni	Deactivation of catalysts depends on feed chemical properties, pore structure, active metal loading, surface area, pellet size of catalyst, and position along the packed-bed reactor. Dominant deactivation by metals at upper part of the first reactor
Callejas et al. (2001)	Residue from Maya crude	Trickle-bed reactor	$P(H_2)$ = 100 bar, LHSV = 1.1/h g_{cat}, H_2/oil ratio = 0.2 kg/kg, T = 375 °C	NiMo/γ-Al_2O_3, HDM catalyst	Vanadium deposition becomes more diffusion rate limited, resulting in a less deep penetration into the catalyst pellet
Higashi et al. (2002)	Heavy residue with high asphaltene content	Fixed-bed reactor	T_{max} = 400 °C, P = 108 bar, LHSV = 1.0 h^{-1}, T-O-S = 241 days	38/31/31 vol%, HDM/NiO–CoO –MoO_3/NiO –MoO_3 catalysts	Catalyst activity strongly damaged with low H_2/HC ratio and low hydrogen pressure
Marafi et al. (2003a, 2003b)	Kuwait AR, HDM residue, HDM/HDS residue	Fixed-bed reactor systems	T = 360–380, 390 °C, P = 120 bar, LHSV = 0.32 and 2.0 h^{-1}, H_2/oil ratio = 570 and 680 mL/mL	Three catalysts: Cat A: HDM, Cat B: HDS, and Cat C: HDS/HDN	Proper selection of feedstocks is crucial to study the performance and screening of candidate catalysts for graded catalyst systems

(*continued*)

TABLE 5.1 (*Continued*)

Author	Type of Feed	Type of Reactor	Reaction Conditions	Type of Catalyst	Main Findings
Hauser et al. (2005)	Kuwait AR	Fixed-bed reactor	$T = 380\,°C$, $P(H_2) = 120$ bar, LHSV $= 1\ h^{-1}$, H_2/oil ratio = 570 mL/mL, T-O-S = 1–240 h	Mo/Al_2O_3 HDM catalyst	Initial coke formation was described in terms of its chemical composition and structural alterations as a function of TOS
Marafi et al. (2007)	Kuwait AR	Fixed-bed single microreactor	$T = 380\,°C$, $P = 120$ bar, LHSV $= 1.0\ h^{-1}$, H_2/oil ratio = 680 mL/mL, T-O-S = 120 h	Three catalysts: Mo/Al_2O_3, $Ni–Mo/Al_2O_3$, and $Ni–MoP/Al_2O_3$	Physical–chemical properties of the catalysts have a strong influence on catalyst deactivation by coke and metal deposition
Almutairi et al. (2008)	AR from Kuwait export crude and Eocene crude	Two fixed-bed reactors in series	$T = 370–412\,°C$, $P = 120$ bar, LHSV $= 2.0\ h^{-1}$, T-O-S = 3800 and 7500 h	Five catalysts: 2 HDM catalysts and 3 highly active NiMo HDS catalysts	Difference in performance of catalyst system attributed to content and composition of asphaltenes in feed

Most HDT and hydrocracking (HCR) commercial units employ fixed-bed reactors. The exothermic nature of hydroprocessing reactions provokes a temperature rise along the reactor that can exceed 100 °C, when processing heavy feeds. Temperature runaways beyond acceptable limits boost catalyst deactivation, change selectivity toward undesirable products, and sometimes even damage the reactor vessel, which affects process profitability. Therefore, controlling this variable is essential for these processes.

Hydroprocessing of heavy fractions in single-bed reactors would be very unpractical and unsafe due to the huge temperature increase. To obtain a more favorable and controllable temperature profile, the total heat release is limited to smaller portions by dividing the total catalyst volume into several beds, commonly of increasing depth, separated by quench zones for introducing relatively cold hydrogen-rich streams.

It is then anticipated that the study of HDT of heavy oils and residua is a formidable task due to the difficulty for obtaining reliable experimental information as well as the complex nature of those feeds and the more number of reactions occurring. The collection of data over representative long-term experiments can be another insuperable economical barrier for many researchers. Various works reported in the literature have dealt with studies on the kinetics of HDT of heavy feeds; however, absence of the effect of deactivation phenomenon is noticeable. From the industrial point of view, modeling the catalyst performance at long TOS is more valuable than at short TOS. Several HDT processes use fixed-bed reactor for HDT purposes. In that type of reactor, undesirable loss of catalyst activity is observed as a function of TOS.

Traditionally, the deactivation phenomenon has been divided into three stages: early deactivation due to coke deposition (SOR: start-of-run), middle-stage deactivation due to the loss of sites by poisoning and pore plugging by metal sulfide deposits (MOR: middle-of-run), and the total loss of activity by severe diffusional resistances due to almost total pore plugging (EOR), which reduces the activity to such a lower value that leads to unavoidable unit shutdown.

Most of the reported studies have been carried out with commercial catalysts at first hours of TOS, and kinetic parameters are lumped with an initial rapid deactivation factor. Some approaches for modeling the deactivation period use experimental information to determine parameters of semiempirical models, but simulations are only valid within the range of operation variables and parameters have not always physical meaning; hence, limited applications of such models are devised. The end of life of catalyst, that is, EOR, although valuable, is not of enough interest to be modeled with confidence. However, studies on middle stage of catalyst deactivation seem to be useful because during this period is when the catalyst is employed in the industrial practice. Predicting the performance of the catalyst for long TOS by using models with tuning parameters is almost impossible. Thus, it seems that more fundamental modeling of deactivation during middle stage is necessary. Some approaches have been used for predicting catalyst life, but the effect of metal deposits was not taken into account and in some cases very long times were necessary in order to tune the empirical constants (Krishnaswamy and Kittrell, 1981; Ho, 1984).

Among the several reported models for predicting the performance of HDT catalyst in middle stage, the one developed by Oyekunle and Hughes (1987) and

Oyekunle et al. (2005) is a suitable approach for representing the growth of metal sulfide deposits within the catalyst pore due to hydrodemetallization reaction.

Juraidan et al. (2006) used a mathematical model that covers chemical reaction kinetics of various catalytic and noncatalytic HCR and HDT reactions, catalyst deactivation, and catalyst performance in an atmospheric residue hydroprocessing reactor. The apparent rate constants of hydrodesulfurization, removal of vanadium (HDV), removal of nickel (HDNi), and hydrodeasphaltenization reactions, process conditions, catalyst system configurations, feed properties, and maximum values of metal-on-catalyst were considered. A marked improvement was observed during simulation after including the effect of noncatalytic hydrothermal reactions.

Beckman and Froment (1980) have developed deactivation models for coke deposition on the catalyst surface that proceeds following two mechanisms: site coverage and pore blockage, both limiting the diffusion of reactants to enter the pore structure. When dealing with hydrocracking of heavy oil and residua, it is accepted that the deactivation by coke occurs during the first hours of time-on-stream. It has also been accepted that the most important changes in catalyst activity are due to sites coverage and pore blockage. The simplest way to take into account the loss of catalyst activity is by means of average power deactivation function that relates the activity coefficient with time-on-stream. It is generally assumed that a simple function allows for modeling deactivation by coke during first hours of TOS without a distinction between sites coverage and pore blockage. Another assumption frequently made is that final activity reduces to zero, which is not the case for deactivation by coke in hydrocracking reactions according to experimental evidence.

In the case of heavy oils, kinetic studies involving catalyst deactivation are of great importance for proper reactor and process design as well as to establish suitable operating policies to compensate for the loss of catalyst activity during time-on-stream. The few literature reports regarding catalytic and thermal HDT of heavy oils are summarized in Tables 5.2 and 5.3, respectively, which are limited to those works in which reaction orders and activation energies for different HDT reactions were provided. It is well known that coke is formed very rapidly during the first hours of time-on-stream, and reaches a pseudo-steady-state level, while metals in the feed (mainly V and Ni) are converted into their sulfides, which deposit within the pores and irreversibly deactivate the catalyst (Ancheyta et al., 2003).

It has been demonstrated that hydrocracking of crude oils and residua can be also properly modeled by the continuous kinetic lumping approach (Elizalde et al., 2009; Elizalde and Ancheyta, 2011). The main attractiveness of such a model is the use of fixed reaction order and accurate predictions of the entire boiling point curve with few parameters. Also, because the use of modern chromatographic apparatuses for simulated distillation allows for obtaining almost the entire distillation curve of hydrocracked products, it is desirable to use a kinetic model compatible with those experimental results.

Coke and metal deposits cause loss of catalytic surface area and decrease in mean pore diameter and volume. The effective diffusivity shows a stronger correlation with coke content because the more the deposited coke, the more difficult the access to

TABLE 5.2 Studies on Catalytic Hydrotreating of Heavy Oil

Author	Reactor	Operating Conditions	Feed API Gravity	Catalysts	Reaction Order (Activation Energy, kcal/mol)					
					HDS	HDNi	HDV	HDAsph	HDN	HDCCR
Scamangas et al. (1982)	TBR	350–400 °C, 5–10 MPa, 0.3–2.5 h^{-1}, 0.08 std m^3/h H_2 flow	10.7	CoMo/γ-Al_2O_3	2 (29)					
Papayannakos and Marangozis (1984)	Batch-recycle TBR	325–400 °C, 3–10 MPa, 0.5–9 mol% H_2S, 14.8 g cat., 200 g feed	9.0	CoMo/γ-Al_2O_3	2.5 (36.1)					
Ammus and Androutsopoulos (1987)	SBR	385–395 °C, 3–7 MPa, 33 cm^3 cat, 500 g feed	9.0	Two different CoMo/γ-Al_2O_3	1.9/2.3 (32/35)					
Chang et al. (1998)	FBR	370–405 °C, 12–16.8 MPa, 0.15–1 h^{-1}, 400–1000 std m^3/m^3 H_2/oil	13.9–14.7	Three different HDS catalysts	1 (18.3–24)					
			10.4–16.1	Three different HDM catalysts		1 (20–34)	1 (18–35)			
Callejas and Martínez (2000a, 2000b)	CSTR	375–415 °C, 10–15 MPa, 0.25–7.1 h^{-1}, 1781 std m^3/m^3 H_2/oil	18.15	NiMo/γ-Al_2O_3	0.5 (35.6)	0.5 (47.7)	0.5 (45.3)	0.5 (41.5)	0.5 (39.3)	0.5 (66.3)
Callejas and Martínez (1999)	CSTR	375–415 °C, 12.5 MPa, 2.7–7.1 h^{-1}, 1781 std m^3/m^3 H_2/oil	18.15	NiMo/γ-Al_2O_3	2 (68.6)	1 (95.1)	1 (53.6)		0.5 (43.3)	

(continued)

TABLE 5.2 (Continued)

Author	Reactor	Operating Conditions	Feed API Gravity	Catalysts	Reaction Order (Activation Energy, kcal/mol)					
					HDS	HDNi	HDV	HDAsph	HDN	HDCCR
Marafi et al. (2003a, 2003b)	Multi-FBR system	360–420 °C, 12 MPa, 0.5–4 h^{-1}, 120 h TOS, 570 std m^3/m^3 H_2/oil	12.27	HDM, HDS, and HDS/HDN	2 (25–29)	1.5 (12.5–28.9)	2 (23.4–62.2)	2 (15.1–31.5)	1–1.5 (29.2–31)	2 (20–24.5)
Alvarez and Ancheyta (2008)	Multi-FBR system	380–420 °C, 9.81 MPa, 0.25–1 h^{-1}, 891 std m^3/m^3 H_2/oil	5.4	HDM, HDS, and HCR	1.17 (24.85)	0.55 (20.4)	1.56 (23.5)	0.75 (27.9)	2 (22.5)	
Oleck and Sherry (1977)	Batch	357–400 °C, 13.8 MPa, 15 g cat., 300 g feed, 80 min reaction time	20	CoMo/γ-Al_2O_3	2 (45.1)		2 (38.2)			
Philippopoulos and Papayannakos (1988)	TBR	350–425 °C, 5 MPa, 0.14–2.72 h^{-1}, 0.04 std m^3/h H_2 flow	21.4*	Two different CoMo/Al_2O_3				2 (32.6/48)		
Trejo and Ancheyta (2005)	FBR	380–420 °C, 6.8–9.8 MPa, 0.33–1.5 h^{-1}, 891 std m^3/m^3 H_2/oil	20.9	NiMo/Al_2O_3				2.1 (10.35)		
Trasobares et al. (1999)	CSTR	375–415 °C, 10–15 MPa, 0.12–0.56 h^{-1}, 1781 std m^3/m^3 H_2/oil	HDT Maya residue	NiMo/γ-Al_2O_3						1 (36.3)
Martínez et al. (1998a, 1998b)	Batchwise tubular reactor	425–475 °C, 15 MPa, 10–40 min reaction time	Asphaltenic coal residue	NiMo/γ-Al_2O_3						4.7 (21.9)

TABLE 5.3 Studies on Noncatalytic Hydrotreating of Heavy Oil

Author	Reactor	Operating Conditions	API Gravity	Inert Material	HDT Reaction	Reaction Order	E_a (kcal/mol)
Khorasheh et al. (1989)	CSTR	440 and 450 °C, 13.9 MPa, 1 h^{-1}, 0.066 std m^3/h H_2 flow	12.4	Nonused	HDS	1	45–60
Marafi et al. (2008a, 2008b)	FBR	360–400 °C, 12 MPa, 0.28–1.0 h^{-1}, 680 std m^3/m^3 H_2/oil ratio	10.2	Alumina balls	HDS	1.74	25.42
					HDV	1.79	43.83
					HDNi	1.38	50.64
					HDAsph	1.76	35.27
					HDCCR	1.85	31.44
Martínez et al. (1998a, 1998b)	Batchwise tubular reactor	425–475 °C, 15 MPa, 10–40 min reaction time	Asphaltenic coal residue	Nonused	HDA	2	26.3
					HDS	3.8	32
Ramírez et al. (2011)	FBR	380–420 °C, 9.8 MPa, 0.2–0.6 h^{-1}, 890 m^3/m^3 H_2/oil ratio	6.21	Silicon carbide	HDS	1.9	48.2
					HDM	1.4	42

pores in whose active sites chemical transformation is carried out. As a consequence, a decrease in effectiveness factor is observed.

Continuous lumping approach for hydrocracking of heavy oils is relatively new, and until now it has been applied to experiments conducted at steady-state conditions, but there is no report on its application to hydrocracking of heavy oil experiments affected by catalyst deactivation.

Although various catalyst deactivation models have been reported in the literature for studying the deactivation at SOR and MOR separately, only very few reports deal with modeling the three stages together. Some models that account for the three stages of deactivation do not include the main HDT reactions occurred in heavy petroleum processing but only a single reaction. On the other hand, complex deactivation models have been developed mainly for model molecules, and some conclusions drawn from these approaches cannot be straightly extrapolated to deactivation of catalysts employed for hydrotreatment of real complex feedstocks.

It is often argued that if the catalyst activity decay is due to coke formation and coke formation depends on the concentration of reacting species, then the catalyst activity cannot be a simple function of time. Thus, catalyst deactivation as a function of concentration of the agent that is indeed causing it would be the proper way for describing the deactivation process; however, during hydrocracking of heavy oils not only coke is responsible of the loss of catalyst activity but also metal deposition, both following different complex mechanisms. This situation together with the need of conducting a series of experiments (unit preparation, catalyst loading, catalyst activation, adjustment of reaction conditions, recovery of liquid products for characterization, and catalyst unloading) to obtain various samples of deactivated catalyst for further characterization, which can be time consuming and costly particularly for small experimental setups, make impractical the use of this approach. Hence, expressing catalyst deactivation as a function of time-on-stream is preferred to simplify data acquisition.

5.2 EFFECT OF REACTOR CONFIGURATION ON THE CYCLE LENGTH OF HEAVY OIL FIXED-BED HYDROPROCESSING

5.2.1 Experimental Procedure

5.2.1.1 Kinetic Measurements A series of bench-scale hydroprocessing experiments were carried out to obtain kinetic data. A bench-scale plant consists of two fixed-bed reactors in series (~500 cm^3 of catalyst in each reactor), which were loaded with a system of commercial size catalysts: a front-end HDM catalyst, a midsection catalyst with balanced HDM/HDS activity, and a tail-end highly active catalyst for HDS and HCR. The catalysts were *in situ* activated by sulfiding with straight-run gas oil (1.46 wt% sulfur) containing 1.0 wt% dimethyl disulfide. The feedstock was an atmospheric residue (343 °C+) from HCO (13° API). Physical and chemical properties of the feed (AR1) are presented in Table 5.4.

Before the tests, the fresh catalyst system was stabilized at low temperature (360 °C) by processing the feed for 100 h. After this initial deactivation period, the

TABLE 5.4 Physical and Chemical Properties of the Feedstock

Property	AR1	AR2
Density at 15.6 °C (g/cm^3)	1.0326	1.0475
API gravity	5.4	3.2
Kinematic viscosity at 121 °C (cSt)	1637.9	1832.4
S (wt%)	5.74	6.21
Metals (wppm)		
Ni	102	117
V	620	578
Ni+V	722	695.6
Asphaltenes in C_7 (wt%)	21.77	25.10
ASTM D1160 distillation		
IBP (°C)	296	380.3
5 vol% (°C)	372	415.4
10 vol% (°C)	401	447
15 vol% (°C)	438	474
20 vol% (°C)	475	504
30 vol% (°C)	521	551
40 vol% (°C)	541	–
Vol% recovery at 538 °C	35.25	26.90

experiments were conducted by varying the liquid hourly space velocity (LHSV) (0.25–1.0 h^{-1}) and temperature (380–420 °C). Pressure and H_2/oil ratio were kept constant at 9.81 MPa and 891 std m^3/m^3, respectively. For each experiment, operating conditions were adjusted in a period of 5 h, followed by a stabilization period to obtain representative products. Under steady operation, two consecutive mass balance runs were performed by collecting the product during 3 h. The tests were carried in less than 250 h of operation so as to minimize the influence of catalyst deactivation on the kinetic analysis. After the experiments, catalyst activity was monitored with a check-back test using the conditions of the first experiment.

Physical and chemical properties of the collected liquid samples from the bench-scale plant were analyzed with standard methods.

5.2.1.2 Catalyst Aging Catalyst stability was evaluated under process conditions in a 5.2-month-long test. The study was carried out in another bench-scale plant with two fixed-bed reactors in series (~900 cm^3 of catalyst in each reactor). The total reactor volume was loaded with the same catalyst system as that used for generating the kinetic data. The feedstock was a similar atmospheric residue coming from another batch of 13°API HCO, whose properties are also exhibited in Table 5.4 (AR2).

One important aspect of catalyst life testing is the selection of the operation mode: fixed performance or fixed temperature. The former mode, in which reaction temperature is increased periodically to compensate for catalyst deactivation, is more suitable for representing commercial operations. For this reason, in this work, catalyst aging was studied under those conditions. Unlike most of the residue upgrading processes that typically operate under HDS isoperformance, the objective of this process is to keep the API gravity of the product constant, as this parameter is a good measure of the overall character of oil.

The evaluation was carried out under the following operating conditions: LHSV of $0.25\,h^{-1}$, initial temperature of 380 °C, pressure of 9.81 MPa, and H_2/oil ratio of 891 std m^3/m^3. The mass balance runs were performed consecutively by recovering the products every 12 h. Interreactor samples were taken every 24 h in order to monitor the behavior of the first reactor. The sample size was kept at less than ~2 % of the total feed rate so as to reduce the disturbance of the system. This small amount was sufficient enough to carry out some analyses, particularly to determine the metal content in the products for estimating the metals uptake in the first reactor.

5.2.2 Modeling Approach

The performance model is a three-phase plug-flow reactor model that incorporates mass transfer at the gas–liquid and liquid–solid interfaces, in accordance with the two-film theory, and uses correlations to estimate mass-transfer coefficients and fluid properties at process conditions. The feedstock and products are represented by chemical lumps (S, N, Ni, V, asphaltenes, and 538 °C+ residue) defined by overall elemental and physical analyses. Thus, the model accounts for the corresponding hydroprocessing reactions such as HDS, HDN, HDM (HDNi+HDV), HDAs, and HCR. The following assumptions were considered to develop the model equations:

- The reactor operates in plug-flow mode; there is no axial or radial dispersion.
- Liquid velocity is constant through the reactor, since liquid expansion caused by hydrocracking is between 3 and 5 vol% under the experimental conditions.
- Vaporization of the feed does not take place.
- Gas velocity changes along the reactor due to hydrogen consumption.
- Pressure is constant along the reactors.
- Coke buildup occurs during the first 100 hours-on-stream and then reaches equilibrium, whereas metals are deposited during the entire operation period.
- Since catalyst deactivation is relatively slow with respect to the duration of the mass balance runs, a quasi-steady-state approximation can be applied to simulate the performance of the process.

5.2.2.1 ***Mass and Heat Balances*** The model incorporates reactant mass-transfer phenomena at the gas–liquid and liquid–solid interphases. Hydrogen is dissolved in the liquid and then travels to the catalyst surface in order to react, whereas the gaseous products, such as H_2S, NH_3, and light hydrocarbons (constituted mainly by CH_4), are released to the gas phase. Thus, the change in molar flow of gaseous compounds along the reactor is equal to the gas–liquid transport rate:

$$\frac{dN_i^G}{dz} = -A_S k_i^L a_L \left(\frac{P}{H_i} \frac{N_i^G}{\sum N_i^G} - C_i^L \right) \tag{5.1}$$

where $i = H_2$, H_2S, NH_3, and CH_4.

The change in the concentration of gaseous compounds in the liquid phase is owed to gas–liquid equilibrium and mass transfer to the solid phase:

$$\frac{dC_i^L}{dz} = \frac{1}{u_L} \left(k_i^L a_L \left(\frac{P}{H_i} \frac{N_i^G}{\sum N_i^G} - C_i^L \right) - k_i^S a_S (C_i^L - C_i^S) \right) \tag{5.2}$$

where $i = H_2$, H_2S, NH_3, and CH_4.

The chemical species are transferred from the liquid bulk to the catalyst surface; thus, the mass balance equation for these compounds is as follows:

$$\frac{dC_i^L}{dz} = -\frac{1}{u_L} k_i^S a_S (C_i^L - C_i^S) \tag{5.3}$$

where i = S, N, Ni, V, Asph, and 538 °C+ residue.

The compounds transported between the liquid–solid interphase are either consumed or produced by chemical reaction:

$$k_i^S a_S (C_i^L - C_i^S) = \pm r_j \tag{5.4}$$

where $i = H_2$, H_2S, NH_3, CH_4, S, N, Ni, V, Asph, and 538 °C+ residue, and j = HDS, HDN, HDNi, HDV, HDAs, and HCR. The "−" sign is for the reactants, while the "+" sign is for the products.

The following heat balance is included for representing adiabatic operation of the commercial-scale reactors:

$$\frac{dT}{dz} = [(-\Delta H_R) r_{HDS}] \frac{1}{u_G \rho_G c_P^G + u_L \rho_L c_P^L} \tag{5.5}$$

The heat balance incorporates an overall heat of reaction for the atmospheric residue HDS process, reported in the literature (Shah and Paraskos, 1975). This is a fitting parameter originated from several heat balances of similar processes, which accounts for the contribution of all reactions (HDS, hydrogenation, etc.).

It is also necessary to include equations for describing the quench zones. The following heat balance can be employed to estimate the quench rate (q) for a desired cooled mixture temperature (T_{in}):

$$\int_{T_{out}}^{T_{in}} l_{out} Cp^L dT + \int_{T_{out}}^{T_{in}} g_{out} Cp^G dT + \int_{T_Q}^{T_{in}} q Cp^Q dT = 0 \tag{5.6}$$

Once the quench rate is known, the actual rates of gas and liquid entering the following bed ($l_{in} + g_{in}$) can be determined by mixing the preceding bed effluent ($l_{out} + g_{out}$) with the quench fluid:

$$q + l_{out} + g_{out} = l_{in} + g_{in} \tag{5.7}$$

5.2.2.2 ***Reaction Kinetics*** Kinetic rates of each chemical lump were represented by power-law expressions of *n*th order:

$$r_j(z, t) = k_{t,j}^{app}(T, z, t)(C_i^S(z, t))^{n_j} \tag{5.8}$$

In the case of HDS, it was further considered that the reaction is inhibited by H_2S adsorption:

$$r_{HDS}(z, t) = k_{t,HDS}^{app}(T, z, t)\frac{(C_S^S(z, t))^{n_{HDS}}(C_{H_2}^S(z, t))^{0.5}}{(1 + K_{H_2S}^S C_{H_2S}^S(z, t))^2} \tag{5.9}$$

Equations 5.8 and 5.9 represent the local (with respect to the axial position z along the reactor) rates of reaction j at time t, where $k_{t,j}^{app}$ is the time-evolving local apparent rate coefficient of reaction j, with an Arrhenius-type temperature dependence, C_i^S is the local concentration of lump i at the catalyst particle at time t, and n_j is the order of reaction j. The time dependence of $k_{t,j}^{app}$ is given by the activity of the catalyst, which can be represented as follows:

$$k_{t,j}^{app}(T, z, t) = k_j^{app}(T)\varphi(z, t) \tag{5.10}$$

where k_j^{app} is the temperature-dependent apparent rate coefficient of reaction j and ϕ_j is the local deactivation function of reaction j at time t. It is important to notice that by definition k_j^{app} is a rate coefficient distorted by hydrodynamic and mass-transfer phenomena, which are typically present when obtaining kinetic data from bench-scale trickle-bed reactors. Therefore, it is necessary to correct these apparent coefficients in order to predict industrial-scale performance, as it is discussed in the following section.

H_2 consumption and H_2S generation are accounted for through overall stoichiometric coefficients, whereas CH_4 is considered to be produced exclusively from residue HCR (Sánchez et al., 2005).

The rate parameters were estimated with the Levenberg–Marquardt's algorithm from the kinetic data, as presented in previous work (Alvarez and Ancheyta, 2008).

5.2.2.3 ***Scale-Up of Kinetic Data*** The accuracy when scaling-up reaction kinetics strongly depends on the quality of the experimental data. The experience indicates that bench-scale trickle-bed reactors produce apparent kinetics because they suffer from several deviations from ideal plug-flow behavior, as a result of low liquid velocities (Bej, 2002). Keeping this in mind, apparent kinetics can be translated to intrinsic kinetics by considering these effects as follows:

$$k_j^{app}(T) = \eta_0 \eta_{CE} k_j^{in}(T) \tag{5.11}$$

Equation 5.11 correlates the experimentally obtained apparent rate coefficient of reaction j with the intrinsic rate coefficient k_j^{in}, through intraparticle diffusion and reactor hydrodynamics, represented by the catalyst effectiveness factor (η_0) and the external catalyst wetting efficiency (η_{CE}), respectively. When the experiments are carried out using commercial size catalysts, η_0 is already included in the apparent rate coefficients. This allows for describing industrial performance, but only if the catalyst used in the tests is exactly the same as the one in the commercial unit. Considering that in the present work, the experiments were carried out using commercial size catalyst, η_0 can be grouped with k_j^{in}, and thus Equation 5.11 becomes

$$k_j^{app} = \eta_{CE} k_j^{p} \tag{5.12}$$

where k_j^p is the "particle" rate coefficient.

This simple expression correlates bench-scale and industrial reactor performances through a single scale-up parameter. The term η_{CE} is a measure of the extent of catalyst utilization, which tends to become complete ($\eta_{CE} \sim 1$) at industrial scale (Mederos et al., 2009). In this work, this effect was handled with the empirical approach proposed by Bondi (1971):

$$\frac{1}{k^{app}} = \frac{1}{k^{p}} + \frac{A}{G_L{}^{B}} \tag{5.13}$$

Equation 5.13 is a relationship in which the apparent rate coefficient approaches that of particle kinetics as superficial liquid mass velocity (G_L) goes to infinity, with fitting parameters A and B. This approach is based on the fact that k^{app} tends to increase with the liquid rate in trickle-bed reactors. By plotting $1/k^{app}$ versus $1/G_L{}^B$, it is possible to approximate the value of the particle rate coefficient at infinite G_L, which is the intercept on the y-axis ($1/k^p$). For this reason, it is necessary to perform the tests at various liquid rates and if possible, as Bondi (1971) recommended, using two reactor scales in order to reduce the uncertainty of the extrapolated k^p. In this case, wetting efficiency was estimated from the LHSV bench-scale experiments and also from semi-industrial-scale data (10 bpd plant), using the same feedstock, operating conditions, and catalysts.

5.2.2.4 Catalyst Deactivation During hydroprocessing of heavy feeds, coking and metal deposition reduce catalytic activity with time-on-stream. Our previous experimental observations indicate that coke formation is the main cause of the substantial activity loss during the few first hours of operation (generally the first 100 h), but in long term, metals buildup is more relevant. It was also noticed that coking reaches equilibrium after this initial stage, hence, the assumption that coke formation occurs only during the first 100 h.

The deactivation function ϕ_j is defined as the ratio between the local apparent rates of reaction j at time t and at $t = 0$ (fresh catalyst):

$$\phi_j(z,t) = \frac{r_j(z,t)}{r_j(z,t=0)} \tag{5.14}$$

The values of ϕ_j range between 1 (fresh catalyst) and 0 (totally spent catalyst). This function is essentially similar to the concept of active surface area of the catalyst, employed in other deactivation models (Kodama et al., 1980; Toulhoat et al., 2005). The formulation of the deactivation function requires measuring r_j at $t = 0$; however, in these cases this is not possible because the liquid product is accumulated during several hours to obtain enough sample to carry out all the required analyses. For practical reasons, it has been proposed to assume that the initial activities are those obtained from the first sample (Marafi et al., 2008a, 2008b), which is collected after the first 10–12 h of operation. In this study, however, initial activities are obtained from the SOR aging curve by extrapolation, which seems to be more accurate.

As already mentioned, it is considered that catalyst aging is provoked by coking reactions at SOR conditions and metal deposition during the whole cycle. The contribution of these two processes to the overall catalyst deactivation can be expressed as follows:

$$\phi_j(z,t) = \phi_j^{\text{Coke}}(t) + \phi_j^{\text{Metals}}(z,t) \tag{5.15}$$

where ϕ_j^{Coke} and ϕ_j^{Metals} stand for the deactivation functions for coking at time $t \leq 100$ h and local metals buildup (Ni+V) at time t, respectively.

The following empirical relationship is proposed for describing the deactivation process:

$$\phi_j(z,t) = \frac{1}{(1+\alpha_j \cdot t)^{\beta_j}} - (x_{\text{MOC}}(z,t))^{\gamma_j} \tag{5.16}$$

The first term of Equation 5.16 represents the initial activity decline (ϕ_j^{Coke}) during the first 100 h, as a function of time t with fitting parameters α_j and β_j for reaction j. At $t = 100$ h, it is considered that this initial period is terminated and the value of ϕ_j^{Coke} is kept constant for the rest of the operation cycle.

It must be stressed out that the proper way of describing this process is to link this function to a coking agent instead of time; unfortunately, there was no experimental information available that provided insights into this mechanism and, therefore, there was no other possible way to represent this behavior. Hence, application of such a functionality is restricted to the experimental operating conditions. Nevertheless, SOR data obtained with several heavy crudes and residues at different operating conditions indicated that the HDS and HDM activity decay in this period interestingly follows the same pattern even in such a wide range of conditions. This behavior is represented in Figure 5.2 as the change in the deactivation functions of HDM and HDS with respect to the change in time ($\Delta\phi_j/\Delta t$). It is also observed that in all cases

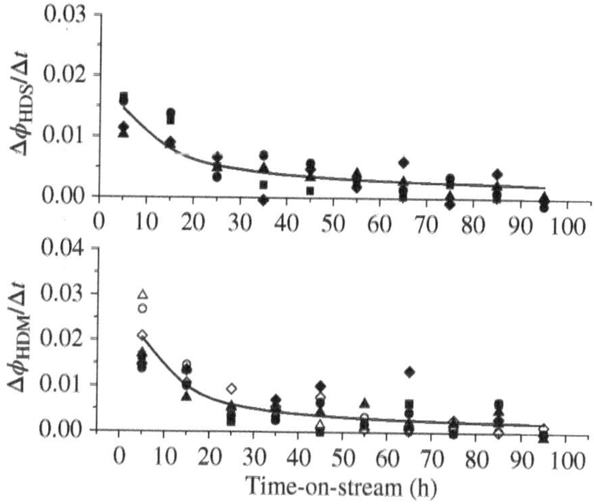

Figure 5.2 HDS and HDM deactivation curves at SOR conditions. Heavy crude oil (13°API) at LHSV = 0.5 h^{-1}, P = 6.86 MPa, and H_2/oil = 891 std m^3/m^3: (●) 380 °C, (▲) 400 °C, (◆) 420 °C; (■) atmospheric residue (5.4°API) at 360 °C, LHSV = 0.25 h^{-1}, P = 9.81 MPa, and H_2/oil = 891 std m^3/m^3; Maya crude oil (21°API) at LHSV = 0.5 h^{-1}, P = 6.86 MPa, and H_2/oil = 891 std m^3/m^3: (○) 380 °C, (△) 400 °C, (◇) 420 °C.

the catalyst system is rapidly stabilized at t = 100 h. On this basis, it can be considered that it is reasonable to use this approach and it can be assumed that the shape of the SOR curve is not affected by the variable temperature profile when running the adiabatic simulations.

The second term of Equation 5.16 describes ϕ_j^{Metals}. In this case, this function is directly linked to the deactivating agent: the amount of metals buildup, with fitting parameter γ_j for reaction j. The quantity of metal deposition is expressed in terms of x_{MOC}, defined as a ratio between the local concentration of MOC at time t and the specific maximum metals uptake of each catalyst. Such a functionality allows for establishing a deactivation profile with respect to time t and axial position z, which is a determining factor for the catalyst cycle life. MOC is obtained by solving the metals mass balance equation:

$$\frac{d\text{MOC}(z,t)}{dt} = a_{MOC} r_{HDM}(z,t) \tag{5.17}$$

where MOC is expressed in weight percent, r_{HDM} is the local rate of HDM at time t, and a_{MOC} is a unit conversion factor.

5.2.2.5 Solution Technique The reactor model consists of a set of coupled ordinary differential and algebraic equations, which were coded in Fortran. The mass balances for the gas and liquid phases, as well as the heat balance, are solved simultaneously using the fourth-order Runge–Kutta method, whereas the solid-phase equations are solved with the Newton–Raphson method, in order to obtain the concentration and temperature profiles along the reactor. With the concentration profiles, the local amount of metals buildup is estimated and then the deactivation functions for each reaction are evaluated. The time step is increased and all the calculations are updated. The simulation is terminated when a MAT of 420 °C is reached at any point of the reactor length.

5.2.3 Results and Discussion

5.2.3.1 Bench-Scale Unit Simulations The reactor performance model was applied to simulate the behavior of the bench-scale reaction system at various LHSV and temperatures under steady catalyst activity (kinetic tests). The results presented in Figure 5.3 show that the evolution of the various chemical lumps along the reactor is well predicted. The model is also able to describe H_2S and CH_4 yields and H_2 consumption along the reactor. In general, the model predictions are in good agreement with the experimental observations.

Figure 5.4 illustrates the levels of HDS, HDM, and HCR of 538 °C+ residue along the catalyst stability test. It can be observed that the reaction temperature was

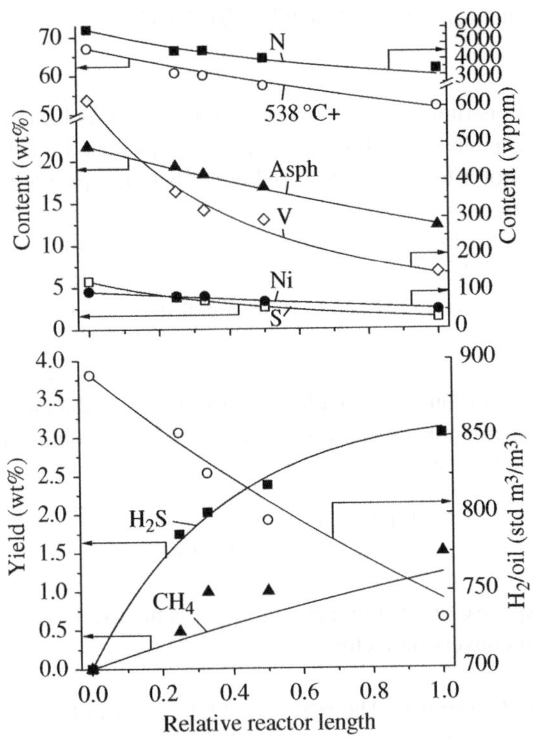

Figure 5.3 Evolution of the chemical lumps and gas yields along the reactor at $T = 380\,°C$ and LHSV = $0.25\ h^{-1}$: (symbols) experimental, (—) simulated.

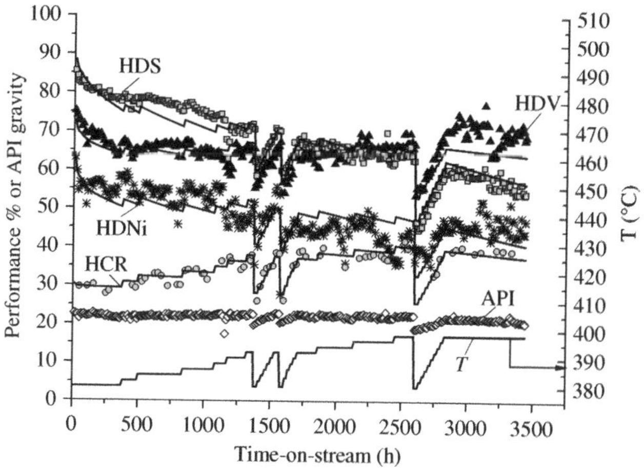

Figure 5.4 Process performance during time-on-stream: (symbols) experimental, (—) simulated.

increased periodically to keep the target oil quality at 23°API (the upgraded oil is obtained by blending the bypassed light fraction with the reactor liquid effluent). The deactivation curves followed different patterns under the selected operation mode, for instance, HDS and HDNi decreased progressively with time-on-stream, whereas HDV and HCR responded well to the gradual temperature increase. Such a behavior is owed to diverse factors such as the properties of the catalysts, activation energies, feed type, and the thermal effect on the overall performance. There are two discontinuities at around 1500 h of time-on-stream and one close to 2500 h, which are owed to operational problems caused by high pressure drops between the reactors and the separation system. In those cases, the reaction temperature was decreased and the heavy feed was switched to light gas oil in order to wash off the plugs. This represented a 297 h interruption (in total) of normal operation; however, these periods are not included in Figure 5.4. Once the pressure drop was within acceptable limits, the feed was switched back to residue and the temperature was increased again until the target performance was achieved. The test demonstrated that IMP HDT technology is capable of keeping the product within the specification during the 3440 hours-on-stream (~5 months of operation). The activity levels at the end of the test also indicated that the catalyst was still active; however, the substantial loss in HDS activity implied that at this stage the process is mainly driven by thermal effect. The operation was terminated due to the high temperature level (398 °C) in order to avoid excessive coking, which eventually would produce even more severe fouling problems.

The aging data were employed to fit the deactivation parameters in order to simulate the performance of the process during time-on-stream. From Figure 5.4, it can be observed that the model predicts reasonably well the deactivation curves. The parity plot shown in Figure 5.5 indicates that most of the data falls into the ±10% interval. Nickel removal exhibits the highest deviations, which is owed to the wide dispersion of the experimental data.

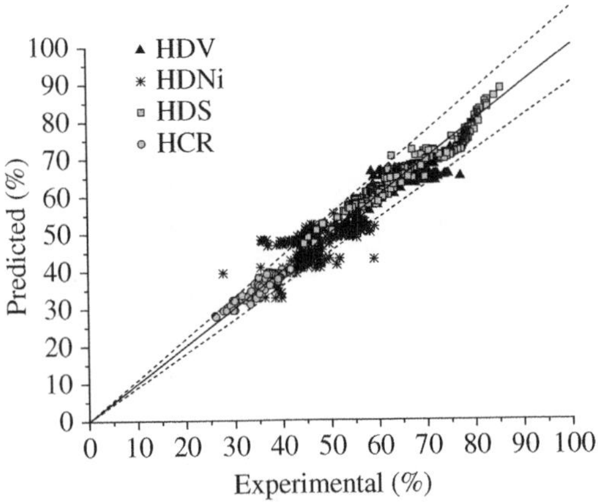

Figure 5.5 Comparison between experimental data and model predictions, (---) ±10%.

As mentioned in the experimental section, interreactor samples were taken to monitor the behavior of the first reactor. The HDM performance of the first reactor, along with the global HDM level, is presented in Figure 5.6. The performance data of the first reactor are quite dispersed, which is why the model describes only the general trend. The quality of the data is mostly owed to the fact that the interreactor samples are not completely representative. As for the overall HDM, the model works sufficiently well. The same figure shows the evolution of the MOC in both reactors. It is clear that MOC accumulation follows a linear behavior. Apparently, there is a good match between the model results and the observed MOC, which was estimated from the metal balances between the feed and products. It is noteworthy that metals

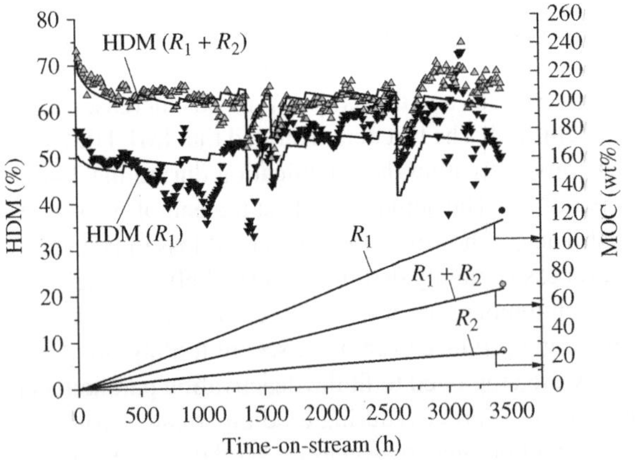

Figure 5.6 HDM performance and metals-on-catalyst (MOC): (symbols) experimental, (—) simulated, R_1 (first reactor), R_2 (second reactor). The experimental values (°) of MOC at the end of the run were estimated based on the metal balance between feed and products.

buildup in the first reactor is as high as ~120 wt% (in fresh-catalyst basis), whereas in the second reactor is only ~20 wt%, which indicates that the front-end catalyst is indeed protecting downstream catalysts.

The model is also capable of describing the evolution of axial MOC profiles, as illustrated in Figure 5.7. It can be observed that initially (500 h) the MOC profile is relatively uniform along the whole reactor length. As time-on-stream increases, the MOC profile acquires the typical descending shape, particularly in the first reactor. Clearly, it can be noticed that the front-end HDM catalyst, located in the first 60% of the first reactor (0–0.6), exhibits the highest metal accumulation, followed by the mid-end catalyst, which is distributed in the remaining space of the first reactor and the first 20% of the second reactor (0–0.2). Such a layered catalyst arrangement reduces significantly the metal content in the stream entering the highly active HDS catalyst bed located in the tail end of the second reactor.

5.2.3.2 Scale Effect Analysis Ideally, in trickle-bed processes sustaining liquid-limited reactions, the whole catalyst bed contributes to the overall conversion because every catalyst particle is covered by a flowing film of liquid. However, the behavior of bench-scale reactors deviates significantly from ideal behavior as a result of low liquid velocities that cause incomplete catalyst wetting. For this reason, it is essential to estimate wetting efficiency to determine the extent of catalyst utilization and thus correcting kinetic data.

Figure 5.8 provides the analysis of the scale effect on catalyst wetting efficiency, defined earlier as the ratio between the apparent and particle rate coefficients. The analysis is based on the bench-scale and semi-industrial-scale data at 380 °C. The shape of the curve indicates that wetting efficiency is a weak function of superficial liquid mass velocity in this system. In fact, the values of η_{CE} for all experiments fall in the range of commercial operation (0.7–1.0); for instance, at the lowest

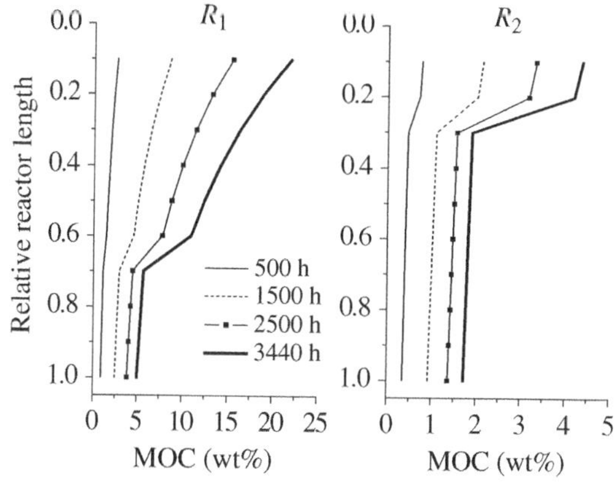

Figure 5.7 Simulated axial MOC profiles in the first (R_1) and second (R_2) reactors. The MOC units are referred to the total amount of fresh catalyst in each reactor.

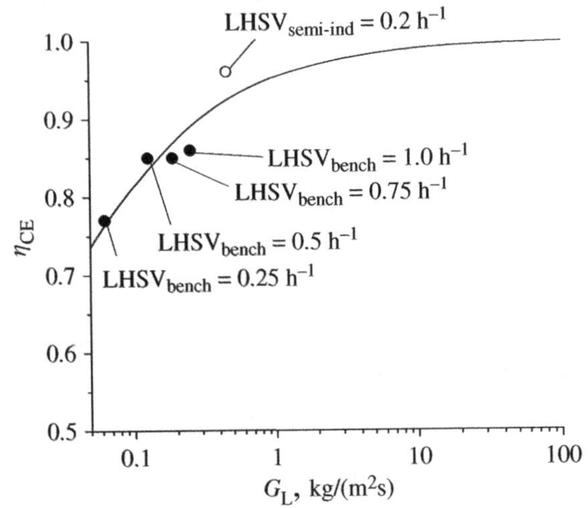

Figure 5.8 Catalyst wetting efficiency as a function of superficial liquid mass velocity: (●) bench-scale, (○) semi-industrial scale, (—) theoretical.

G_L (~0.06 kg/(m^2 s)), η_{CE} is already higher than 0.7 (more than 70% catalyst utilization), whereas for the semi-industrial test η_{CE} is very close to 1, as this plant has roughly 230 times more processing capacity than the bench-scale unit operating at LHSV = 0.25 h^{-1}. Apparently, this indicates that the semi-industrial-scale reactor behaves ideally in terms of catalyst utilization, and, therefore, this test can be considered representative of commercial operations.

The semi-industrial-scale data are compared against model predictions in Figure 5.9. From the previous analysis, it was determined that a wetting efficiency of 0.7 at those operating conditions correlates sufficiently well bench-scale and semi-industrial reactor performances. It is evident that the semi-industrial scale outperforms the bench-scale because the latter uses only 70% of its catalyst inventory.

5.2.3.3 ***Industrial Reactor Simulations*** The performance model was applied to analyze the effect of reactor design on the behavior of the process. The following reactor arrangements were studied (Figure 5.10):

- *Case 1*: Two single-bed reactors of equal length, with one hydrogen quench.
- *Case 2*: One reactor with two beds of equal length followed by a second single-bed reactor, with two hydrogen quenches.
- *Case 3*: One reactor with four beds of increasing length followed by a second reactor with two beds of increasing length, with two hydrogen quenches and three heat exchangers.

It can be noticed from Figure 5.10 that the configuration of Case 3 uses a combination of hydrogen quenching and heat exchange. The amount of hydrogen

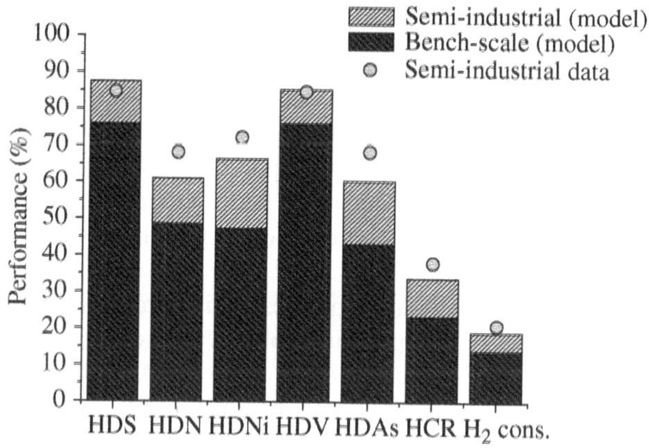

Figure 5.9 Comparison of semi-industrial plant data with model predictions, under the following conditions: LHSV = 0.2 h^{-1}, T = 380 °C, P = 9.81 MPa, and H_2/oil = 890 std m^3/m^3.

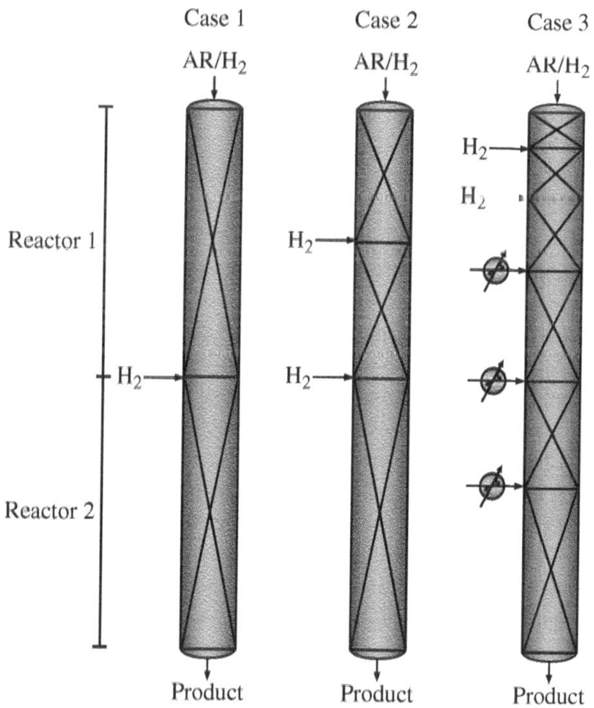

Figure 5.10 Proposed industrial reactor configurations.

quenching in this system is restricted by a low H_2/oil ratio, which is meant to reduce compression requirements. In previous studies (Alvarez and Ancheyta, 2008; Alvarez et al., 2009), it was determined that in this process it is not possible to quench with hydrogen the total heat release and at the same time to keep the required average H_2/oil ratio (891 std m^3/m^3) along the reactors. In fact, no more than two hydrogen quenches are allowed to keep the required H_2/oil ratio, because each extra hydrogen quench decreases significantly this ratio, especially in the first reactor. Having this in mind, the amount of hydrogen quenching in Cases 1 and 2 was also optimized to meet this criterion.

The time-evolving axial temperature profiles are illustrated in Figure 5.11. It can be observed that in all cases the temperature profile is progressively displaced upward because the feed temperature is increased so as to keep API gravity of the product constant (HCR at ~35%). The configuration of Case 1 is characterized by a large and continuous heat release in both reactors. The considerable ΔT produces high temperature zones at the outlet of each reactor. The temperature rise in the second reactor slowly decreases with process time because the tail-end catalyst is less tolerant to metal deposition and loses activity more rapidly, whereas the catalysts in the first reactor are more stable. Eventually, the run would have to be terminated earlier, as the axial temperature at the outlet of the first reactor reaches the permissible limit (420 °C) at 2748 h (~3.8 months). It is evident from this simulation that the drawback of the design of Case 1 is the absence of quenching in the first reactor where it is necessary. In Case 2, the hydrogen quench in the first reactor slightly increases the time to reach MAT to 2964 h (~4.1 months) of operation; however, now the high temperature zone is shifted to the outlet of the second reactor. In both cases, a substantial amount of catalyst is underused and the cycle length is lower than that of the aging test. In addition, the high temperature zones at the outlet of both reactors may induce premature thermal cracking leading to pressure drop problems as the ones observed during the aging test. Case 3 overcomes these disadvantages by the use of multiple catalyst beds of increasing length. In this case, there is a better temperature distribution along the reactor: the temperature levels in all catalyst beds are similar and there are no high temperature zones. With such an arrangement, it is possible to operate 4500 h (~6.3 months) without reaching MAT at any point of the reactor length.

Figure 5.12 shows the axial MOC pattern of each case at 2700 h. Clearly, the MOC profile of Case 3 has the typical descending shape. This means that the majority of the metals are deposited in the front-end HDM catalyst as it should be. An improper reactor design as those of Cases 1 and 2 shifts metal deposition toward the mid and tail-end section. In those cases, the feed temperature is considerably lower than in Case 3, which reduces the HDM rate in the front-end section, whereas higher temperatures in the remaining reactor length increase metal accumulation and consequently deactivation in the mid-end and tail-end catalysts. This also explains why the temperature rise in the second reactor decreases more rapidly with time in Cases 1 and 2 than in Case 3.

These results indicate that appropriate reactor configuration is essential for maximizing the cycle length and for making better use of the catalyst inventory. For highly exothermic processes, the key factor is to limit the total heat release into smaller

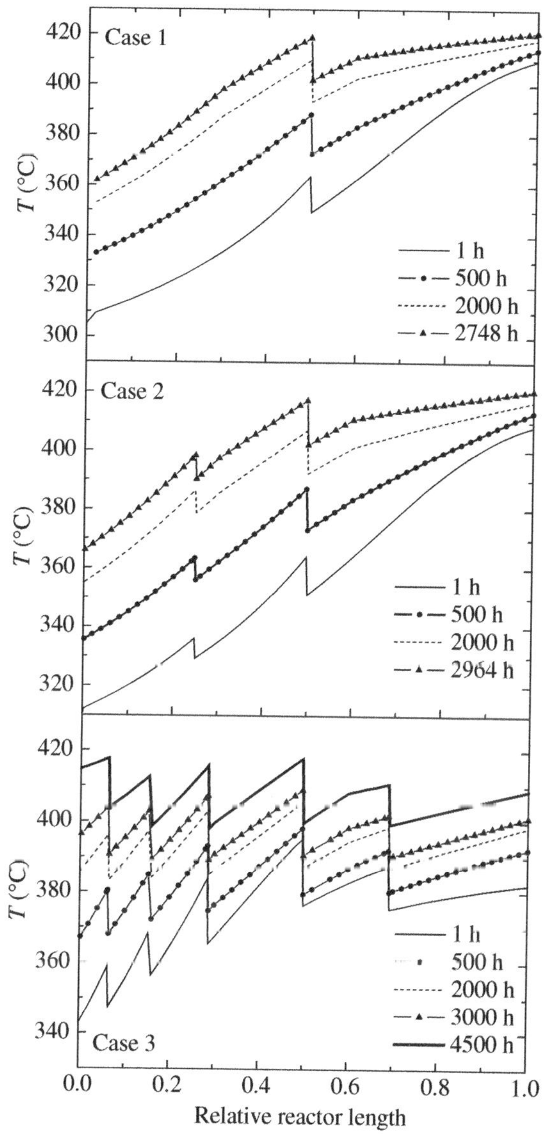

Figure 5.11 Evolution of the axial temperature profiles.

portions (i.e., small ΔTs) in order to achieve a more even temperature distribution along the reactor length. To achieve this goal, it is preferable to use multiple beds of increasing length rather than equal bed lengths because the heat release decreases with the bed depth. Quench rate must also be carefully selected to obtain the desired temperature profile. It is possible to increase the number of beds in order to flatten even more the temperature profile, thus reducing the dependency on temperature

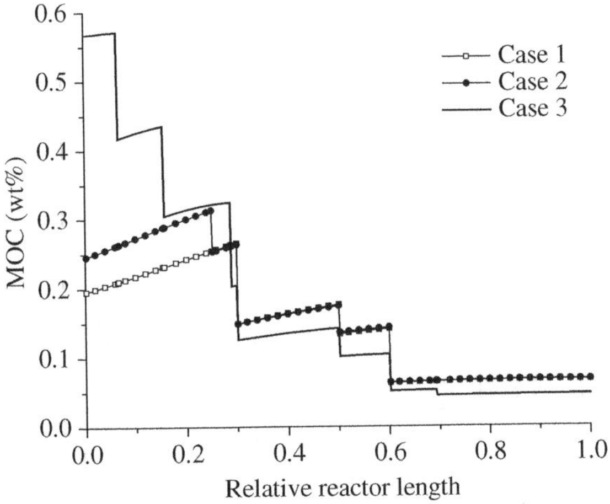

Figure 5.12 Metals deposition profiles at 2700 h.

constraints; however, this increases the height of the reactor vessel and consequently internal hardware investment.

5.2.4 Conclusions

The time-evolving uneven axial temperature and activity profiles add a new dimension to the design of industrial reactors. A hydroprocessing reactor model coupling catalyst deactivation was developed in order to analyze the influence of reactor configuration on the cycle length. The model showed to predict reasonably well the deactivation curves obtained from a long-term bench-scale test. The industrial reactor simulations revealed that for this kind of process, single beds and double beds are inadequate reactor designs due to two main reasons: (1) they exhibit wide axial temperature differences as a result of large bed length, leading to high temperature zones, which are eventually the main cause of premature shutdown; (2) the uneven temperature distribution adversely displaces HDM reactions from the front end toward the mid- and tail-end sections. The configuration with multiple beds of increasing length is characterized by a more regular temperature profile and, therefore, corrects the deficiencies of other arrangements, thereby increasing significantly the cycle length.

5.3 EFFECT OF DIFFERENT HEAVY FEEDSTOCKS ON THE DEACTIVATION OF A COMMERCIAL CATALYST

5.3.1 Experimental Procedure

5.3.1.1 Catalyst and Feedstock A commercial nickel–molybdenum alumina-supported catalyst (tetralobe extrudate shape, 175 m^2/g of surface area, 0.56 cm^3/g of pore volume (PV), mean pore diameter of 127 Å and molybdenum and nickel contents of 10.7 and 2.9 wt%, respectively) was used for all experiments.

TABLE 5.5 Properties of the Feedstocks and Their Asphaltenes

Properties	HCO	ARHCO	EHCO
Crude oil			
Specific gravity at 60 °F	0.9264	1.0219	0.9845
API gravity	21.24	6.97	12.23
Sulfur (wt%)	3.44	4.73	5.25
Asphaltenes in nC_7 (wt%)	10.91	17.74	19.23
Nickel (wppm)	54.7	87.6	87.1
Vanadium (wppm)	298.8	411.5	448.0
Asphaltenes			
C (wt%)	83.86	83.22	83.91
H (wt%)	7.58	7.52	7.57
S (wt%)	7.11	7.55	7.15
H/C atomic ratio	1.0770	1.0766	1.0749

Three different heavy feedstocks (HCO: heavy crude oil, ARHCO: atmospheric residue from heavy crude oil, EHCO: extra heavy crude oil) were used for deactivation studies. Their main physical and chemical properties are reported in Table 5.5. The atmospheric residue was obtained from HCO distillation by fractionation in a 40 L glass column according to ASTM D1160 modified method. All feeds are characterized by having high amount of metals (350–535 ppm Ni+V) and asphaltenes (10–19 wt%), which make them suitable for studying deactivation due to metal deposition.

5.3.1.2 Experimental Setup and Procedure The catalyst deactivation experiments were carried out in a fixed-bed bench-scale reactor, whose details have been described in the previous sections. The main characteristics of this unit are (1) fixed-bed reaction system, (2) continuous downflow reactor, and (3) isothermal operation.

A 100 mL of catalyst was loaded to the reactor. The pressure was increased from atmospheric to 110 kg/cm^2 using nitrogen and kept constant for 3 h in order to check the tightness in all sections of the bench-scale unit.

The catalyst was *in situ* activated by sulfiding with hydrodesulfurized diesel containing 2.5 wt% DMDS (dimethyl disulfide) at the following conditions: pressure of 28 kg/cm^2, H_2-to-oil ratio of 2000 ft^3/bbl, and LHSV of 2 h^{-1}. The sulfiding temperature was programmed in two steps: (1) at 260 °C during 3 h and (2) at 320 °C for 8 h.

5.3.1.3 Reaction Conditions For each feedstock, the experiments were conducted under the following conditions: 70 kg/cm^2 of total pressure, LHSV of 0.5 h^{-1}, and 5000 ft^3/bbl of H_2-to-oil ratio. Reaction temperature was varied at 380, 400, and 420 °C. The hydrotreated product quality was monitored every 10 h during 240 h time-on-stream.

5.3.1.4 Characterization of Samples Chemical and physical properties of hydrotreated liquid products were determined according to the following methods: specific gravity using a pycnometer by ASTM D70 method, total sulfur content by energy-dispersive X-ray fluorescence spectrometry with ASTM D4294 method, nickel and vanadium contents by atomic absorption using ASTM D5863 method, and asphaltene content as *n*-heptane insolubles by ASTM D3279 method.

Spent catalyst samples were collected after reaction according to procedures reported elsewhere (Ancheyta et al., 2002, 2003). For the case of spent catalysts using EHCO, the samples were unloaded from three sections of the catalytic bed (top, middle, and bottom).

Nickel and vanadium contents on spent catalysts were analyzed by atomic absorption in a Spectrometer Thermoelectron Model Solaar AA. Carbon content was determined by combustion in a PerkinElmer 2400 analyzer.

5.3.2 Results and Discussion

To simulate the reactor, the previously described model was used.

5.3.2.1 Removal of Impurities Figure 5.13 shows the effect of temperature, type of feed, and time-on-stream on sulfur, asphaltene, and metal removals, respectively. The influence of reaction temperature on impurities removal was the expected one, that is, the higher the temperature, the higher the conversion of impurities. The effect of the type of feed is different depending on the reaction severity, at temperatures higher than 400 °C the trend in impurities content is ARHCO > EHCO > HCO. According to the content of sulfur, metals, and asphaltenes in the three feeds, the frequently found behavior would be that observed at 380 °C (EHCO > ARHCO > HCO), because the lightest crude oil, that is, HCO, would have less amount of impurities, which presumably are less complex than those contained in heavier crude oil, that is, EHCO. Thus, the differences are surely due to the type of compounds present in each feed. Since ARHCO was obtained from HCO fractionation, the complexity of compounds in both feeds is expected to be the same, which is confirmed by their almost similar H/C atomic ratio of asphaltenes (Table 5.5). The very small differences may be due to the fractionation of HCO, in which part of the maltenes adhered to asphaltene molecules were released. Hence, the explanation of the behavior of catalyst deactivation with respect to each feed can be reduced only to the comparison of HCO (or ARHCO) and EHCO feedstocks. HCO possesses lower content (almost a half) of asphaltenes than does EHCO (10.91 vs 19.23 wt%). According to the H/C atomic ratio (Table 5.5), asphaltenes contained in EHCO have lower value of this ratio than those of the HCO and ARHCO feeds (1.0749 vs 1.0770 and 1.0766) respectively. This implies that EHCO asphaltenes are more aromatic in nature and most of their sulfur and metal (~24%) atoms are surrounded by aromatic cores with low number of heteroatom bridges and alkyl chains compared with HCO and ARHCO asphaltenes.

At low temperatures, that is, 380 °C, the reaction is more hydrogenation-dominated (Ancheyta et al., 2005a, 2005b). When increasing the temperature, hydrocracking

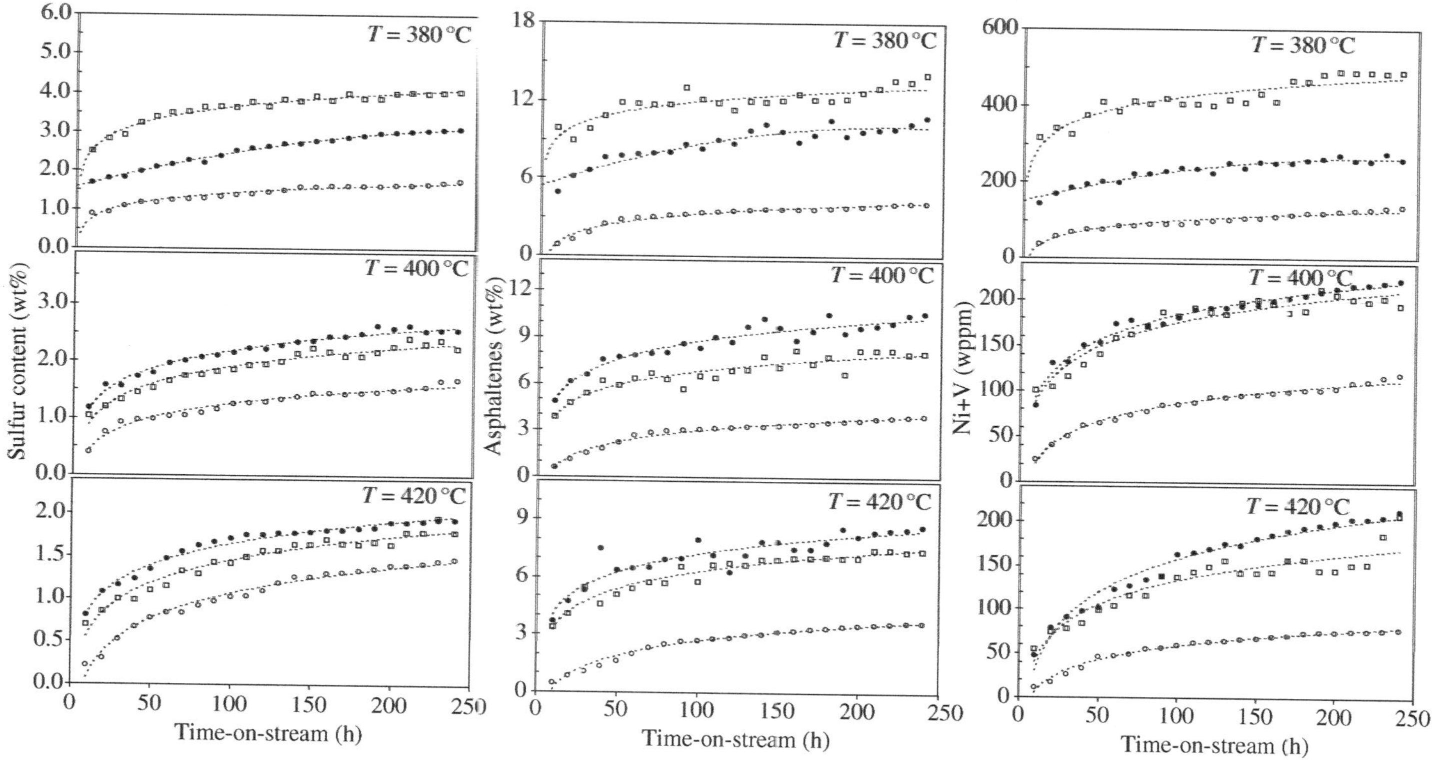

Figure 5.13 Total sulfur, asphaltene, and metal contents in hydrotreated products at different reaction temperatures and feedstocks: (○) HCO, (●) ARHCO, (□) EHCO.

is the predominant reaction, which breaks alkyl chains and C–C bonds, saturates and cracks hydrocarbons, leading to polyaromatic rings to concentrate. Having asphaltenes of EHCO lower H/C ratio than those of HCO and ARHCO means that at 380 °C of HDT temperature asphaltenes of EHCO are more prone to be hydrogenated, that is why at low temperature this feed (i.e., HCO) showed the lowest impurities conversion. However, at temperatures higher than 400 °C, the aromatic carbons of EHCO asphaltenes are broken, which allows for the heteroatoms allocated in the inner part of the aromatic cores to be released, and thus increasing the conversion of impurities in such an extent that it becomes higher than the ARHCO feed.

5.3.2.2 Deposition of V and C on Spent Catalysts When working with the same catalyst and under identical reaction conditions and time-on-stream, it is expected that the more content of metals and asphaltenes in the feed would cause higher vanadium and carbon deposits on the spent catalyst. This indeed happens if the feeds come from the same source and are of similar nature. However, the feeds used in the present work strongly differ in chemical and physical properties, particularly in the properties of their asphaltenes as discussed before. The composition and structure of the asphaltenes then influence the vanadium and carbon depositions.

The effect of temperature on V and C deposits is the expected one, that is, at higher temperatures more metals and carbon are accumulated on the spent catalyst as can be seen in Figure 5.14. For both types of deposits, linear behavior was obtained.

For the spent catalyst samples collected at different reactor positions, the results of analyses show that vanadium deposit is higher at the top of the reactor while carbon

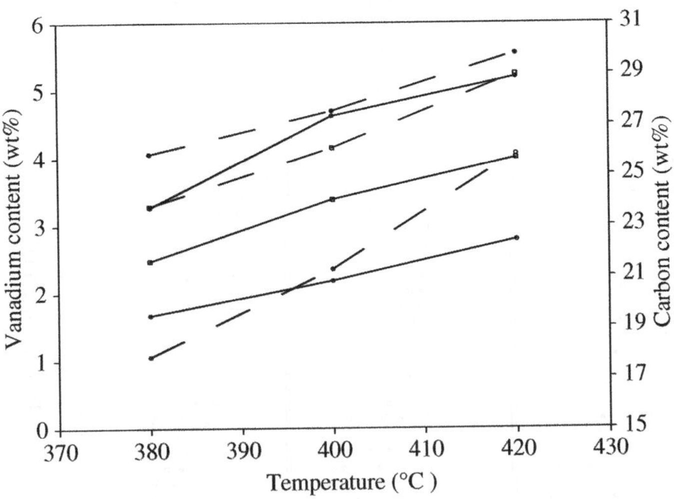

Figure 5.14 Effect of temperature on V and C deposits on the catalyst for different feedstocks: (O) HCO, (●) ARHCO, (□) EHCO, (—) V, (---) C.

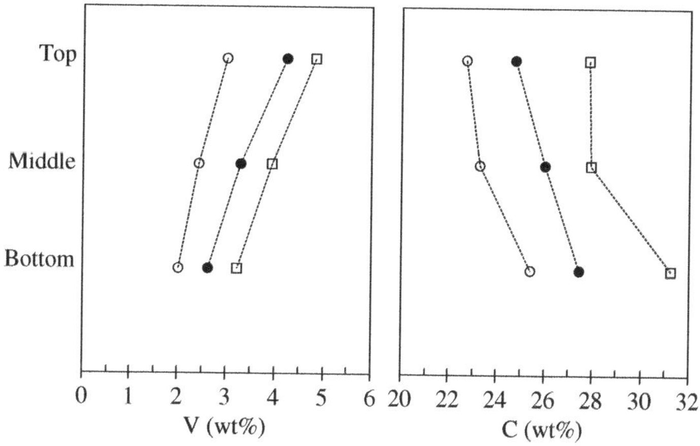

Figure 5.15 Vanadium and carbon contents in the spent catalyst for EHCO feedstock at different positions of catalytic bed.

exhibits the opposite deposition pattern, that is, it is lowest at the top of the reactor (Figure 5.15). The way in which vanadium is deposited along the reactor implies that the first section of the catalytic bed is catching more amount of metals, so that less amount of metals passes through the next sections. Hence, in commercial operations graded catalysts are used, being the front-end catalyst more tolerant to metal deposition, which in turn protects the following catalysts designed to accomplish other reactions, for example, hydrodesulfurization and hydrocracking. The ascending carbon deposition pattern can be explained as follows: at the end of the reactor, the temperature is higher than that at the beginning due to the exothermal nature of the reactions taking place. Since conversion of asphaltenes is more controlled by temperature than by the catalytic reaction, thus at the bottom of the reactor, thermal cracking of asphaltenes is expected and so does the carbon deposition on the catalyst.

5.3.2.3 ***Reactor Simulations*** A set of experimental data was employed to fit the deactivation curves in order to analyze the behavior of the process during time-on-stream. Figure 5.16 exhibits the HDM and HDS activity profiles with HCO and ARHCO feeds. It can be observed that the model predicts reasonably well the overall trend of deactivation curves in the range of experimental conditions.

Apparently, during the whole run, there was no substantial effect of the feedstock type on the shape of the activity curves. This is exemplified in Figure 5.17a for the case of HDS. It is clearly seen that the HDS deactivation function decreases at a similar rate in all of the cases. However, this is only valid for SOR conditions, as the length of run of the experiments is really small compared with the length observed in commercial practice. Certainly, it could be expected that the rate of deactivation will increase proportionally to feedstock heaviness with long time-on-stream and it

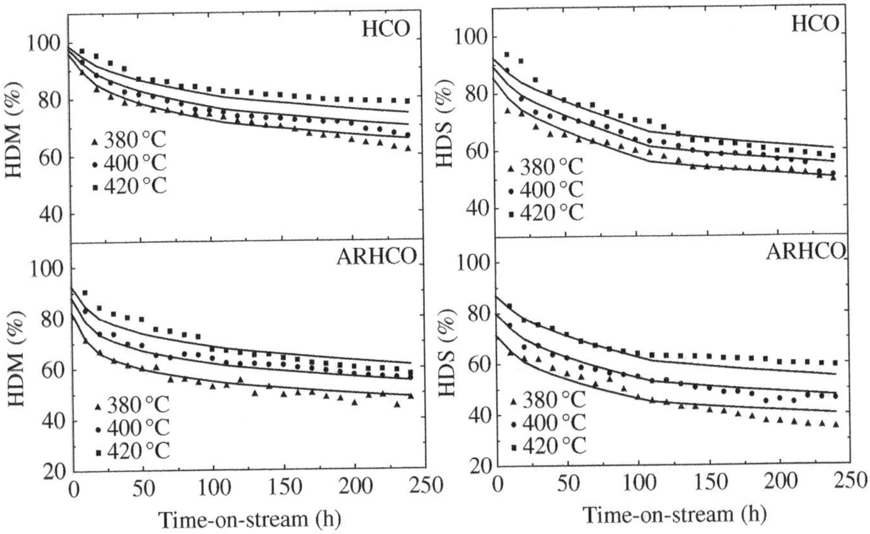

Figure 5.16 HDM and HDS performance with HCO and ARHCO feeds during time-on-stream: (symbols) experimental, (—) simulated.

is possible to capture this effect with the model, as it is directly linked to the amount of metal deposits. As for the effect of temperature, for each reaction and feed, the curves were fitted with a single set of parameters for the whole reaction temperature range (for hydrodesulfurization reaction: HCO, $\alpha = 1.09 \times 10^{-2}$, $\beta = 0.3$, $\gamma = 0.21$; ARHCO, $\alpha = 2.04 \times 10^{-2}$, $\beta = 0.19$, $\gamma = 0.26$; EHCO, $\alpha = 8.83 \times 10^{-2}$, $\beta = 0.32$, $\gamma = 0.72$). This means that the activity profiles during the experimentation period are temperature independent, which is reflected in the overlapped profiles illustrated in Figure 5.17b. Once again, at longer periods of operation, the deactivation rate will increase progressively with the reaction temperature as a result of higher HDM rates. For this particular study, this analysis justifies the choice of using a simple time-dependent function for describing catalyst aging during the SOR.

Figure 5.18 shows the profiles of metal accumulation during time-on-stream, which were determined from the metal balances between the feed and products. Clearly, MOC buildup follows a linear behavior in the whole temperature range. At the end of the experiment, the MOC for the HCO feed is between 3.5 and 4.5 wt%, depending on the reaction temperature, whereas for ARHCO it is between 4.5 and 6 wt%. For both of the feeds, the MOC ranges are very narrow, which explains why there is no substantial effect of temperature on the deactivation rates shown in Figure 5.17b. Also the small difference between both feeds supports the discussion on the effect of feed type in Figure 5.17a.

It is also possible to track the evolution of the axial MOC profiles, as presented in Figure 5.19. Initially (10 h), the MOC profile is relatively flat along the whole reactor length. But as time increases, the MOC profile converges in the typical descending shape, which means that most of the metals tend to accumulate in the front end of the reactor.

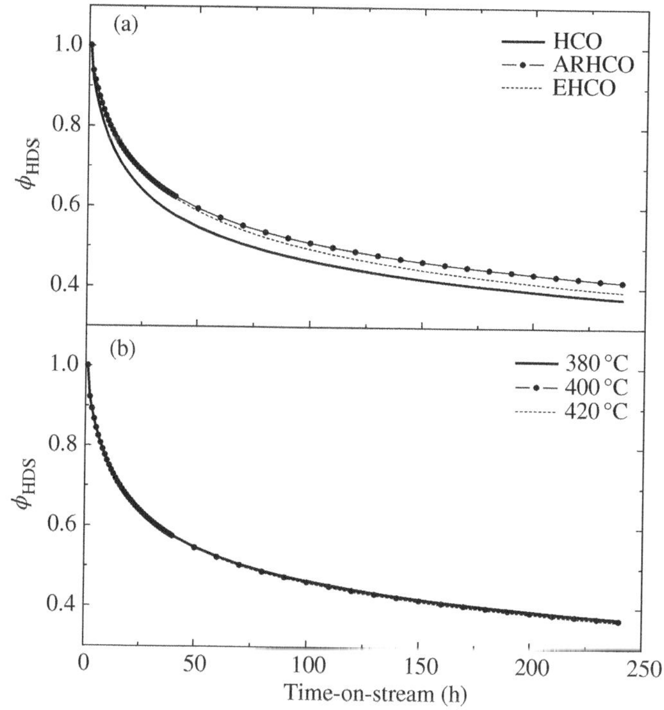

Figure 5.17 Profiles of the HDS deactivation function. (a) Effect of feedstock type at 380 °C. (b) Effect of temperature for HCO feed.

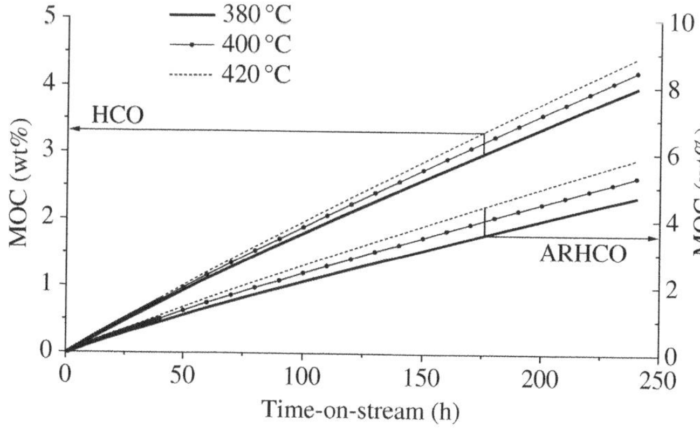

Figure 5.18 MOC accumulation with HCO and ARHCO feeds.

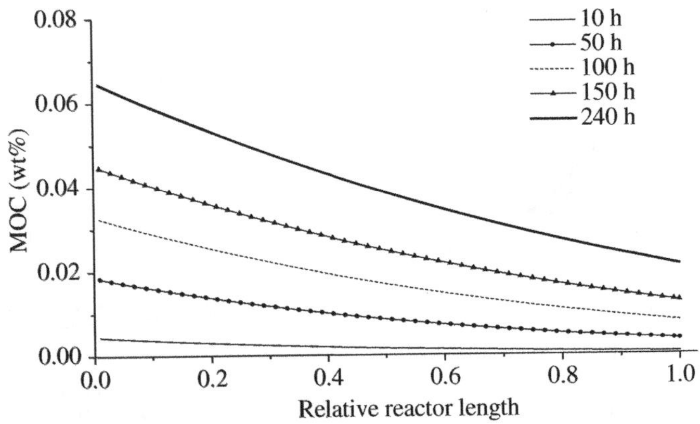

Figure 5.19 Evolution of axial MOC profiles with HCO feed at 380 °C.

5.3.3 Conclusions

The following conclusions can be drawn:

- The effect of the type of feedstock on catalyst deactivation is different depending on reaction temperature.
- Not only asphaltene content in the feed but also properties of asphaltenes have a great impact on catalyst deactivation.
- Vanadium deposit is more at the top of the reactor, while carbon deposit is low at the same position.
- Carbon deposition is more at the bottom of the reactor due to the enhanced asphaltenes thermal cracking.
- Catalyst deactivation can be modeled by a two-term equation, each accounting for carbon and metal depositions.

5.4 MODELING THE DEACTIVATION BY METAL DEPOSITION OF HEAVY OIL HYDROTREATING CATALYST

5.4.1 The Model

5.4.1.1 Mass Balances The model employed to describe the pore plugging of the catalyst due to deposit of metal sulfide species is the one developed by Oyekunle and Hughes (1987) and Oyekunle et al. (2005). The following assumptions were made:

- Irreversible HDM reaction (HC-Metal → HC-free metal)
- Power-law kinetic model
- Constant hydrogen concentration due to its presence in excess

- No resistances to mass transfer between fluid and catalyst particle
- Cylindrical pores
- Loss of activity due to reduction of pore size
- Isothermal particle with constant size and shape
- Plug-flow reactor model
- Negligible pressure drop
- No vaporization of feed due to its heaviness.

The mass balance within the pore (Figure 5.20) is as follows:

$$-N_A|_z S + N_A|_{z+\Delta z}\ S + k\overline{C_A^n}\ S\Delta z = -\frac{\partial C_A}{\partial t} S\Delta z \tag{5.18}$$

where the over-bar on concentration means the average reactant concentration in the volume element.

After dividing by the cross-flow section (S) and Δz and taking the corresponding limit:

$$\frac{\partial N_A}{\partial z} + kC_A^n = -\frac{\partial C_A}{\partial t} \tag{5.19}$$

Fick's first law is assumed:

$$-D\frac{\partial C_A}{\partial z} = N_A \tag{5.20}$$

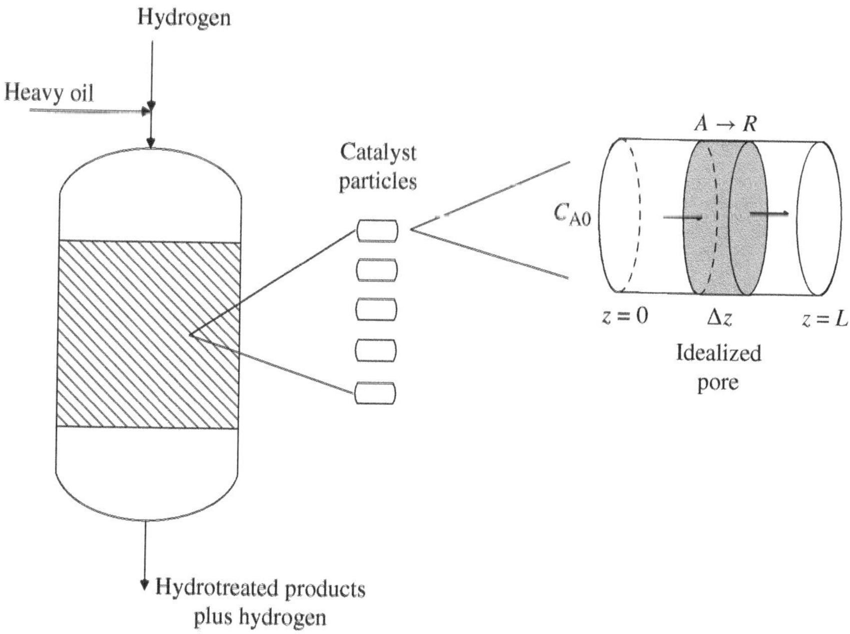

Figure 5.20 Internal structure of reactor and idealized pore catalyst.

Substitution of Equation 5.20 into Equation 5.19 produces

$$D\frac{\partial^2 C_A}{\partial z^2} - kC_A^n = \frac{\partial C_A}{\partial t} \tag{5.21}$$

The following dimensionless variables are used:

$$\chi = \frac{C_A}{C_{A0}} \quad \zeta = \frac{z}{L} \quad \tau = \frac{tD}{L^2} \quad \phi = L\sqrt{\frac{kC_{A_0}^{n-1}}{D}}$$

After proper application of chain rule, Equation 5.21 is transformed to

$$\frac{\partial \chi^2}{\partial \zeta^2} - \phi^2 \chi^n = \frac{\partial \chi}{\partial \tau} \tag{5.22}$$

The size of molecule bearing metal sulfide compounds can restrict the diffusivity of species, so that this restriction was taken into account by using the following equation:

$$D_r = D(1-\lambda)^4 \tag{5.23}$$

where $\lambda = \frac{r_{mol}}{r_p}$.

Hence, the mass balance (Eq. 5.22) takes the following form:

$$\frac{\partial \chi^2}{\partial \zeta^2} - \phi_r^2 \chi^n = \frac{\partial \chi}{\partial \tau_r} \tag{5.24}$$

With initial conditions:

$$\text{At } \tau_r = 0 \quad \chi = \chi_{max} \quad \forall_\zeta | \chi \in (0,1], \quad \chi = 1, \chi = 0$$

And boundary conditions:

$$\text{At } \tau_r > 0 \quad \zeta = 0 \quad \chi = 1$$

$$\zeta = 1 \quad \frac{\partial \chi}{\partial \zeta} = 0$$

Rajagopalan and Luss (1979) and Oyekunle and Hughes (1987) used the following relationship to model the metal deposits in pores:

$$\frac{\partial \gamma}{\partial t} = -\frac{\varepsilon K C_0^n M_{ms}}{\rho_{ms}} \frac{\chi^n}{r_0} \tag{5.25}$$

where $\gamma = \frac{r_p}{r_{p0}}$.

Equation 5.25 is transformed to dimensionless form and the final relationship is given by

$$\frac{\partial \gamma}{\partial \tau} = -\Lambda \chi^n \tag{5.26}$$

with

$$\Lambda = L^2 \frac{\epsilon K C_0^n M_{ms}}{\rho_{ms} D_r} \frac{1}{r_p}$$

And initial conditions:

$$\tau = 0, \quad \gamma = 1$$

The thickness of metal sulfide deposits can be calculated with

$$\delta = r_{p0} - r_p = r_{p0}(1 - \gamma) \tag{5.27}$$

5.4.1.2 *Estimation of Properties*

Kinetic Parameters Reaction rate constant and order were calculated by using the data obtained during the first 200 h of TOS with the following equation:

$$\frac{\varepsilon_L (n+1) k}{\text{LHSV}} = C_p^{1-n} - C_f^{1-n} \tag{5.28}$$

To calculate the output concentration of the reactor system at longer values of TOS, the following expression was employed:

$$\frac{\varepsilon_L (n+1) k}{\text{LHSV}} \varphi = C_p^{1-n} - C_f^{1-n} \tag{5.29}$$

where $\varphi = \frac{\eta_t}{\eta_0}$.

For calculating the effectiveness factor (η), the following relationship reported by Oyekunle et al. (2005) was used:

$$\eta = \int_0^1 \chi^n \gamma \; d\zeta \tag{5.30}$$

Catalytic Textural Properties It is widely accepted that initial textural catalyst properties (i.e., pore volume, specific surface area (SSA), and pore distribution) change during short TOS due mainly to coke deposits. So that in order to simulate the pore plugging due to metal deposits textural properties of the spent catalyst at 200 h TOS were assumed to be the initial values for this latter deactivation mechanisms.

Average Boiling Point of the Feed The correlation developed by Riazi (1997) was used to predict the entire boiling point and to calculate the average boiling temperature of oil.

Transport Properties Correlation presented by Beggs and Robinson (1975) was used to estimate the viscosity of oil, while diffusivity was calculated with Tyn and Calus (1975) correlation.

Other Properties Density of oil was estimated with Standing–Katz correlation (Korsten and Hoffmann, 1996). The characteristic size of particle (L) was calculated by using the formulae derived from Ancheyta et al. (2005a, 2005b). Correlation of Haughey and Beveridge (1969) was used to estimate the bed void fraction, while liquid holdup was calculated with the correlation of Goto and Smith (1975). A summary of the correlations used is given in Table 5.6.

The number of metal sulfide molecules per molecule of reactant (ϵ) was assumed to be two. According to other reports, average molecular weight of metal compounds can be assumed to be the same as that of the feed (Rodríguez et al., 2012), and it can be determined from the average boiling point distillation. Metal sulfide density value (ρ_{ms}) was taken from the literature (Rajagopalan and Luss, 1979).

5.4.1.3 Algorithm of Solution The average feed boiling temperature was calculated from Riazi's (1979) correlation. Given the properties of the feed and of the catalyst, the other transport and related properties were computed.

Equation 5.24 is solved for a short range of dimensionless time and profiles of dimensionless concentration are substituted into Equation 5.26, which is integrated with that information. After that new value of pore radius is obtained and hence ϕ_r parameter is updated. The procedure is repeated until pore plugging is complete.

TABLE 5.6 Correlations for Thermodynamic and Transport Properties

$$\frac{T-T_0}{T_0} = \left[\frac{A_1}{B_1}\ln\left(\frac{1}{1-\text{cwt}}\right)\right]^{1/B_1}$$

$$\left(\frac{T-T_0}{T_0}\right)_{av} = \left(\frac{A_1}{B_1}\right)^{1/B_1}\Gamma\left(1+\frac{1}{B_1}\right)$$

$$\mu = 10^{(X*T^{-1.163})} - 1, \quad X = 10^{(3.0324-0.02023°\,\text{API})}$$

$$D = 8.93\times10^{-8}\frac{v_L^{0.267}}{v_i^{0.433}}\frac{T}{\mu}$$

$$\rho_{oil} = \rho_0 + \rho_1 + \rho_2, \text{ where}$$

$$\rho_1 = [0.167 + 16.181\times10^{-0.425\rho_0}]\left[\frac{P}{1000}\right] - 0.01[0.299 + 263\times10^{-0.0603\rho_0}]\left[\frac{P}{1000}\right]^2$$

$$-\rho_2 = [0.0133 + 152.4(\rho_0+\rho_1)^{-2.45}][T(\circ\text{R}) - 520]$$
$$-[8.1\times10^{-6} - 0.0622\times10^{-0.764}(\rho_0+\rho_1)]\ \ldots\ [T(\circ\text{R}) - 520]^2$$

$$L = \frac{V_p}{S_p}\ V_p = 4\left[\pi\left(\frac{d_p}{4.828}\right)^2 L_p\right] - 2.944\times10^{-2}d_p^2L_p\ S_p = 4\left[2\pi\left(\frac{d_p}{4.828}\right)^2 + 2\pi\left(\frac{d_p}{4.828}\right)L_p\right]$$
$$+5.888\times10^{-2}d_p{}^2 - 2\pi\left(\frac{d_p}{4.828}\right)L_p$$

$$\varepsilon_0 = 0.39 + 0.07\frac{d_p}{d_t} + 0.54\left(\frac{d_p}{d_t}\right)^2$$

$$\varepsilon_L = A_2(v_L)^{1/3}(100\mu)^{1/4}$$

To resolve Equation 5.24, the method of orthogonal collocation was used. The second derivative of concentration with respect to position from mass balance was substituted by a matrix of values (B). To calculate matrix B, Legendre polynomial roots were used with $w = 1$; due to this condition, collocation points are near the surface of catalyst, where the steepest gradients are present. To generate the polynomials, the following hypergeometric function was used:

$$F(-i, i + \alpha/2, \alpha/2, x^2) \tag{5.31}$$

where $\alpha = 1$, 2, or 3 for flat plate, cylindrical, or spherical geometry, respectively.

Using the corresponding weights derived from standard collocation method and dimensionless concentration from Equation 5.24, integral in Equation 5.30 was accurately computed. A detailed algorithm is shown in Figure 5.21.

5.4.2 Experimental Procedure

Experimental data were obtained from a series of two trickle-bed bench-scale reactors (volume of 900 cm^3 each), both of which are having a diameter of 2.94 cm. The reactors were loaded with a system of commercial size catalyst: a front-end HDM catalyst, a mid-section catalyst with balanced HDM/HDS activity, and a tail-end highly active catalyst for HDS and HCR. The main properties of the feed are 3.2°API, 6.21 wt% sulfur, 117 wppm Ni, 578 wppm V, and 25.1 wt% asphaltenes. Operating conditions were average temperature of 385 °C, 9.8 MPa of initial pressure, hydrogen-to-oil ratio of 5000 scf/bbl, and LHSV of 0.25 h^{-1}. Metal contents on feed and products were determined by atomic absorption. ASTM D1160 method was used to obtain the boiling point distillation curve of feed and products.

Two tests were conducted, one at 200 h TOS and other one at 2200 h TOS. The short-term test data were used to determine the reaction kinetics, initial effectiveness factor, and properties of the spent catalyst at minimal metal deposition, while the long-term test was employed to validate the extrapolations of the deactivation model.

BET method was used to obtain the SSA, average pore diameter (APD) and its distribution, and pore volume (PV) of fresh and spent catalysts.

In the test of 2200 h of TOS after stabilization period, liquid samples were collected each 12 h and characterized by metal content. More details of experimental setup, catalyst activation, and sample characterization can be found elsewhere (Rana et al., 2004; Ancheyta et al., 2001).

5.4.3 Results and Discussion

Properties of fresh and spent (200 h of TOS) catalysts are compared in Figure 5.22. Properties were normalized taking the property values of the fresh catalyst as reference. It is observed that the SSA of the catalyst diminishes about 40% after 200 h. Such a reduction is attributed to initial coke deposits in the pores as was evidenced recently (Elizalde and Ancheyta, 2014). Also, a diminution of pore volume occurred

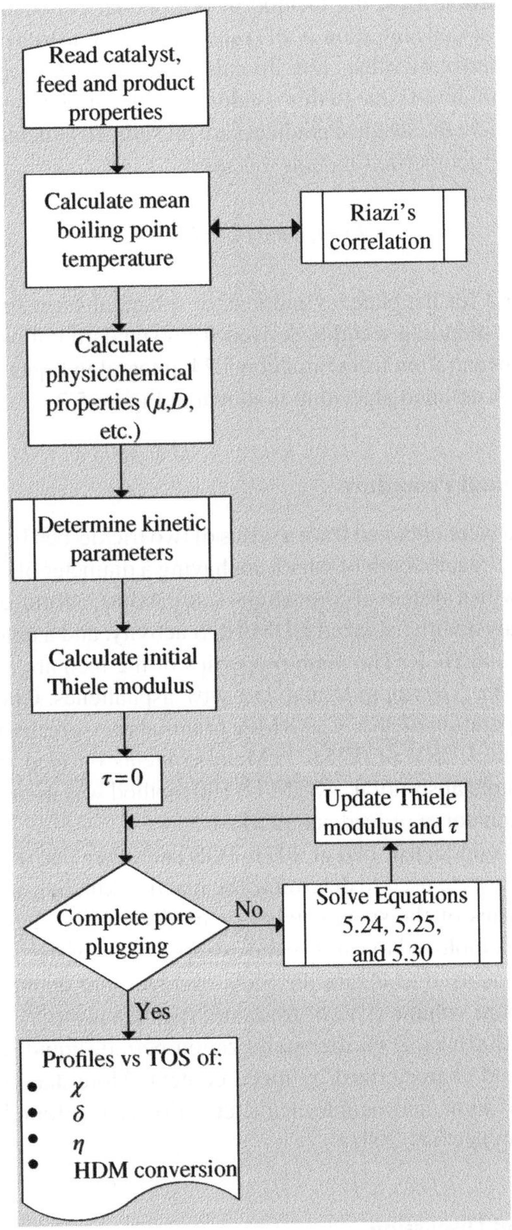

Figure 5.21 Algorithm for solution of deactivation model.

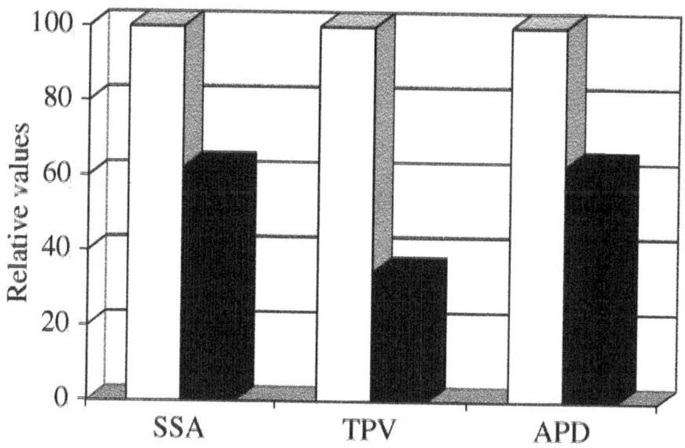

Figure 5.22 Changes in textural properties of fresh (□) catalyst and spent (■) catalyst (200 h TOS).

because of micropores plugging. The percentage of macropores also diminishes due to plugging of some pore mouths.

Results of kinetic modeling have shown that second-order kinetics is adequate to represent our experimental data. This observation agrees with experimental evidence reported by others (Marafi et al., 2010; Rodríguez et al., 2012) and theoretical predictions for removing metal compounds from heavy petroleum fractions.

Figure 5.23 shows initial boiling point distillation curves from experimental data and from simulation obtained by linear and nonlinear parameter estimations from Riazi's (1997) correlation. The experimental points agree quite well with nonlinear regression fitting, whereas linear parameter estimation is not a good choice because it deviates considerably.

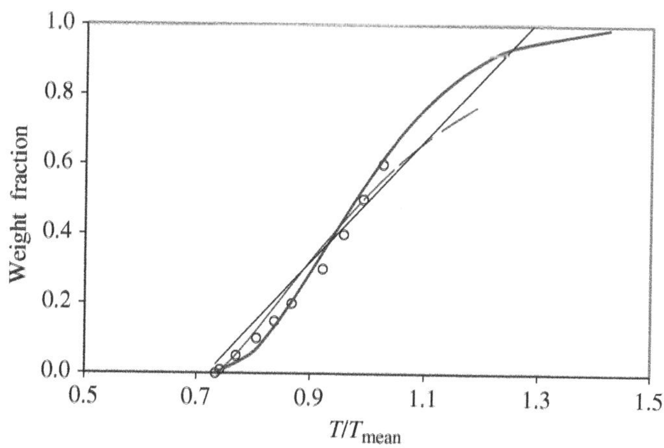

Figure 5.23 Distribution of normal boiling points of feed. (O) Experimental, (.....) linear regression, (—) nonlinear regression.

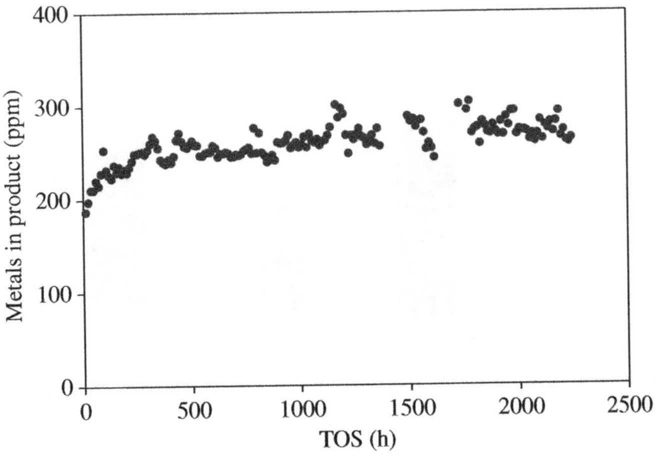

Figure 5.24 Metals concentration in product as a function of TOS.

Figure 5.24 reports metal content in the product as a function of TOS. Initial rapid increase in metal concentration in the product is observed due to the loss of activity attributed mainly to deposition of carbon on the catalyst surface. After a short interval of TOS (200 h), a low rate of deactivation results in small changes in metal removal of the feed. Figure 5.25 depicts the dimensionless concentration (χ) as a function of dimensionless time (τ) and intraparticle position (ζ). It is clearly appreciated that reactant concentration diminishes as species moves from the surface to the center of the particle due to the removal of these metal compounds. Also, at the beginning of HDM reaction ($\tau = 0$), the maximum concentration within the pore at each position ($0 < \zeta < 1$) is reached. As time-on-stream increases, concentration of species within the pore ($\zeta \rightarrow 1$) is reduced because their access is each time more restricted. It is predicted that at higher values of TOS ($\tau = 3500$), utilization of the internal surface of catalyst drastically diminishes and only the external surface is available for HDM reaction. This, of course, provokes low efficiency of catalyst and its replacement is unavoidable.

The thickness distribution (δ) at the pore mouth as a function of dimensionless time is shown in Figure 5.26. Initially, a linear trend is observed, that is, the deposits in catalyst pores increase proportionally as time increases. Evidence of that fact has been observed experimentally by many researchers. At longer TOS, an accelerated increase in thickness is found and exponential change in metal deposits is predicted.

The profiles of the ratio of instantaneous pores to initial pores are represented in Figure 5.27. Near the catalyst surface, steeper gradients of thickness are predicted as expected, whereas at the center of particle, no appreciable changes are observed. Thus, a significant portion of pores are unutilized because the pore mouth size reduction restrict the diffusion of species.

Effectiveness factor as a function of TOS is a good mean of following the catalyst utilization during HDM reactions (Figure 5.27). As can be expected, this factor

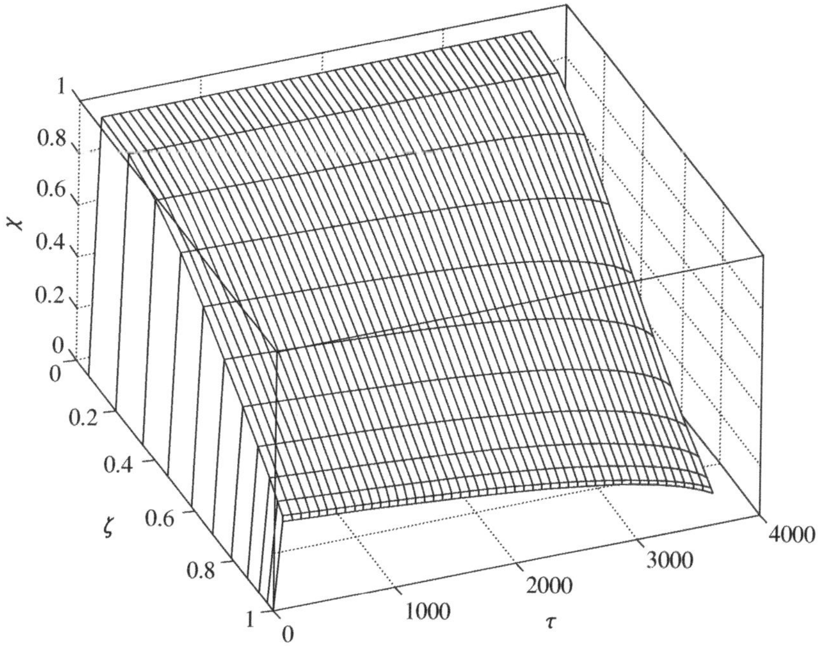

Figure 5.25 Dependence of dimensionless concentration (χ) on dimensionless time (τ) and intraparticle position (ζ).

decreases monotonically. For most of the TOS, the decrease in effectiveness factor is linear but at longer times it reduces in accelerated way indicating that the life of catalyst is close to EOR conditions.

By using the profile depicted in Figure 5.28 and Equation 5.29, HDM conversions as a function of TOS were calculated and compared with the experimental values. Final results are plotted in Figure 5.29. A reasonable agreement between the predicted and experimental information is observed.

Many models available in the literature are empirical and depend on data available at short times-on-stream and thus extrapolation is not recommendable. The model used here allows for predicting the catalyst activity and catalyst life during considerable period of TOS. Taking as criteria the value of effectiveness factor (Figure 5.28) or the level of conversion (Figure 5.29) as limit to shut down the unit operation, it is possible to predict the catalyst life. No particular calculations were done in this work to predict the life of our catalyst, but by using these figures it is easily done. For instance, for $\tau = 1800$, experimental point and simulation agree vey well as can be seen in Figure 5.29 (last point). On the other hand, at $\tau = 3000$ a mean pore diameter of 10 nm is predicted by the model, which is comparable with the experimental measure of 11 nm.

Further applications of the model by taking into consideration variations in metal concentration at different reactor axial positions are necessary for accurate prediction

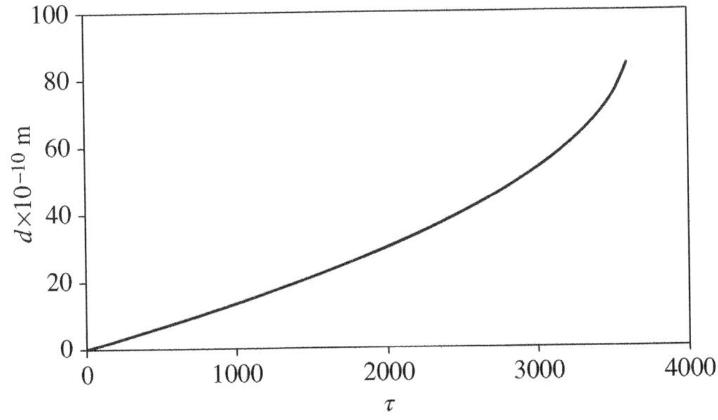

Figure 5.26 Thickness of metal sulfides at pore mouth of catalyst.

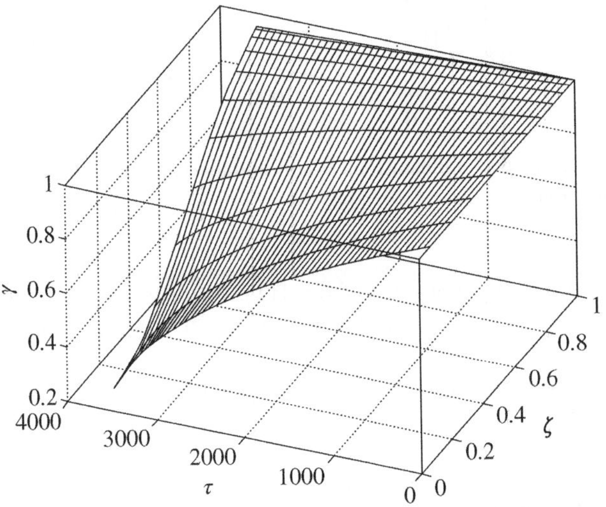

Figure 5.27 Profiles of instantaneous to initial pore radius (γ) ratio as a function of intraparticle position and dimensionless time.

of the catalyst performance. Also, changes of catalyst activity due to variations in catalytic bed temperature are needed to be modeled. However, due to the nature of the feed, unavailability of correlations for external transport properties, absence of heat of HDM reaction, changes in boiling point molecules due to hydrocracking, and so on, the development of a more refined approach requires much more experimental characterization and modeling work. In addition, coupling hydrocracking and HDM reactions as well as other reactions that contribute to catalyst deactivation such as hydrodeasphaltenization could result in a better approach because they provoke loss

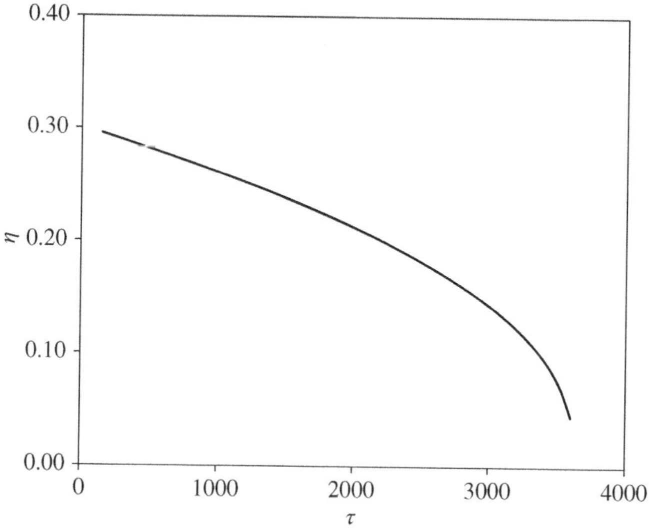

Figure 5.28 Effectiveness factor (η) as a function of dimensionless time (τ).

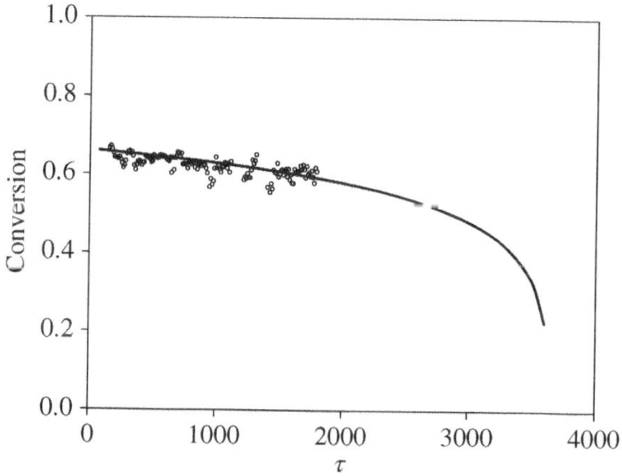

Figure 5.29 Comparison between experimental and predicted HDM conversion as a function of dimensionless time.

of catalytic activity due to coke and metal deposition during hydroprocessing of heavy oils and residua.

5.4.4 Conclusions

Modeling deactivation due to pore plugging caused by metal deposition during HDT of heavy oil was carried out. Kinetics of metal removal was found to be of second order. Experimental data have shown that as TOS increases the utilization of catalyst diminishes due to loss of activity, which can be attributed to pores filling by metal sulfide compounds deposited on it. Although several assumptions have been done to simplify the model, reasonable agreement was found between experimental and simulated result of changes in conversion profiles as function of time-on-stream.

5.5 KINETIC MODEL FOR HYDROCRACKING OF HEAVY OIL IN A CSTR INVOLVING SHORT-TERM CATALYST DEACTIVATION

5.5.1 Experimental Procedure

5.5.1.1 Experimental Setup All experiments were performed in a bench-scale continuous stirred tank basket reactor (CSTBR), which is shown in Figure 5.30. The heavy oil feedstock is preheated and mixed with hydrogen and then passed through the reactor. The reaction temperature is measured inside the reactor with three different thermocouples positioned at the bottom, the middle, and the top part of the basket. Liquid/gas separation is carried out at high and low pressures. Liquid products are collected at the bottom of the separation system, while gases are recovered at the upper part and sent to a chromatograph for analysis. The bench-scale unit has a control system based on distributed PID controllers and remote computerized supervision.

5.5.1.2 Catalyst Loading and Activation The main properties of the catalyst used are given in Table 5.7. A 100 mL of catalyst was loaded into the catalytic basket and activated *in situ* by sulfiding with gas oil containing 1.6 wt% DMDS at the following conditions: 2 $mL_{feed}/(mL_{cat}$ h), 28 kg/cm^2, 2000 std ft^3H_2/bbl of oil, at temperatures of 260 and 320 °C for 3 and 12 h, respectively.

5.5.1.3 Experiments and Product Analysis An atmospheric residue (312 °C+) recovered from an EHCO (13° API) was used as feedstock. Its main properties are presented in Table 5.8. All the experiments were carried out at constant pressure and H_2-to-oil ratio of 100 kg/cm^2 and 5000 std ft^3/bbl, respectively. At 380 and 400 °C,

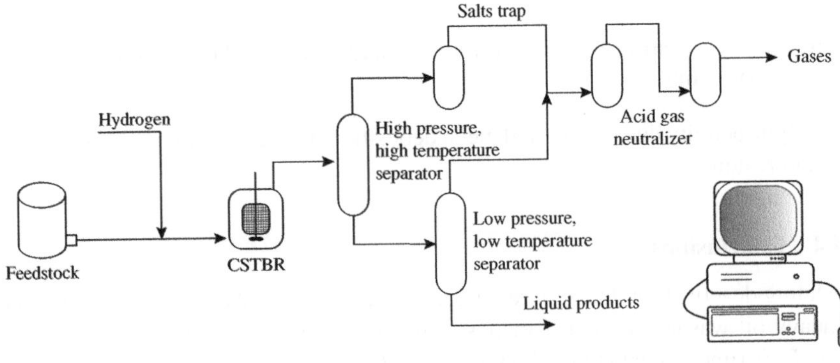

Figure 5.30 Continuous stirred tank basket reactor pilot plant.

TABLE 5.7 Properties of the Catalyst

Nominal size (in.)	1/18
NiO/Ni (dry basis) (wt%)	0.73/0.58
MoO_3/Mo (dry basis) (wt%)	3.27/2.18
Physical properties	
ABD (g/mL)	0.45
CBD (g/mL)	0.55
SSA (m^2/g)	197.2
TPV (mL/g)	0.85
APD (Å)	172.6
Pore size distribution (vol%)	
<50 Å	2.32
50–100 Å	14.86
100–200 Å	49.44
200–500 Å	30.34
500–1000 Å	2.26
>1000 Å	0.78

ABD, average bulk density; CBD, compacted bulk density; SSA, specific surface area; TPV, total pore volume; APD, average pore diameter.

LHSV was studied at 0.5, 0.75, 1.00, and 1.25 h^{-1}, while at 410 °C, LHSV was varied at 0.75 and 1.25 h^{-1}, and at 420 °C, at 0.5 and 1.0 h^{-1}. The products and feedstock were analyzed by simulated distillation following the ASTM D7169 method.

5.5.2 Results and Discussion

5.5.2.1 Mass Transfer Limitations To ensure that the experimental data were collected in the true kinetic regime, the absence of external and internal mass-transfer gradients were investigated using a commercial size catalyst at the following operating conditions: 380 °C, 100 kg/cm^2, 5000 std ft^3 H_2/bbl of oil, and 0.5 $mL_{feed}/(mL_{cat}$ h), increasing the stirring rate from 750 to 1250 rpm. The sulfur conversion obtained every 10 h was followed up to establish the optimal stirring rate. The results are presented in Figure 5.31. The common initial deactivation period is first observed, after which sulfur conversion stabilizes. A stability period of 40 h was considered after changing the stirring rate. At 750 rpm, the conversion was 44.5% on average. When the stirring rate was increased to 1250 rpm, the sulfur conversion also increased to about 45.9%. After that, the stirring rate was set to 1000 rpm and a conversion of 45.8% similar to that obtained at 1250 rpm was achieved. Based on this fact, it was established that external mass-transfer limitations are significantly diminished at 1000 rpm.

To minimize intraparticle mass-transfer gradients, experimental runs with catalyst crushed at different mesh sizes were conducted. It was first necessary to modify the original basket design to be able to keep smaller catalyst particles inside it. Nevertheless, it was not possible to perform proper experiments since the fine particles

TABLE 5.8 Properties of the Feedstock

Specific gravity, 60/60 °F	1.028
API gravity	6.15
Elemental analysis (wt%)	
Carbon	82.51
Hydrogen	10.05
Sulfur	5.80
Nitrogen	1.09
CCR (wt%)	20.83
Metals (ppm)	
Ni	106
V	558
Kinematic viscosity (130 °C) (cSt)	330.12
Simulated distillation ASTM D7169 (°C)	
IBP	206.0
10	344.7
30	462.3
50	568.3
70	644.7
90	700.9
FBP	744.6

produced during the grinding of the catalyst passed through the smaller mesh causing important loss of the catalyst mass. Therefore, the commercial size catalyst was chosen.

5.5.2.2 *Kinetic Modeling* The kinetic model (see Figure 5.32) consists of the following five lumps: gases, naphtha (IBP-204 °C), distillates (204–343 °C), vacuum gas oil (343–538 °C), and unconverted residue (538 °C+). As mentioned before, experiments were performed in a disguised kinetic regime due to the presence of internal diffusion limitations; this made the introduction of the effectiveness factor (η_{EF}) necessary since internal gradients were indeed present. η_{EF} was calculated using the following expression:

$$\eta_{EF} = \frac{I_1(2\phi)}{\phi I_0(2\phi)} \tag{5.32}$$

The generalized Thiele modulus was obtained as follows:

$$\phi = \frac{V_p}{S_p}\left[\left(\frac{n+1}{2}\right)\frac{k_{i,s}^{n-1}}{D_e}\right]^{0.5} \tag{5.33}$$

Some catalyst and feedstock properties were calculated using correlations reported in the literature (Froment et al., 2010; Chen et al., 2005; Ancheyta et al., 2005a, 2005b; Korsten and Hoffmann, 1996). The effectiveness factors obtained for each

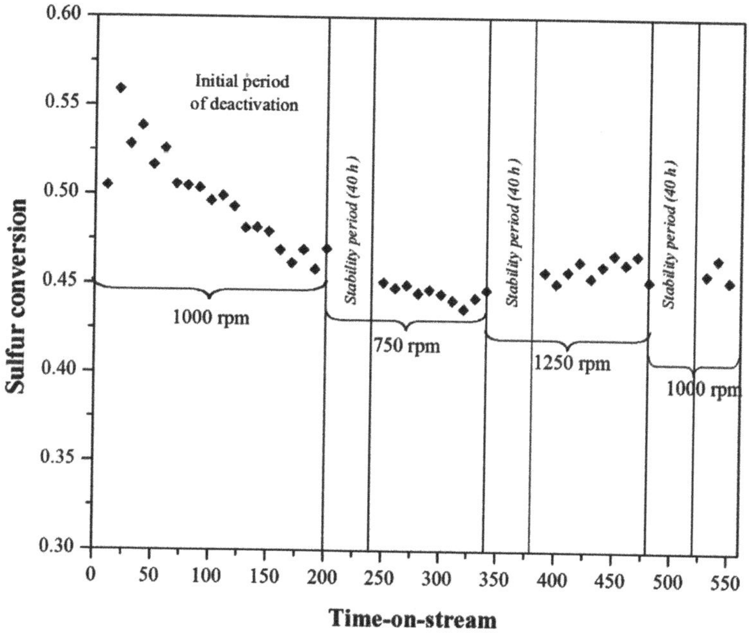

Figure 5.31 Variation of sulfur conversion with stirring rate.

TABLE 5.9 Effectiveness and Contact Efficiency Factors

LHSV (h^{-1})	$\eta_{EF\,380°C}$	$\eta_{EF\,400°C}$	$\eta_{EF\,410°C}$	$\eta_{EF\,420°C}$	η_{CE}
0.50	0.9778	0.9702	–	0.9586	0.7349
0.75	0.9749	0.9682	0.9590	–	0.9224
1.00	0.9708	0.9644	–	0.9405	0.9814
1.25	0.9666	0.9541	0.9508	–	0.9977

operating condition are given in Table 5.9. It can be observed that the effectiveness factor decreases with the increasing reaction temperature and LHSV. When the temperature is increased, a more strong increase in the rate constant is observed than in diffusivity; this leads to a more pronounced diffusion limitation since it becomes the limiting step and therefore to smaller effectiveness factors. When LHSV is increased at constant temperature, the decrease in the effectiveness factor can be attributed to the increased reaction mixture viscosity that is obtained at higher LHSV; since viscosity is directly related to reactants diffusivity, an increase in viscosity leads to a decrease in effectiveness factor.

To take into account the deactivation rate, the following function was used:

$$\varphi = \frac{1}{(1 + k_d t)^m} \tag{5.34}$$

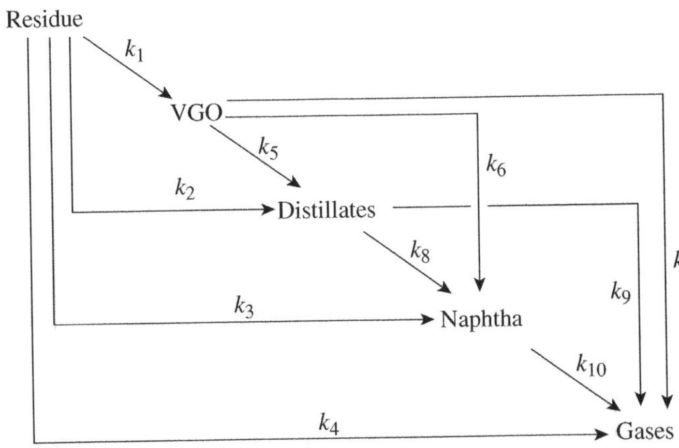

Figure 5.32 Five-lump kinetic model.

Simple stirred tank reactors are often used for kinetic studies due to efficient heat and mass transfers, which is true only for experiments using crushed or powdered catalysts (Pitault et al., 2005). The use of large pellets could drastically reduce liquid–solid transfer in this reactor and the presence of basket wall would increase this phenomenon by increasing the pressure drop, therefore decreasing the fluid velocities (Magnico and Fongarland, 2006). As LHSV increases, different fluid patterns will be obtained inside the reactor, affecting somehow the contact between the catalyst and reactants and hence the reaction rates. To take into consideration these phenomena, a contact efficiency factor (η_{CE}) was introduced, which becomes less significant as the LHSV is increased, that is, its value approximates to unity, and was considered to be independent of temperature (Table 5.9).

For each reaction, a kinetic expression (r_i) was formulated as a function of the product composition (y_i), kinetic constant, effectiveness factor, and contact efficiency factor. Product compositions were determined from bench-scale mass balances and simulated distillation curves. The hydrocracking of residue was considered to follow second reaction order, while the remaining reactions were assumed to be first order as reported in the literature (Sánchez and Ancheyta, 2005). On the basis of these considerations, the reaction rates of the proposed model are as follows:

Residue

$$r_R = -\varphi \eta_{EF} \eta_{CE} k_0 y_R^2 \tag{5.35}$$

where $k_0 = k_1 + k_2 + k_3 + k_4$.

VGO

$$r_{VGO} = \varphi \eta_{EF} \eta_{CE} [k_1 y_R^2 - (k_5 + k_6 + k_7) y_{VGO}] \tag{5.36}$$

Distillates

$$r_D = \varphi \eta_{EF} \eta_{CE} [k_2 y_R^2 + k_5 y_{VGO} - (k_8 + k_9) y_D] \tag{5.37}$$

Naphtha

$$r_N = \varphi\eta_{EF}\eta_{CE}[k_3 y_R^2 + k_6 y_{VGO} + k_8 y_D - k_{10} y_N] \tag{5.38}$$

Gases

$$r_G = \varphi\eta_{EF}\eta_{CE}[k_4 y_R^2 + k_7 y_{VGO} + k_9 y_D + k_{10} y_N] \tag{5.39}$$

The kinetic model was incorporated into the CSTBR model. The following mass balance was used to evaluate the product composition from a set of kinetic constants for each temperature:

$$\frac{W_{Cat}}{m_{T_0}} = \frac{y_{i,0} - y_i}{(-r_i)} \tag{5.40}$$

A sequential method for determining the parameter values was employed, which is reported elsewhere (Ancheyta and Sotelo, 2000). The original kinetic model was divided into two-, three-, and four-lump kinetic models to simplify parameters estimation, as presented in Figure 5.33. From this approach, it is possible to obtain some kinetic parameters, which become constant when solving the five-lump kinetic model, thus reducing the number of parameters to be estimated simultaneously.

The minimization of the objective function, based on the sum of square errors between the experimental and calculated product compositions, was applied to find the best set of kinetic parameters. This objective function was solved using the least-squares criterion with a nonlinear regression procedure based on Marquardt's algorithm.

The values of the kinetic parameters, deactivation constants, and activation and deactivation energies are summarized in Table 5.10. The Arrhenius plots for all the kinetic constants are presented in Figure 5.34. The first part of Table 5.10 shows the estimated parameters of the global hydrocracking of residue (Residue → Hydrocracked products). Reaction order was confirmed to be two, which agrees with literature reports (Sánchez and Ancheyta, 2005). Catalyst deactivation order was found to be close to 0.19. This value was kept constant for parameter estimation of the other lumps. Curiously, activation and deactivation energies were same (50 kcal/mol), this, in principle, means that there is not a predominant effect of the deactivation process over the global hydrocracking reaction.

The activation energies for the individual hydrocracking pathways of residue were in the range of 41–57 kcal/mol. The average value of the four activation energies is quite similar to the global activation energy (50 kcal/mol). This implies that the reaction kinetics when grouping various hydrocarbon components in a single lump (Prod = VGO + D + N + G) behaves in similar way to that of the individual lump reaction pathways.

For hydrocracking of both residue and VGO, the lower activation energies dictate that distillates are preferentially obtained over naphtha and gases. Hydrocracking of distillates toward naphtha and gases happens more quickly than the other reactions, while the reaction of naphtha toward gases is the least favored.

It can be observed that residue hydrocracking has a higher selectivity toward VGO production than the other lumps in the following order:

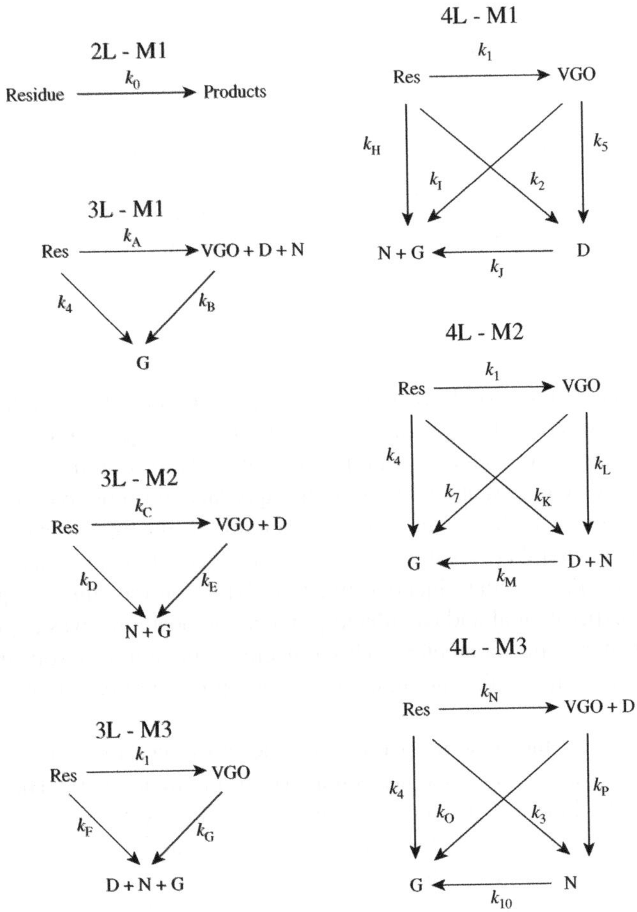

Figure 5.33 Two-, three-, and four-lump kinetic models.

VGO > distillates > naphtha > gases. For the hydrocracking of VGO, distillates are more preferentially obtained than naphtha and gases: distillates > naphtha > gases, the distillates formation is approximately 8 and 11 times faster than the formation of naphtha and gases, respectively. Hydrocracking of distillates was found to be more selective toward naphtha than gases, being the naphtha formation approximately three times faster than that of gases. Higher amount of gas production comes from the residue and has a lower contribution from VGO, distillates, and naphtha. It can be noticed that as temperature is increased, the reaction rates of residue to naphtha and gases and VGO to naphtha and gases become very competitive, that is, their reaction rate coefficients approach, and that naphtha conversion toward gases is the more favored reaction when temperature is above 400 °C.

TABLE 5.10 Kinetic Parameters

Parameters	Reaction Path	Temperature (°C)				E_a (kcal/mol)
		380	400	410	420	
k_0	Res → Prod	3.059	9.174	16.928	27.841	49.99
n_R		2.000	2.000	2.000	2.000	
k_d		0.071	0.276	0.418	0.643	50.00
m		0.188	0.188	0.188	0.188	
k_1	Res → VGO	1.099	3.902	7.459	13.038	55.89
k_2	Res → D	0.981	2.427	4.330	6.078	41.84
k_3	Res → N	0.651	1.613	2.723	4.673	44.02
k_4	Res → G	0.327	1.233	2.416	4.052	57.22
k_5	VGO → D	5.024E−03	1.506E−02	2.124E−02	3.090E−02	40.96
k_6	VGO → N	5.410E−04	1.844E−03	2.816E−03	4.422E−03	47.32
k_7	VGO → G	3.370E−04	1.173E−03	2.247E−03	3.924E−03	55.45
k_8	D → N	6.325E−02	1.329E−01	1.538E−01	1.774E−01	23.53
k_9	D → G	1.672E−02	3.363E−02	4.088E−02	6.137E−02	28.42
k_{10}	N → G	8.894E−03	3.915E−02	1.058E−01	1.472E−01	65.70

A comparison between experimental product composition and those determined by solving Equations 5.34–5.40 with k_i values given in Table 5.10 is shown in Figure 5.35. It can be observed that the product composition is predicted quite well for the different lumps. A random distribution around the horizontal axis with a quite balanced number of positive and negative residual values is presented in Figure 5.36. Intercept, slope, number of positive and negative residuals, and absolute error are summarized in Table 5.11. It can be noted that slope and intercept are close to unity and zero, respectively. The higher absolute percentage error was 5.26% and corresponds to the prediction of naphtha composition.

5.5.3 Conclusions

Kinetic parameters of a five-lump kinetic model for hydrocracking of a heavy oil taking into account catalyst deactivation were obtained in a CSTBR in the range of reaction temperature of 380–420 °C and LHSV of 0.5–1.25 $mL_{feed}/(mL_{cat}\ h)$. Hydrocracking of residue, VGO, and distillates indicates a high selectivity toward the heavier lumps at the studied temperatures. At higher temperatures, the reaction rates of residue to naphtha and gases and VGO to naphtha and gases become very competitive.

A notorious increase in the reaction rate of naphtha toward gases is observed at temperatures above 400 °C. Activation energies show that the reaction pathway of naphtha toward gases is the least favored.

The composition of the selected lumps can be predicted with absolute average errors less than 5.0 %.

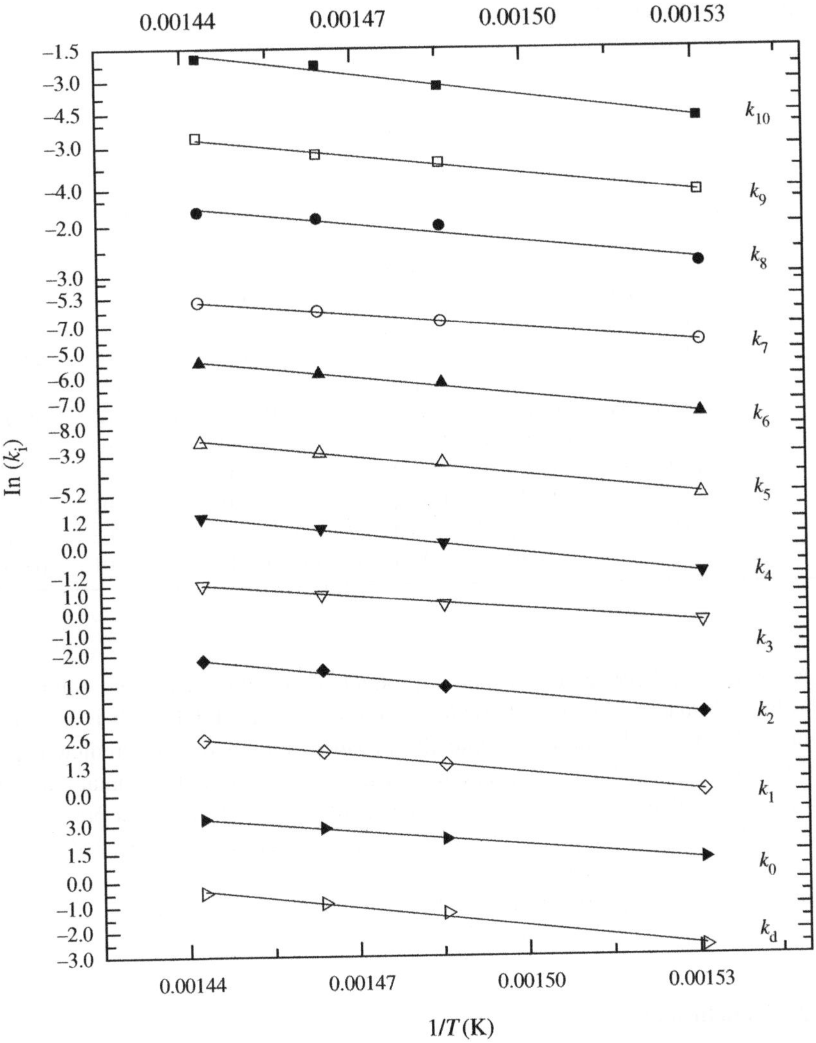

Figure 5.34 Arrhenius plot for the different kinetic parameters.

5.6 MODELING THE KINETICS OF PARALLEL THERMAL AND CATALYTIC HYDROTREATING OF HEAVY OIL

5.6.1 The Model

5.6.1.1 Model Assumptions The main HDT reactions considered in this work are HDS, HDV, HDNi, HDAs, hydrodenitrogenation, and removal of Conradson carbon residue (HDCCR). Since a commercial size catalyst was used for performing all the experiments, the effectiveness factor was included in the kinetic expression

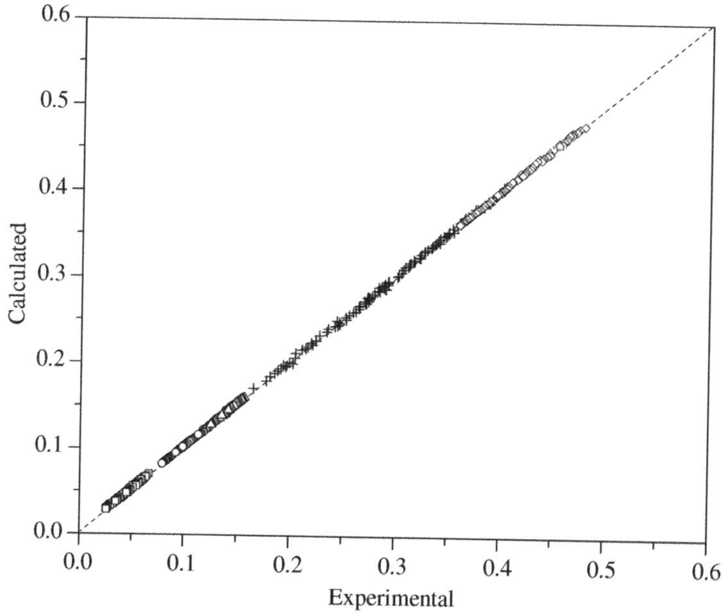

Figure 5.35 Comparison between experimental and calculated product compositions: (+) residue, (◇) VGO, (Δ) distillates, (□) naphtha, (O) gases.

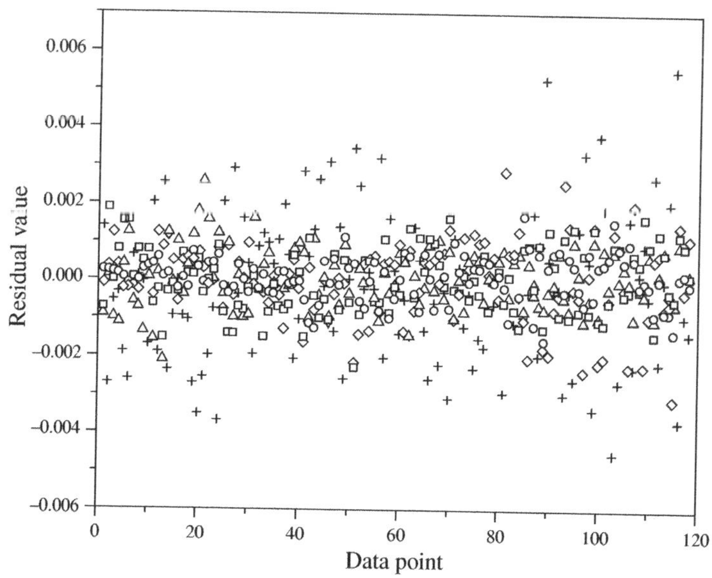

Figure 5.36 Residual values obtained: (+) residue, (◇) VGO, (Δ) distillates, (□) naphtha, (O) gases.

TABLE 5.11 Statistical Analysis

Lump	Slope	Intercept	Residual (+)	Residual (−)	Maximum Absolute Percentage Error
Residue	1.0019	0.0044	53	65	2.63
VGO	1.0034	−0.0014	61	57	0.74
Distillates	1.0052	−0.0008	59	59	1.84
Naphtha	0.9891	0.0005	57	61	5.26
Gases	0.9968	0.0003	68	50	1.38

for evaluating the intraparticle resistances to mass transfer at each temperature and space velocity. In each reaction pathway, parallel thermal and catalytic reactions were included. The general HDT reactions were expressed as shown in Figure 5.37.

The following assumptions were made for deriving the kinetic expressions:

- Parallel thermal and catalytic reactions occur during the HDT process.
- Thermal HDT reactions are not affected by catalyst deactivation.
- Mass-transfer limitation through the catalyst pores affects only the catalytic reaction pathway.
- Thermal, catalytic, and deactivation rate constants are well described by the Arrhenius law.

5.6.1.2 *Model Formulation* Based on the above-listed assumptions, the following kinetic expressions were formulated for each HDT reaction:

HDS

$$(-r_{\mathrm{HDS}}) = \O \eta_{\mathrm{S}} k_{C_{\mathrm{S}}} \frac{C_{\mathrm{S}}^{n_{C_{\mathrm{S}}}} C_{\mathrm{H_2}}^{n_{C_{\mathrm{H_2}}}}}{(1 + K_{\mathrm{H_2S}} C_{\mathrm{H_2S}})^2} + k_{T_{\mathrm{S}}} C_{\mathrm{S}}^{n_{T_{\mathrm{S}}}} \tag{5.41}$$

The influence of temperature on the adsorption-equilibrium constant for the hydrogen sulfide was calculated according to the van't Hoff equation (Korsten and Hoffmann, 1996):

$$K_{\mathrm{H_2S}} = A_{\mathrm{H_2S}} \exp\left(\frac{\Delta H_{\mathrm{ads}}}{RT}\right) \tag{5.42}$$

HDNi

$$(-r_{\mathrm{HDNi}}) = \O \eta_{\mathrm{Ni}} k_{C_{\mathrm{Ni}}} C_{\mathrm{Ni}}^{n_{C_{\mathrm{Ni}}}} + k_{T_{\mathrm{Ni}}} C_{\mathrm{Ni}}^{n_{T_{\mathrm{Ni}}}} \tag{5.43}$$

HDV

$$(-r_{\mathrm{HDV}}) = \O \eta_{\mathrm{V}} k_{C_{\mathrm{V}}} C_{\mathrm{V}}^{n_{C_{\mathrm{V}}}} + k_{T_{\mathrm{V}}} C_{\mathrm{V}}^{n_{T_{\mathrm{V}}}} \tag{5.44}$$

HDAsph

$$(-r_{\mathrm{HDAsph}}) = \O \eta_{\mathrm{Asph}} k_{C_{\mathrm{Asph}}} C_{\mathrm{Asph}}^{n_{C_{\mathrm{Asph}}}} + k_{T_{\mathrm{Asph}}} C_{\mathrm{Asph}}^{n_{T_{\mathrm{Asph}}}} \tag{5.45}$$

HDS:

$$R\text{–}S + H_2 \xrightarrow[\text{Catalytic}]{\text{Thermal}} RH + H_2S$$

HDM:

$$R\text{–}M + H_2 \xrightarrow[\text{Catalytic}]{\text{Thermal}} RH + M$$

HDAsph:

$$R\text{–}Asph + H_2 \xrightarrow[\text{Catalytic}]{\text{Thermal}} RH + Asph\text{–}R \text{ (smaller hydrocarbons)}$$

HDCCR:

$$R\text{–}CCR + H_2 \xrightarrow[\text{Catalytic}]{\text{Thermal}} RH + CCR\text{–}R \text{ (smaller hydrocarbons)}$$

HDN:

$$\text{Non-basic nitrogen} \xrightarrow[\text{Catalytic}]{\text{Thermal}} \text{Basic nitrogen} \xrightarrow[\text{Catalytic}]{\text{Thermal}} RH + NH_3$$

Asph = Asphaltenes
CCR = Conradson carbon residue
M = Metals
R = Hydrocarbon molecule

Figure 5.37 Hydrotreating reaction pathways.

HDCCR

$$(-r_{\text{HDCCR}}) = \O\eta_{\text{CCR}} k_{C_{\text{CCR}}} C_{\text{CCR}}^{n_{C_{\text{CCR}}}} + k_{T_{\text{CCR}}} C_{\text{CCR}}^{n_{T_{\text{CCR}}}} \tag{5.46}$$

HDNNBN

$$(-r_{\text{HDN}_{\text{NBN}}}) = \O\eta_{\text{NBN}} k_{C_{\text{NBN}}} C_{\text{NBN}}^{n_{C_{\text{NBN}}}} + k_{T_{\text{NBN}}} C_{\text{NBN}}^{n_{T_{\text{NBN}}}} \tag{5.47}$$

HDNBN

$$(-r_{\text{HDN}_{\text{BN}}}) = \O\eta_{\text{BN}} k_{C_{\text{BN}}} C_{\text{BN}}^{n_{C_{\text{BN}}}} + k_{T_{\text{BN}}} C_{\text{BN}}^{n_{T_{\text{BN}}}} - \O\eta_{\text{NBN}} k_{C_{\text{NBN}}} C_{\text{NBN}}^{n_{C_{\text{NBN}}}} - k_{T_{\text{NBN}}} C_{\text{NBN}}^{n_{T_{\text{NBN}}}} \tag{5.48}$$

The effect of temperature on kinetic constants was represented by the Arrhenius law:

Catalytic kinetic constant

$$k_{c_i} = A_{c_i} e^{\left(-\frac{E_{a_{C_i}}}{RT}\right)} \tag{5.49}$$

Thermal kinetic constant

$$k_{T_i} = A_{T_i} e^{\left(-\frac{E_{a_{T_i}}}{RT}\right)} \tag{5.50}$$

Deactivation function

$$\text{Ø}_i = \frac{1}{(1 + k_{d_i} t)^{m_i}} \tag{5.51}$$

$$k_{d_i} = A_{d_i} e^{\left(-\frac{E_{d_i}}{RT}\right)} \tag{5.52}$$

5.6.1.3 CSTR Model The following mass balance was used to evaluate the product composition from a set of kinetic constants for each temperature:

$$\frac{1}{\text{WHSV}} = \frac{C_{i,0} - C_i}{(-r_i)} \tag{5.53}$$

$$\text{WHSV} = \frac{m_{T_0}}{W_{\text{Cat}}} \tag{5.54}$$

5.6.1.4 Model Solution Equations 5.41–5.52 were coupled to the reactor model (Eqs 5.53 and 5.54) and the following function that relates the operating variables to the concentration of the different species was obtained:

$$\frac{1}{\left(1 + \left[A_{d_i} e^{\left(-\frac{E_{d_i}}{RT_i}\right)}\right] t\right)^{m_i}} \eta_i \left[A_{C_i} e^{\left(-\frac{E_{a_{C_i}}}{RT_i}\right)}\right] C_i^{n_{C_i}} + \left[A_{T_i} e^{\left(-\frac{E_{a_{T_i}}}{RT_i}\right)}\right] C_i^{n_{T_i}} - \frac{m_{T_0}}{W_{\text{Cat}}} (C_{i,0} - C_i) = 0 \tag{5.55}$$

Equation 5.55 was solved by using the Newton–Raphson method to obtain the calculated concentration (C_i^{Exp}) value. The minimization of the following objective function, based on the sum of square errors between the experimental and calculated concentrations of the species, was applied to find the best set of kinetic parameters along with the effectiveness factors:

$$\text{SSE} = (C_i^{\text{Exp}} - C_i^{\text{Calc}})^2 \tag{5.56}$$

This objective function was solved using the least-squares criterion with a nonlinear regression procedure based on Marquardt's algorithm.

5.6.2 Experimental Procedure

5.6.2.1 Experimental Setup A scheme of the bench-scale CSTBR that was used in all of the HDT experiments is shown in Figure 5.30. The operation of this reactor under minimal external mass-transfer regime has been verified.

5.6.2.2 Experiments and Product Analysis A 312 °C+ residue recovered from an EHCO (13°API) was employed as feedstock. The main properties of the catalyst and feedstock used are presented in Tables 5.7 and 5.8, respectively. All the experiments were carried out at constant pressure and H_2/oil ratio of 9.8 MPa and 891 std m^3/m^3, respectively. At 380 and 400 °C, WHSV was studied at 0.98, 1.49, 2.0, and 2.56 h^{-1}, while at 410 °C, WHSV was varied at 1.49 and 2.56 h^{-1}, and at 420 °C, at 0.98 and 2.56 h^{-1}. In all experiments, 54 g of catalyst was loaded to the reactor. For each condition, the time allowed during the run was 200 h.

Energy-dispersive X-ray fluorescence spectrometry technique was used for sulfur content measurements (ASTM D4294) with a model SLFA-2100 HORIBA spectrometer. Coke-forming tendency of the feedstock was measured as Conradson Carbon (ASTM D189), which determines the amount of carbon residue left after evaporation and pyrolysis of oil. The amount of asphaltenes was measured as the difference between *n*-pentane and toluene insolubles according to the ASTM D2007 and ASTM D4055 methods. Atomic absorption was used for determining Ni and V contents with a model AA Series Solar spectrometer. The total nitrogen content was determined following the ASTM D4629 procedure using the technique of combustion/chemiluminescence. Basic nitrogen content was determined by the color indicator titration method (UOP 313-70). The nonbasic nitrogen content was determined from the difference between the total and basic nitrogen contents.

5.6.3 Results and Discussion

The discussion of results is organized in several sections. Because the tests were conducted with commercial size of catalyst, the results of effectiveness factors calculated for each reaction are analyzed. Second, the deactivation phenomenon is discussed, which has been assumed to affect the catalytic reaction but not the thermal reaction. Finally, the kinetic parameters for all reactions, either thermally or catalytically influenced, are explained and comparisons of model predictions with experimental data are provided.

5.6.3.1 Effectiveness Factor The effectiveness factors obtained for all reactions at each operating condition are given in Table 5.12. The value of η decreases for increasing reaction temperature and WHSV. This effect becomes more notorious at the highest temperature and WHSV (420 °C and 2.56 h^{-1}, respectively). Some authors have also reported the same trend of the effectiveness factor with temperature (Chang et al., 1988; Mahinsa et al., 2012). This behavior is due to the fact that when temperature is increased, a stronger increase in rate constant is achieved compared with diffusivity, thus leading to a more pronounced diffusion limitation, which becomes the limiting step, and therefore leading to smaller effectiveness factors. On the other hand, the reduction of effectiveness factor due to the increase of WHSV implies that internal diffusion does not depend of the flow rate (Macías and Ancheyta, 2004), but it can be attributed to the change in the reaction mixture viscosity, since this property is directly related to reactants' diffusivity, higher viscosity causes a decrease in effectiveness factor. In other words, low severity conditions (high

TABLE 5.12 Effectiveness Factors

HDT Reactions	WHSV (h^{-1})	η (380 °C)	η (400 °C)	η (410 °C)	η (420 °C)
HDS	0.98	0.882	0.775	–	0.535
	1.49	0.789	0.739	0.494	–
	2.09	0.734	0.677	–	0.454
	2.56	0.594	0.438	0.404	–
HDNi	0.98	0.962	0.516	–	0.182
	1.49	0.946	0.518	0.187	–
	2.09	0.919	0.515	–	0.154
	2.56	0.643	0.354	0.147	–
HDV	0.98	0.920	0.495	–	0.309
	1.49	0.778	0.484	0.275	–
	2.09	0.709	0.454	–	0.265
	2.56	0.649	0.303	0.224	–
HDAsph	0.98	0.898	0.552	–	0.381
	1.49	0.865	0.544	0.352	–
	2.09	0.757	0.515	–	0.197
	2.56	0.467	0.206	0.177	–
HDCCR	0.98	0.844	0.467	–	0.423
	1.49	0.795	0.443	0.424	–
	2.09	0.773	0.418	–	0.217
	2.56	0.531	0.198	0.175	–
HDNNBN	0.98	0.922	0.880	–	0.777
	1.49	0.746	0.595	0.551	–
	2.09	0.656	0.456	–	0.375
	2.56	0.444	0.331	0.318	–
HDNBN	0.98	0.889	0.853	–	0.788
	1.49	0.832	0.779	0.766	–
	2.09	0.773	0.758	–	0.667
	2.56	0.707	0.671	0.638	–

WHSV) cause high values of viscosity of the reaction mixture, so that effectiveness factor diminishes.

At the highest temperatures tested (410 and 420 °C), the HDNi, HDV, and HDAsph for all the WHSV and the HDCCR and HDNNBN at 2.09 and 2.56 h^{-1} were more diffusion-limited reactions with an effectiveness factor in the range of 0.14–0.4. The value of the effectiveness factor obtained at 400 °C for the HDNi, HDV, HDAsph, and HDCCR were half those values obtained at 380 °C. This implies that the reaction rate became greater than the rate of diffusion above 400 °C, resulting in a significant decrease in the effectiveness factor value.

Regarding the effect of WHSV, for all the HDT reactions, except for HDNBN, at 380 and 400 °C, the more significant change of the effectiveness factor value is obtained at 2.56 h^{-1}. At this condition, the diffusion rate significantly decreases while the reaction rate remains constant, resulting in a more diffusion-limited operation. In the case of HDNBN, a smooth decrease in the effectiveness factor is observed (from 0.89 to 0.63) in all ranges of operating conditions. This means that the basic nitrogen removal is the least diffusion-limited reaction.

5.6.3.2 ΔH_{ads} *of* H_2S There are no reports on the adsorption energy of hydrogen sulfide on commercial NiMo/γ-Al_2O_3 catalysts used in hydrodesulfurization of heavy oils in the open literature. Only a few works, dealing with model compounds, were found. Nakamura et al. (2008) reported an adsorption energy of 3.3 kcal/mol, while Kabe et al. (2001) reported a value of 16.1 kcal/mol. The discrepancy between these values is because the former used a reactant mixture that contained 0.05% of H_2S-generating compound (dimethyldisulfide) and 0.13% of NH_3-generating compound (tri-*n*-butyl amine), while the latter varied only the partial pressure of H_2S from 0 to 0.088 MPa. When NH_3 and H_2S are present in the reactant mixture, the adsorption of NH_3 is more favored on the NiMo/γ-Al_2O_3 catalyst due to its higher adsorption energy (16.7 kcal/mol), suppressing the adsorption of H_2S.

The adsorption energy of the H_2S that resulted from our experimental data was 10.6 kcal/mol. This result agrees with the combined effect of the adsorption of NH_3 and H_2S on the NiMo catalyst since the feedstock had about six times more H_2S-generating compounds than NH_3-generating compounds.

The reaction order of the hydrogen concentration in Equation 5.41 was estimated to be 0.40, which is near the theoretical value (0.5) that has been reported for HDS reaction and attributed to the dissociation of H at the catalyst sites.

5.6.3.3 Catalyst Deactivation Parameters It is well documented that the coke deposition on the catalyst surface affects not only its activity but also its selectivity (Corella, 2004). Therefore, the catalyst deactivation may affect the conversion of the HDT reactions in different ways. This difference in the variation with time-on-stream of conversion of each product, owed to catalyst deactivation, can be modeled with a selective deactivation kinetic model. If it is assumed that catalysts have a uniform surface and that the blocking of the catalyst pores due to coke formation is the dominant effect of catalyst deactivation on reactivity, then each reaction pathway must be characterized by a different catalyst decay exponent (Bollas et al., 2007). This results in a reactant-oriented catalyst deactivation (ROCD), which is the manner that Equation 5.51 was considered. The optimized values of the catalyst deactivation parameters (deactivation order and deactivation energies) are summarized in Table 5.13.

The highest deactivation energy was obtained for hydrodenitrogenation of the non-basic nitrogen (33.8 kcal/mol), while the deactivation energies for the other reactions were in the range of 20–27 kcal/mol. The deactivation orders vary between 0.19 and 0.85. These different deactivation orders and deactivation energies affect the extent of each reaction during time-on-stream. A decrease in catalyst activity was observed as reaction temperature increased (Figure 5.38a); this trend is expected as temperature is increased, increasing deposition of coke and metals on the catalyst is observed, causing the deactivation proceeds more rapidly. Figure 5.38b shows the ROCD for

TABLE 5.13 Kinetic and Deactivation Parameters

HDT Reaction	m	E_d (kcal/mol)	n_C	E_{a_C} (kcal/mol)	n_T	E_{a_T} (kcal/mol)
HDS	0.228	26.057	1.346	27.007	0.938	61.480
HDNi	0.192	21.452	2.406	35.336	0.650	89.443
HDV	0.287	25.752	1.290	30.227	0.513	71.882
HDAsph	0.211	20.552	1.503	26.507	0.205	58.809
HDCCR	0.607	23.742	1.527	26.314	1.004	78.360
HDNBN	0.305	26.526	2.154	24.037	1.364	48.822
HDNNBN	0.849	33.822	1.792	23.429	1.137	57.530

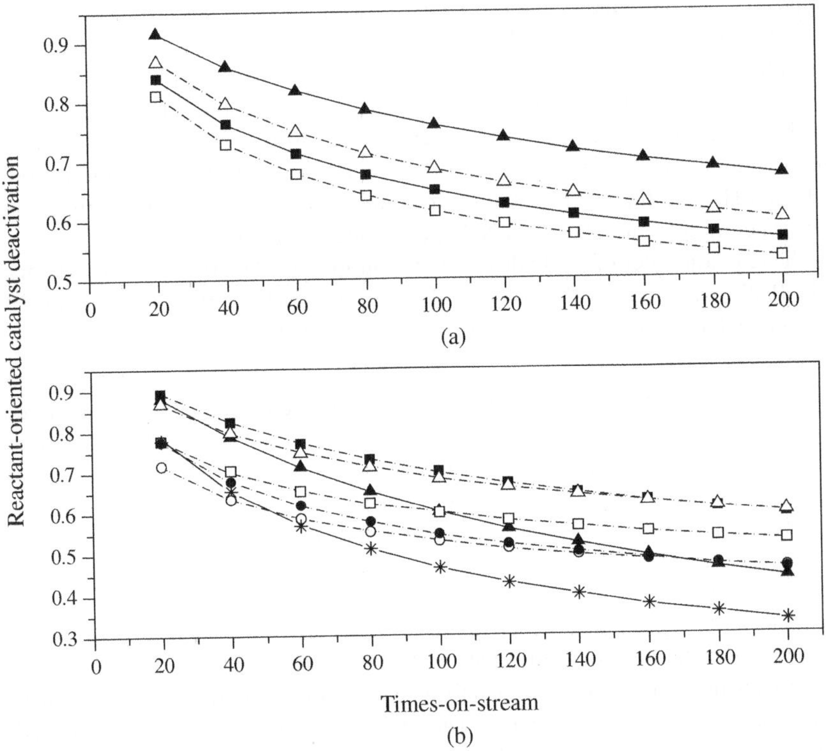

Figure 5.38 Reactant-oriented catalyst deactivation. (a) HDS reaction: (▲) 380 °C, (△) 400 °C, (■) 410 °C, and (□) 420 °C. (b) Hydrotreating reactions at 400 °C: (■) HDNBN, (△) HDS, (▲) HDNNBN, (□) HDNi, (●) HDV, (○) HDAsph, and (*) HDCCR.

all the HDT reactions at 400 °C. It can be noted that HDNNBN and HDCCR reactions are more affected by catalyst deactivation than the other HDT reactions; this can be explained by the high deactivation orders of these reactions: 0.849 and 0.607, respectively. It is important to remark that the HDNNBN reaction shows higher value of ROCD than the HDCCR reaction in spite of having higher deactivation order;

this is because the HDCCR reaction has lower deactivation energy, 23.7 kcal/mol for HDCCR versus 33.8 kcal/mol for HDNNBN. The remaining reactions showed an ROCD in the following order:

$$\text{HDNBN} > \text{HDS} > \text{HDNi} > \text{HDV} > \text{HDAsph}$$

5.6.3.4 Kinetic Parameters for Thermal and Catalytic Reactions It can be observed in Table 5.2 that a wide variety of residues and HCOs (5–21°API) have been hydroprocessed and studied by different research groups. Fractional reaction orders have been reported for the kinetics of residues and asphaltenic fractions, while for the HCOs the values lie in the range of 1–2.5. The given activation energies are in the range of 10–95 kcal/mol for the different HDT reactions. Discrepancies between these values can be attributed to the differences in feed origin, catalyst type, reactor type, and operating conditions.

Asphaltenes present in the residues and heavy oils are the most critical components during the initial stages of hydroprocessing. Their presence is evidenced by the loss of catalyst activity due to fouling. In addition, most of V and Ni are associated to their structure, what makes its disintegration necessary to allow for HDM reactions to proceed. The activation energy values reported in the literature for removal of asphaltenes, nickel, and vanadium vary considerably: 15–42, 12–95, and 18–62 kcal/mol for HDAsph, HDNi, and HDV, respectively. Such differences reflect the different temperature sensitivities of the HDT reactions to the different types of feed and catalyst. The content of CCR is commonly associated to the tendency of the feed to form coke; however, it cannot be expected a direct relationship between the energy required for the CCR (20–67 kcal/mol) and asphaltenes removal since coke formation comes not only from the asphaltenes but also from the polyaromatic structure.

The kinetics of HDN reactions using petroleum residues has been less studied than that of HDS reactions. Nevertheless, the same discrepancies in the values of the reaction orders and activation energies as in the case of the other reactions can be observed, that is, reaction orders vary in the range of 0.5–2 and activation energies between 22 and 43 kcal/mol.

For thermal HDT reactions, the discrepancies in the values for the kinetic parameters become even more evident. The reaction orders ranges from 1 to 3.8, while the activation energy is in the range of 26–60 kcal/mol for different heavy oils and residues (Table 5.3). It is then evident that activation energies for thermal HDT are, as expected, higher than those reported for catalytic HDT. Nonetheless, these parameters are still feed-, catalyst-, and reaction conditions-dependent.

The lowest values of activation energy for the catalytic reactions were obtained for the basic and nonbasic nitrogen removals, which are near the activation energy value of about 26 kcal/mol obtained for the HDAsph reaction. These similar activation energy values suggest that most of the nitrogen compounds may come from the asphaltenic fraction. On the other hand, the activation energies for the HDS, HDNi, and HDV reactions are higher than that of the HDAsph reaction, which implies that conversion of the asphaltenes fraction is necessary to proceed before the removal of

S, Ni, and V contained in the asphaltene structure. The activation energies obtained for the thermal reactions are considerably higher than those of the corresponding catalytic reactions; nevertheless, the similar behavior described earlier for the catalytic reactions was observed.

The reaction orders for the different catalytic reactions (1.5–2.5) were considerably higher than those for the thermal reactions (0.2–1.37); this evidences that HDT reactions occur preferentially via the catalytic reaction pathway. Nonetheless, an important contribution of the thermal reaction is noticed at high temperatures. This effect is illustrated in Figure 5.39. It can be seen that at the lowest severity condition (Figure 5.39a: 380 °C and 2.56 h^{-1}), when reactions are mostly catalytic, the conversion values are between 9% and 50%. At this condition, the HDT reactions present the following conversion sequence:

HDNBN > HDV > HDS ~ HDNi > HDAsph ~ HDNNB > HDCCR

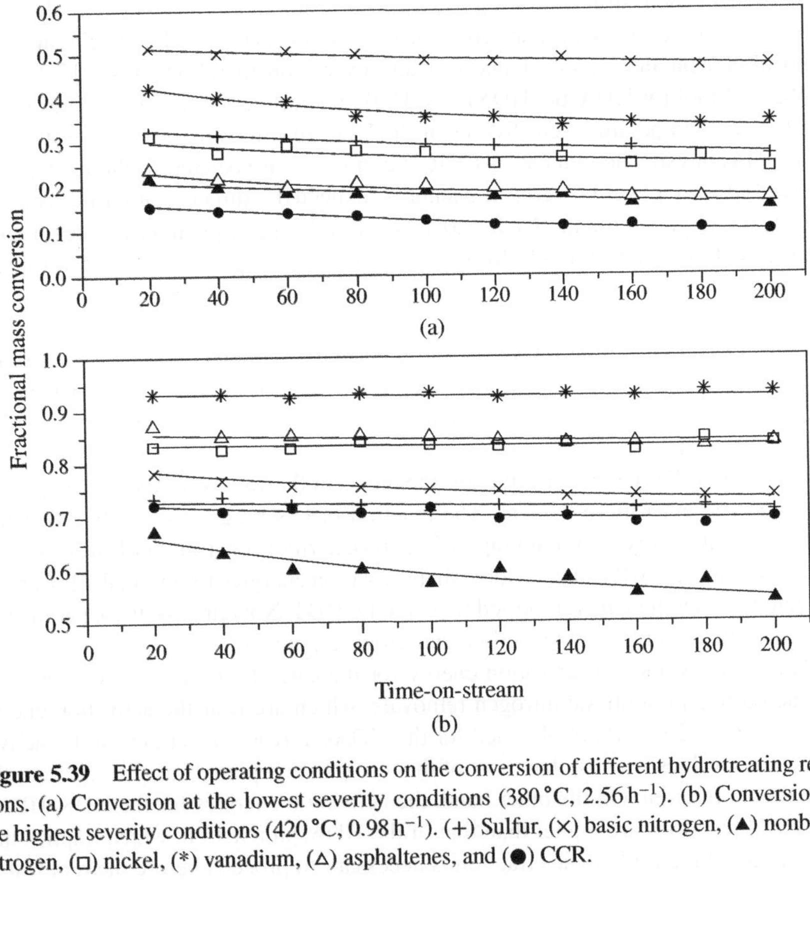

Figure 5.39 Effect of operating conditions on the conversion of different hydrotreating reactions. (a) Conversion at the lowest severity conditions (380 °C, 2.56 h^{-1}). (b) Conversion at the highest severity conditions (420 °C, 0.98 h^{-1}). (+) Sulfur, (×) basic nitrogen, (▲) nonbasic nitrogen, (□) nickel, (*) vanadium, (△) asphaltenes, and (●) CCR.

While at the highest severity conditions (Figure 5.39b: 420 °C and 0.98 h^{-1}), nickel, vanadium, and asphaltenes removals are the more favored reactions due to the thermal contribution. In this latter case, the conversions are in the range of 55–95% with the following sequence:

$$\text{HDV} > \text{HDAsp} \sim \text{HDN} > \text{HDNBN} > \text{HDS} > \text{HDCCR} > \text{HDNNB}$$

It can be noted from Figure 5.39 that at the highest severity conditions, the non-basic nitrogen conversion is 3.2 times higher than that obtained at the lowest severity conditions, while the basic nitrogen conversion increases only 1.5 times. Also at the lowest and at the highest severity conditions, the conversion of nonbasic nitrogen is much lower than that of basic nitrogen compounds, which agrees with the results obtained by Bej et al. (2001). Mitra-Kirtley et al. (1993) analyzed several petroleum asphaltenes for determining the type of nitrogen present in the asphaltenes. It was reported that about 76% of the nitrogen compounds correspond to nonbasic nitrogen (pyrrole forms). This means that at low severity conditions, most of the asphaltenes and nonbasic nitrogen removals will proceed at the same rate (Figure 5.39a), but at a higher severity, hydrocracking of asphaltenes will be the more dominant reaction, its rate being higher than that of nonbasic removal (Figure 5.39b).

5.6.3.5 Comparison Between Experimental and Calculated Conversions A comparison between experimental and calculated mass conversions is shown in Figure 5.40. It can be observed that the mass conversion is predicted quite well for the different components. A random distribution around the horizontal axis with a quite balanced number of positive and negative residual values is presented in Figure 5.41. Intercept, slope, number of positive and negative residuals, and absolute error are summarized in Table 5.14. It has been noted that the slope and intercept are close to unity and zero, respectively, which indicates a good agreement between the experimental and calculated values. The highest absolute percentage error was less than 5%.

5.6.4 Conclusions

Kinetics of parallel thermal and catalytic reactions for hydroprocessing of heavy oil taking into account catalyst deactivation were obtained in a CSTBR in the range of reaction temperature of 380–420 °C and WHSV of 0.98–2.56 h^{-1}.

The HDNNB and HDCCR reactions were more affected by catalyst deactivation than the HDT reactions. The reaction orders for the catalytic reactions were higher than those for the thermal reactions, which corroborated the fact that HDT reactions occur preferentially via catalytic route.

The activation energy of the thermal reactions was higher than that of the catalytic reactions, maintaining a similar sequence of values.

The mass conversion of the HDT reactions can be predicted with the developed model with absolute average errors less than 5%.

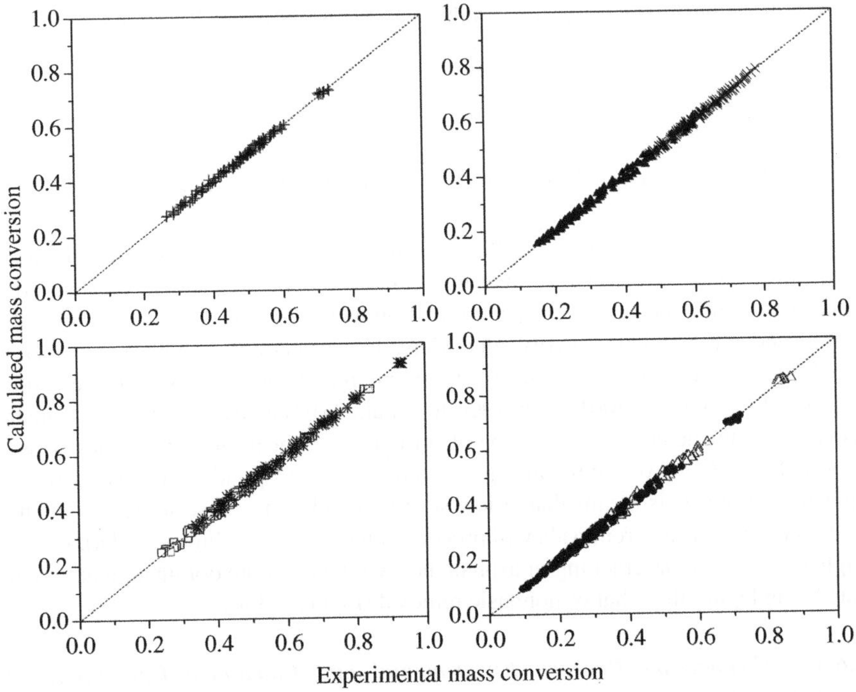

Figure 5.40 Comparison between experimental and calculated mass conversions. (+) Sulfur, (×) basic nitrogen, (▲) nonbasic nitrogen, (□) nickel, (*) vanadium, (△) asphaltenes, and (●) CCR.

5.7 MODELING CATALYST DEACTIVATION DURING HYDROCRACKING OF ATMOSPHERIC RESIDUE BY USING THE CONTINUOUS KINETIC LUMPING MODEL

5.7.1 The Model

5.7.1.1 Kinetic Model The rate of reaction for species with reactivity "k" can be written as follows (Laxminarasimhan et al., 1996):

$$-r_{\mathrm{A}} = k \cdot c(k,\tau) - \int_{k}^{k_{\max}} [p(k,x) \cdot x \cdot c(x,\tau) \cdot D(x)]dx \tag{5.57}$$

The first term on the right-hand side of Equation 5.57 represents the rate of hydrocracking of species with reactivity "k", whereas the second term accounts for the species that being longer produces the species in question by hydrocracking. τ is the reciprocal of space velocity.

The yield species distribution ($p(k, K)$) is given by

$$p(k,K) = \frac{1}{S_0\sqrt{2\pi}}[e^{-[\{(k/K)_0^a - 0.5\}/a_1]^2} - A + B] \tag{5.58}$$

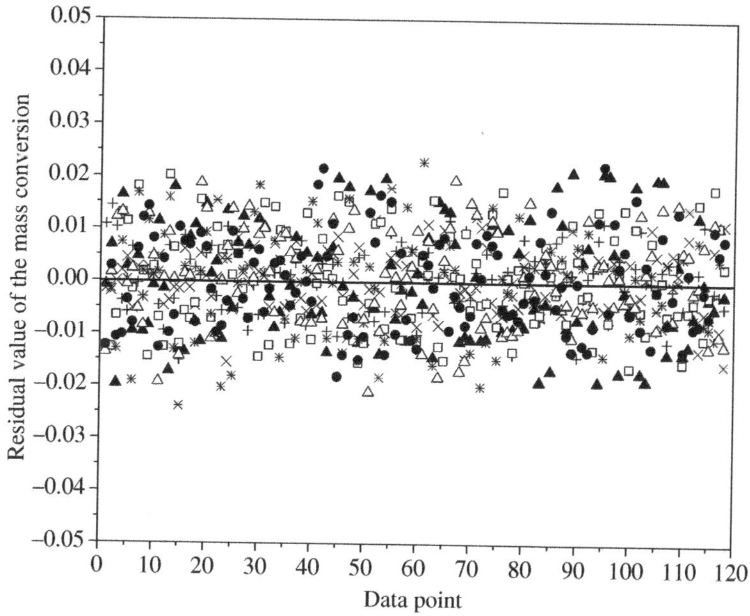

Figure 5.41 Residual values of the mass conversion. (+) Sulfur, (×) basic nitrogen, (▲) non-basic nitrogen, (□) nickel, (*) vanadium, (△) asphaltenes, and (●) CCR.

where A and B are given by the following expressions:

$$A = e^{-(0.5/a_1)^2} \tag{5.59}$$

$$B = \delta[1 - (k/K)] \tag{5.60}$$

and S_0 is calculated with

$$S_0 = \int_0^K \frac{1}{\sqrt{2\pi}}[e^{-[\{(k/x)^{a_0} - 0.5\}/a_1]^2} - A + B \cdot D(x) \cdot dx] \tag{5.61}$$

The species-type distribution or change factor from discrete to continuous approach is given by the following relationship:

$$D(k) = \frac{N\alpha}{k_{\max}^{\alpha}} k^{\alpha - 1} \tag{5.62}$$

Such a relationship permits to keep the invariance during transforming discrete mixture to a continuous one. Table 5.15 reports complementary equations of the kinetic model. Description of each variable is given in "Nomenclature". In the continuous kinetic model, α, a_0, a_1, δ, and $k_{\max}$ are the model parameters.

TABLE 5.14 Statistical Analysis

HDT Reaction	Slope	Intercept	Residual (+)	Residual (−)	Maximum Absolute Percentage Error
HDS	0.996	0.002	63.0	55.0	3.862
HDNi	0.998	0.000	61.0	57.0	4.976
HDV	0.994	0.004	67.0	51.0	4.765
HDAsph	0.998	0.000	65.0	53.0	4.962
HDCCR	0.997	0.001	55.0	63.0	4.872
HDNBN	0.996	0.003	57.0	61.0	3.034
HDNNNB	0.992	0.003	55.0	63.0	4.962

TABLE 5.15 Complementary Equations for the Hydrocracking Kinetic Model

$$\theta = \frac{\mathrm{TBP-TBP(l)}}{\mathrm{TBP(h)-TBP(l)}}$$

$$\frac{k_{\mathrm{HDC}}}{k_{\max}} = \theta^{1/\alpha}$$

$$\int_0^K P(x,K)D(x)dx = 1$$

$$S_0 = \int_0^K \frac{1}{\sqrt{2\pi}}[e^{-[\{(x/K)^{a_0}-0.5\}/a_1]^2} - A + B \cdot D(x) \cdot dx]$$

$$C_{1,2} = \int_{k_1}^{k_2} c(x,t) \cdot D(x) \cdot dx$$

$$\frac{1}{N}\int_0^{k_{\max}} D(x) \cdot dx = 1$$

5.7.1.2 Reactor Model The experiments were carried out in a continuous stirred tank reactor (CSTR). A pseudohomogeneous model is used to represent the reactor. Proper care has been taken to minimize the fluid-to-particle mass-transfer resistance. Thus, kinetic information is affected only by intraparticle mass-transfer resistance. Based on this, the CSTR model is used as follows:

$$\tau = \frac{C_{\mathrm{A}_0} - C_{\mathrm{A}_1}}{-r_{\mathrm{A}}|_{C_{\mathrm{A}}=C_{\mathrm{A}_1}}} \tag{5.63}$$

Substituting Equation 5.57 in Equation 5.63:

$$\tau = \frac{c(k,0) - c(k,\tau)}{k \cdot c(k,\tau) - \int_k^{k_{\max}} [p(k,x) \cdot x \cdot c(x,\tau) \cdot D(x)]dx} \tag{5.64}$$

Solving for hydrocracking product concentration, one arrives at (McCoy and Wang, 1994; Khorasheh et al., 2001):

$$c(k,\tau) = \frac{c(k,0) + \tau \int_{k}^{k_{max}} [p(k,x) \cdot x \cdot c(x,\tau) \cdot D(x)]dx}{1 + k \cdot \tau} \tag{5.65}$$

For the heaviest compound ($c(k_{max}, T)$), Equation 5.65 reduces to

$$c(k_{max},\tau) = \frac{c(k_{max},\tau)}{1 + k_{max} \cdot \tau} \tag{5.66}$$

In order to solve Equation 5.65, discretization of integral term was done and after some arrangements, the following expression was obtained:

$$c(k_i,\tau) = \frac{c(k_i,0) + \tau \left[\sum_{j=i+1}^{n+1} c(k_j,\tau_r) I_{1j} + \sum_{j=i+1}^{n} c(k_j,\tau_r) I_{2j} \right]}{1 + \tau[k_i - I_{1i}]} \tag{5.67}$$

where I_{1i}, I_{1j}, and I_{2j} are given by

$$I_{1i} = \int_{k_i}^{k_{i+1}} (k_i, x) \cdot x \cdot \left(\frac{x - k_{i+1}}{k_i - k_{i+1}} \right) \cdot D(x) \ dx \tag{5.68}$$

$$I_{1j} = \int_{k_{j-1}}^{k_j} p(k_i, x) \cdot x \cdot \left(\frac{x - k_{j-1}}{k_j - k_{j-1}} \right) \cdot D(x) \ dx \tag{5.69}$$

$$I_{2j} = \int_{k_i}^{k_{j+1}} (k_i, x) \cdot x \cdot \left(\frac{x - k_{j+1}}{k_j - k_{j+1}} \right) \cdot D(x) \ dx \tag{5.70}$$

After the calculation of $c(k, \tau)$, the following integral was solved in order to obtain the weight fraction of individual pseudocomponent:

$$wt_{1,2}(\tau) = \int_{k_1}^{k_2} c(k,\tau) D(k) \ dk \tag{5.71}$$

And total mass conservation was verified by using the following equation at each residence time:

$$\sum_{i=1}^{n} wt_i = 1 \tag{5.72}$$

Species distribution in feed ($c(k,0)$) can be determined with the following equation:

$$Wt = A(k)c(k,0) \tag{5.73}$$

where

$$A(k) = \begin{pmatrix} a_{1,1} & a_{1,2} & 0 & \cdots & 0 & 0 \\ 0 & a_{2,1} & a_{2,2} & \cdots & 0 & 0 \\ 0 & 0 & a_{3,1} & \cdots & 0 & 0 \\ \cdots & \cdots & \cdots & \ddots & \cdots & \cdots \\ 0 & 0 & 0 & \cdots & a_{n,1} & a_{n,2} \end{pmatrix} \quad (5.74)$$

and

$$c(k,0) = \left(c(k_{1,0})\ c(k_{2,0})\ c(k_{3,0})\ \cdots\ c(k_{n+1,0})\right)^T \quad (5.75)$$

W is a vector containing the experimental information of individual weight fraction of feed.

Also, matrix coefficients in Equation 5.74 were calculated with

$$a_{i1} = \frac{1}{k_i - k_{i+1}} \frac{N\alpha}{k_{\max}^{\alpha}} \left[\left(\frac{k_{i+1}^{\alpha+1}}{\alpha+1} - \frac{k_{i+1}^{\alpha+1}}{\alpha} \right) - \left(\frac{k_i^{\alpha+1}}{\alpha+1} - k_{i+1} \frac{k_i^{\alpha}}{\alpha} \right) \right] \quad (5.76)$$

$$a_{i2} = \frac{1}{k_{i+1} - k_i} \frac{N\alpha}{k_{\max}^{\alpha}} \left[\left(\frac{k_{i+1}^{\alpha+1}}{\alpha+1} - k_i \frac{k_{i+1}^{\alpha}}{\alpha} \right) - \left(\frac{k_i^{\alpha+1}}{\alpha+1} - \frac{k_i^{\alpha+1}}{\alpha} \right) \right] \quad (5.77)$$

5.7.1.3 Deactivation Model To model the deactivation by sites coverage and pore blockage, two functions were chosen: power-law type considering that activity coefficient does not depend on hydrocarbon composition and considering first-order dependence (Corella et al., 1985). According to Fuentes (1985), modeling deactivation as a function of time-on-stream with the above conditions can be done with the following expression:

$$\varphi = \varphi_S + (1 - \varphi_S)e^{-\psi t} \quad (5.78)$$

where φ is the activity, ψ a parameter decay, and φ_S a steady-state catalyst activity.

Because in our experiments some sharp change in activity was observed, two intervals of TOS can be identified. The time that marks such a change was denominated as t_b. According to Corella and Monzón (1988), the proper way for modeling the deactivation kinetics by two simultaneous and different causes is done by multiplying a partial activity due to cause one by the activity due to cause two. During the first hours of time-on-stream, only site coverage due to coke precursors occurs so that only cause one is present. After certain time, simultaneous sites coverage and pore blockage provoke loss of activity. On the basis of these ideas, the following function defined by intervals is proposed:

$$\varphi = \begin{cases} \varphi_1 & 0 \le t \le t_b \\ \varphi_1 \cdot \varphi_2 & t_b < t \end{cases} \quad (5.79)$$

where φ_1 is taken to be equal to Equation 5.78 and φ_2 is defined as

$$\varphi_2 = \varphi_S + (1 - \varphi_S)e^{-\psi(t - t_b)} \quad (5.80)$$

5.7.1.4 Strategy for Estimation of Model Parameters Prior to parameter estimations, dimensionless curves from experimental results were estimated by calculating the normalized temperature for each point taking as reference the maximum and minimum boiling points in the feed and product mixtures and normalizing the weight fractions by adding the gas yields to the liquid product.

The kinetic model was solved first by using simulated distillation data curve at three different temperatures and 20 h of TOS. In order to avoid overparameterization, it was first assumed that α, a_0, a_1, and δ are only functions of temperature while the reactivity coefficient (k_{max}) also depends on TOS. The former model parameters were calculated at each temperature. After that, the values of $k_{max}\varphi$ were determined at all values of TOS. Posterior refinement of model parameters was followed.

Details of solution of Equation 5.57 can be found elsewhere (Elizalde and Ancheyta, 2011).

5.7.2 Experimental Procedure

A CSTR of 1 L equipped with a basket for catalyst was used to carry out the hydrocracking of residue. It was operated under the following conditions: 9.8 MPa of total initial pressure, three temperatures: 380, 400, and 410 °C; LHSV of 0.75 h^{-1}, hydrogen-to-oil ratio (H_2/HC) of 5000 scf/bbl and 1000 rpm. A 100 mL of extruded catalyst was loaded into the reactor basket. The catalyst properties are nominal size of 1/18 in., Ni content of 0.58 wt%, Mo content of 2.18 wt%; SSA of 197.2 m^2/g, pore volume of 0.85 mL/g, and APD of 172.6 Å. The feedstock consists of atmospheric residue of crude oil of 13°API, whose main properties are listed in Table 5.16.

Feed and products were characterized by the simulated distillation method. The duration of the test at each condition was 200 h, and the samples were taken at 20 h intervals.

TABLE 5.16 Properties of Feedstock

API gravity	5.60
Elemental analysis (wt%)	
Carbon	82.51
Hydrogen	10.05
Sulfur	5.80
Nitrogen	1.09
Metals (wppm)	
Ni+V	664
Kinematic viscosity at 110 °C (cSt)	774.41
Distillation (°C)	
IBP/5 (wt%)	268/338
10/30 (wt%)	374/397
37.8 (wt%)	546

5.7.3 Results and Discussion

5.7.3.1 ***Experimental Results*** Figure 5.42 shows the curves of liquid composition versus boiling temperature of hydrocracked products obtained at 380, 400, and 420 °C of reaction temperature and 0.50 h^{-1} of space velocity at different TOS. Only selected curves and points are presented in order to avoid overlapping. The maximum boiling point for products at all conditions is of about 720 °C. Typical trends of distillation curves are observed, that is, for longer TOS the curves are displaced to the right. The displacement of the curves to the right means that for certain cut temperature the weight percent is lower, which is indicative of diminution of desired product yields due to decay on catalyst activity. At 380 °C (Figure 5.42a), the yield of the fraction with high boiling points (>500 °C) undergoes small changes at increased TOS. This behavior can be attributed to the difficulty-to-react of those compounds (mainly resins and asphaltenes) at the moderate reaction conditions studied here. On the other hand, the yield of fraction with lower boiling points is more TOS dependent because marked differences in the trends of the curves are observed. It is believed that deactivation occurs mainly in small pores so that its effect is more pronounced in the yield of lighter boiling point fractions. Also, it is observed that a sharp change in distillation curves occurs between the curve at 80 and 100 h of TOS, which can be attributed to changes in catalyst activity due to pore mouth plugging of significant fraction of total pores. At 400 °C of reaction temperature (Figure 5.42b), all compounds undergo changes at different TOS including those contained in the heavy fraction, probably because at this temperature the reactions begin to be more dominated by hydrocracking of heavy compounds rather by hydrogenation. As hydrocracking occurs in a higher extent, more changes in activity are observed. At 200 h, an apparent steady-state activity is reached. At 420 °C (Figure 5.42c), appreciable changes in distillation curves are noticed for all fractions as TOS increases. Due to severe reaction conditions, hydrocracking affects hydrocracking products obtained at 180–200 h of TOS although that effect is small. It is observed in all figures that some curves are close to each other in certain intervals of TOS, which could be interpreted as pseudo-steady-state activity at those conditions.

A separate plot of distillation curves as a function of reaction temperature for two TOS, namely, 20 and 200 h, was prepared in order to clearly observe the effect of these variables on distillate yields, which is shown in Figure 5.43. As expected, the higher the temperature, the higher the yield of distillates at the shortest TOS studied. It is worth noting that the yields at 200 h TOS were comparable for the three curves obtained at different reaction temperatures. This behavior has been already observed experimentally (Ancheyta et al., 2003) and is attributed to a steady-state coke deposition, which has reached a maximum value and is almost equal for different reaction temperatures.

5.7.3.2 ***Model Results***

Dependence of Model Parameters on TOS Initially, it was assumed no dependence of some kinetic model parameters on TOS, but poor fitting was observed. In order

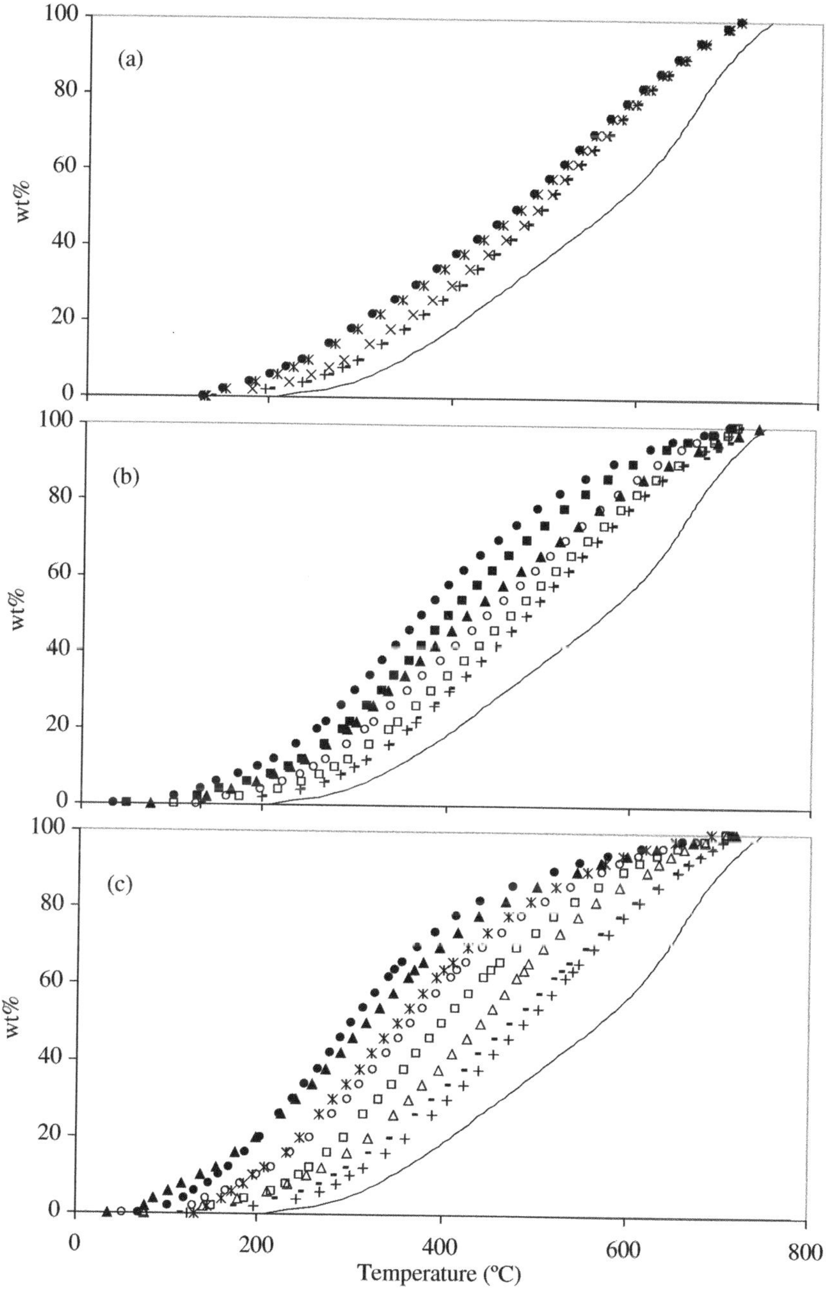

Figure 5.42 Effect of time-on-stream on distillation curves at (a) 380 °C, (b) 400, and (c) 420 °C and LHSV of 0.50 h^{-1}. (—) Feed; (●) 20 h, (■) 40 h, (▲) 60 h, (*) 80 h, (×) 100 h, (○) 120 h, (□) 140 h, (△) 160 h, (♦) 180 h, (+) 200 h of TOS.

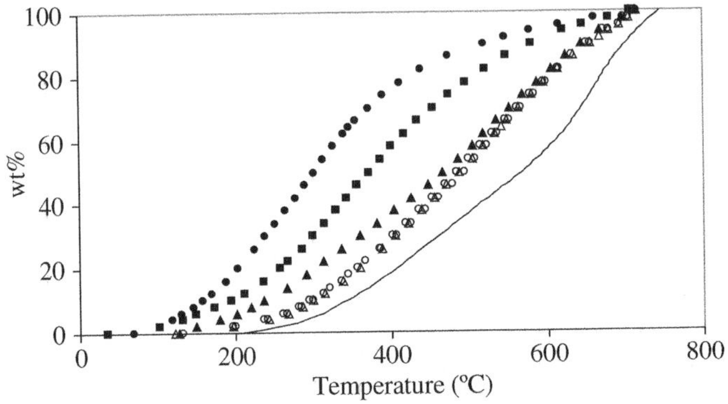

Figure 5.43 Effect of reaction temperature on distillation curves: (●,○) 420 °C, (■,□) 400 °C, and (▲,△) 380 °C. (Full symbols) 20 h of TOS, (void symbols) 200 h of TOS. (—) Feed.

to improve predictions, all kinetic model parameters were calculated at each level of TOS and averaged except k_{max}. However, even with this modification, significant deviations between simulated and experimental points were encountered. The only model parameter that did not deviate was a_1. Ulterior refinement was conducted to formulate linear dependence of the other kinetic model parameters (α, a_0, and δ) on TOS on the basis of their trends. Optimized curves of these parameters are shown in Figure 5.44. It is observed that in general at the lowest reaction temperature, the dependence of model parameters on TOS is low and increases at higher temperatures. Thus, it is confirmed that those model parameters are functions of catalyst activity as suggested by Laxminarasimhan et al. (1996). This connection indeed reduced the deviations between model and experimental results when constant activity was assumed *a priori* in spite of exhaustive parameter optimization. Figure 5.45 shows the dependence of k_{max} on TOS together with the calculated trends by using Equation 5.79. These trends remained unchanged even when other model parameters were kept constants. Notorious changes in activity at 380 and 400 °C of reaction temperature at 80 and 100 h TOS, respectively, are observed (Figure 5.45a and b), while at 420 °C less change is seen at 60 h (Figure 5.45c). With reference to Figure 5.45a and b, the first part of the curve represents the loss of catalyst activity by site coverage and the sharp changes in reactivity can be attributed to some catalyst pore constrictions together with deactivation by coke precursors. No well-defined curve is shown in Figure 5.45c and such a behavior can be attributed to severe reaction conditions, for which the catalytic function is the main cause for hydrocracking but also some thermal cracking could modify the dependence of k_{max} on TOS. Another salient feature that deserves to be mentioned here is that besides the empiricism of expression to describe the loss of activity, the deactivation function given by Equation 5.79 allows for capturing well the changes of reactivity as a function of TOS.

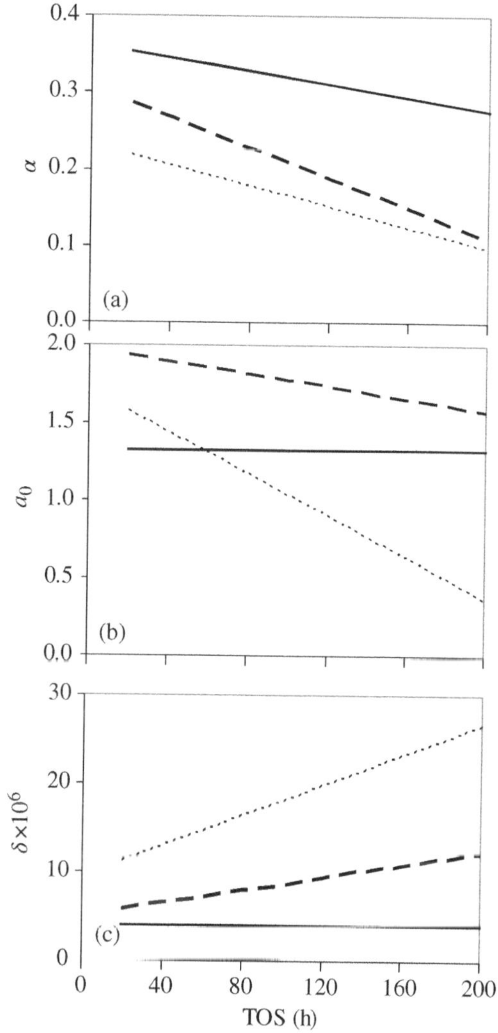

Figure 5.44 Effect of time-on-stream on α, a_0, and δ model parameters at (—) 380 °C, (---) 400 °C, and (···) 420 °C.

Dependence of Model Parameters on Reaction Temperature Derived kinetic model parameters at the shortest TOS are plotted against reaction temperature except a_0 that was constant for all conditions (Figure 5.46). Linear dependence of α and δ on reaction temperature is shown in Figure 5.46a and b, respectively. Some deviation from linearity is found for a_0 (Figure 5.46c). Excellent correlation between reciprocal of absolute reaction temperature and logarithm of k_{max} is seen in Figure 5.46d. These observations perfectly agree with previous reports, in which linear correlation of kinetic model parameters with reaction temperature were derived at short TOS (Elizalde et al., 2009; Elizalde and Ancheyta, 2011).

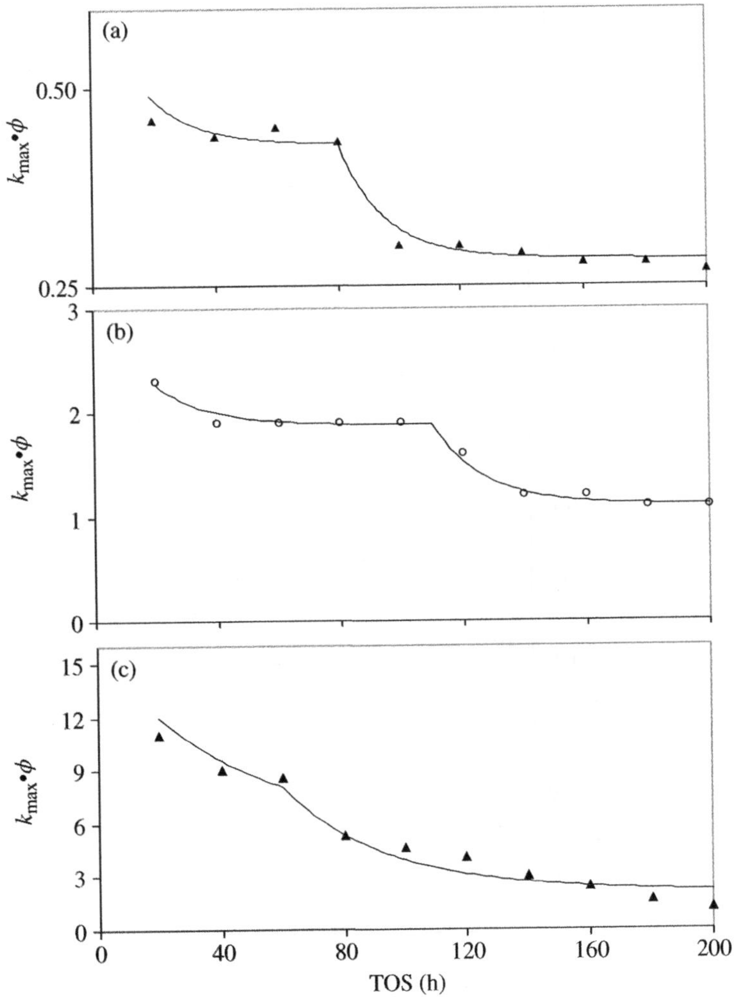

Figure 5.45 Effect of time-on-stream on k_{max} model parameter at (a) 380 °C, (b) 400 °C, and (c) 420 °C.

Comparison of Predicted and Experimental Results The optimized model parameters were used to simulate the dimensionless curves (θ against wt) and they were compared with experimental points in a parity plot shown in Figure 5.47. Good correlation can be observed. Simulation of selected complete composition curves at 410 °C as a function of dimensionless boiling temperature is shown in Figure 5.48 together with experimental results. In general, good correlation is confirmed. The representation of this figure was chosen because plots obtained at different TOS are clearly observed while the simulated and experimental results at the other reaction temperatures are closer than for the case of 410 °C. In Figure 5.49, all predicted curves at

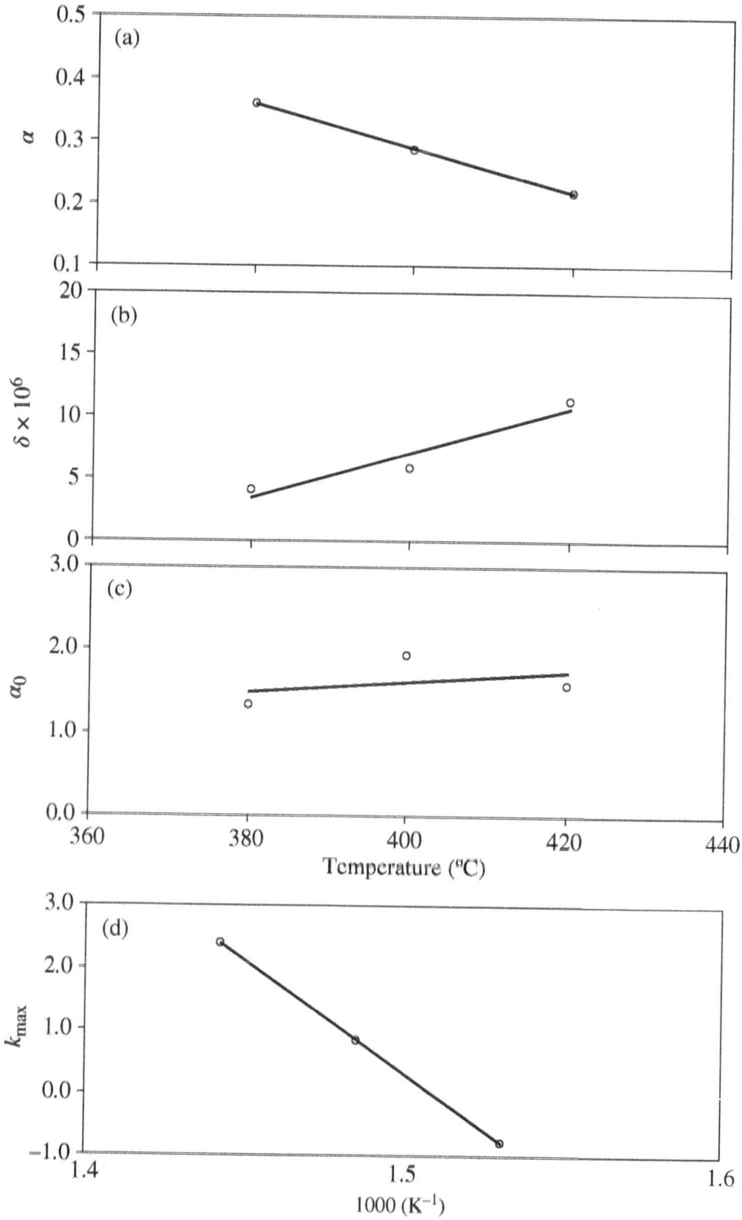

Figure 5.46 Dependence of kinetic model parameters on reaction temperature at 20 h of TOS.

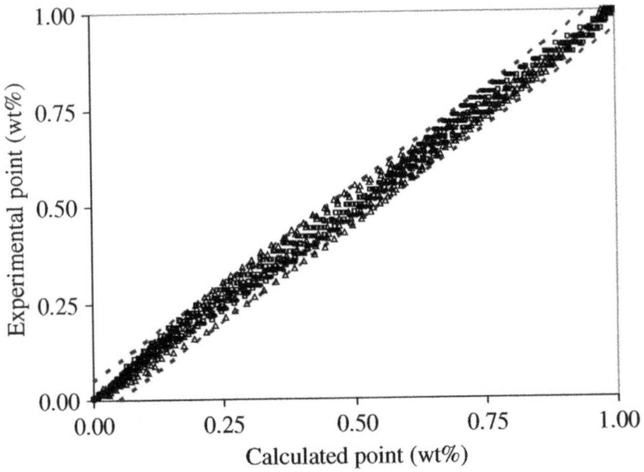

Figure 5.47 Parity plot of simulated and experimental data at all temperatures and all TOS (20–200 h).

380 and 400 °C of reaction temperature are plotted. From this figure, sharp changes are clearly observed to occur at certain TOS in agreement with variations of reactivity commented from Figure 5.46a and b, which can be attributed to changes of mechanism from deactivation of active sites to combined deactivation of sites and pore plugging. Figure 5.49b depicts slight differences in gas yields ($\theta < 0.25$) for the different TOS. This could mean that gases are produced in short TOS and then only small amounts of such a fraction are added due to the loss of ability of catalyst to produce them. A similar behavior was found for fractions containing the heaviest compounds, maybe because at the beginning such a fraction is more prone to react, but at longer TOS less number of active sites able to hydrocrack heavy molecules are available, and thus only a slight increase in heavy compounds yield is observed. Gas formation requires high activity, while heavy compounds need clean pores to enter active sites, and such conditions are only present during the first hours of TOS.

Deactivation Model Parameters It was found for all reaction temperatures that residual activity (φ_s) is present because its value is higher than zero as reported in Table 5.17. It is observed that as reaction temperature increases the corresponding residual activity diminishes. This behavior is anticipated since at the most severe reaction conditions, faster decay of catalyst activity is expected. Parameter t_b, which marks the sharp change in reactivity, is different for experiments conducted at different reaction temperatures. The observed values of t_b can be attributed to relative contributions of reaction mechanisms, that is, hydrocracking, hydrogenation, and thermal cracking as the reaction severity is increased.

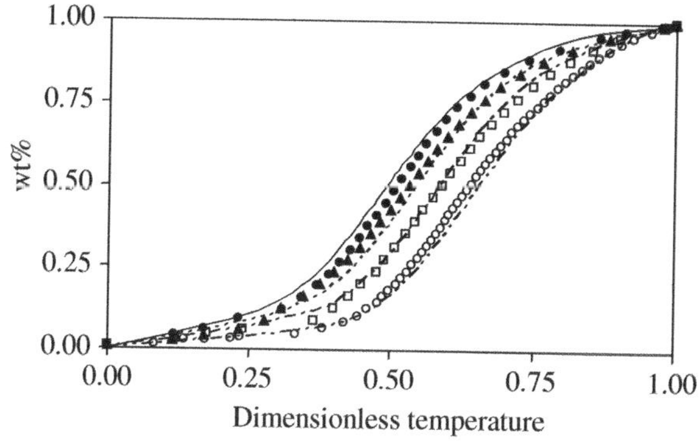

Figure 5.48 Simulated and calculated dimensionless distillation curves as a function of TOS at 410 °C. (●) 20 h, (▲) 60 h, (□) 100 h, (O) 140 h. (Lines) Simulated results.

TABLE 5.17 Model Parameters for Deactivation Function

Model Parameter	380 °C	400 °C	420 °C
k^0_{max} (h^{-1})	0.65	3.2	16
ψ (h^{-1})	0.070	0.061	0.025
ϕ_S	0.660	0.588	0.356
t_b (h)	80	110	60

Final Remarks The use of advanced characterization equipment such as high-temperature simulated distillation (HTSD) is more and more a requirement to properly obtain experimental information of distillation curves that are needed for an accurate description of hydrocracking of heavy oils. HTSD method allows for generating more information of the end part of a distillation curve that represents most complex compounds, whose conversion is the main goal of hydrocracking. If data are not available in this zone, extrapolation of the curves by using any correlation can be used. However, this latter approach is questionable. Regarding the collection of data, much care must be taken because deactivation by coke deposition is indeed present at short TOS, and if catalyst activity is assumed to be constant, it could lead to misinterpretation of kinetics.

From the good agreement between model predictions and experimental information, it is clear that the continuous lumping approach captures some of the main features of activity decay during hydrocracking of heavy oil and reflects the need of more accurate kinetic models and more research in this area.

5.7.4 Conclusions

The effect of temperature and TOS on hydrocracking of residue oil was studied by using the continuous kinetic lumping approach and an empirical deactivation

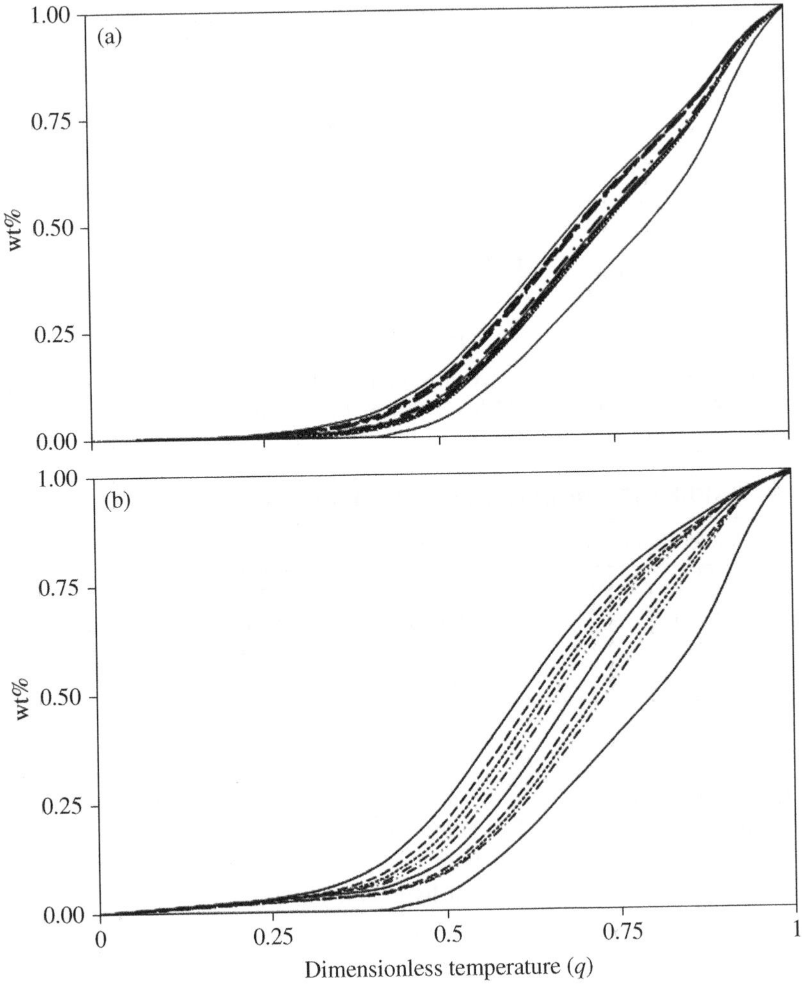

Figure 5.49 Simulated dimensionless curves as a function of TOS at (a) 380 °C and (b) 400 °C.

model that considers two causes: sites deactivation by coverage and pore mouth constrictions. It was observed that kinetic model allows for capturing the general trend of hydrocracking and the derived model parameters undergo changes at different TOS, which confirms the fact that these parameters are functions of catalyst activity. Dependence of model parameters on reaction temperature was also confirmed. Experimental data with simulation results allow for corroborating the fact that during the first hours of TOS accelerated deactivation of catalyst occurs and for longer TOS pseudo-steady-state activity is reached.

5.8 APPLICATION OF A THREE-STAGE APPROACH FOR MODELING THE COMPLETE PERIOD OF CATALYST DEACTIVATION DURING HYDROTREATING OF HEAVY OIL

5.8.1 Deactivation Model

5.8.1.1 Assumptions This model allows accounting for the three stages of catalyst deactivation. It was derived from Idei et al. (1995, 1998, 2003). The following assumptions were considered:

- There are different active sites in the catalyst selective for HDT reactions.
- Deactivation of any type of sites is independent from others.
- Each HDT reaction is carried out in two types of sites: sites type I that are deactivated at a fast rate and sites type II that correspond to those that are deactivated slower than sites type I.
- The rate of loss of any type of sites is proportional to its concentration.
- The concentration of any type of sites is proportional to the reaction rate constant of species that reacts on such sites.
- First reaction order.
- There is no activity profile along the catalyst bed.

5.8.1.2 Development of Deactivation Model: Absence of Diffusional Resistances due to Deposits If N_i is the concentration of any type of active sites in the catalyst, its rate of diminution is proportional to the concentration of corresponding species. For sites type I, it can be written that

$$-\frac{dN_1}{dt} = \alpha_1 N_1 \tag{5.81}$$

A similar equation for loss of sites type II can be written as

$$-\frac{dN_2}{dt} = \alpha_2 N_2 \tag{5.82}$$

Because of assumption II, both Equations 5.81 and 5.82 can be solved independently with the following initial conditions:

$$\text{At } t = 0, N_i = N_i^0$$

Thus, after proper substitution of conditions, the solutions for Equations 5.81 and 5.82 are

$$N_1 = N_1^0 N_1^0 e^{-\alpha_1 t} \tag{5.83}$$

$$N_2 = N_2^0 e^{-\alpha_2 t} \tag{5.84}$$

If $N = N_1 + N_2$ one arrives at

$$N = N_1^0 e^{-\alpha_1 t} + N_2^0 e^{-\alpha_2 t} \tag{5.85}$$

where N represents the total concentration of sites available for any particular reaction.

It is assumed that the initial concentration of sites I and II is, according to Equation 5.85:

$$N^0 = N_1^0 + N_2^0 \tag{5.86}$$

The deactivation function of active sites φ_A is defined as

$$\varphi_A = \frac{N}{N^0} \tag{5.87}$$

Using Equations 5.85 and 5.86, Equation 5.87 can be written as follows:

$$\varphi_A = \frac{N_1^0 e^{-\alpha_1 t}}{N^0} + \frac{N_2^0 e^{-\alpha_2 t}}{N^0} \tag{5.88}$$

According to assumption IV, it is possible to relate the concentration of active sites of any type with the reaction rate constant:

$$\varphi_A = \frac{k_1}{k_0} e^{-\alpha_1 t} + \frac{k_2}{k_0} e^{-\alpha_2 t} \tag{5.89}$$

Multiplying both sides of Equation 5.89 by $(k_0/k_1)e^{-\alpha_1 t}$ and taking logarithms one arrives at

$$\ln(\varphi_A e^{-\alpha_1 t} k_0/k_1) = \ln\left(\frac{k_0}{k_1}\frac{k_1}{k_0} e^{-\alpha_1 t} + \frac{k_0}{k_1}\frac{k_2}{k_0} e^{-\alpha_2 t}\right) \tag{5.90}$$

Simplifying Equation 5.90 and solving for φ_A:

$$\ln(\varphi_A) = \ln\left(1 + \frac{k_2}{k_1} e^{(\alpha_1 - \alpha_2)t}\right) - \left[\ln\left(\frac{k_0}{k_1}\right) + \alpha_1 t\right] \tag{5.91}$$

5.8.1.3 *Deactivation Model Considering Diffusional Resistances due to Deposits*

At MOR, pore diameter reduction due to metal sulfide deposits and the aging of coke provoke diffusional resistances that make the access of molecules to catalyst pores difficult. Also the small pores can be plugged provoking loss of catalyst pore volume. The combination of deactivation of sites type I and II in the (a) absence and (b) presence of diffusion problems by deposits must be accounted for modeling the catalyst deactivation properly. According to the literature (Corella and Monzon, 1988), the deactivation due to two causes (a, b) can be represented by

$$\varphi = \varphi_A \varphi_D \tag{5.92}$$

φ_D can be related to the initial effectiveness factor (η_0) and effectiveness factor at any TOS (η_t) as follows:

$$\varphi_D = \frac{\eta_t}{\eta_0} \tag{5.93}$$

In several situations, the catalyst particle can be considered spherical by using an appropriate characteristic length. Thus, the effectiveness factor is calculated with the following expression:

$$\eta = \frac{3}{\phi^2}[\phi \coth(\phi) - 1] \tag{5.94}$$

This last equation can be approximated without an appreciable loss of accuracy with

$$\eta = \frac{1}{\sqrt{1+\phi^2}} \tag{5.95}$$

Equation 5.95 can be rearranged and after substitution in Equation 5.93, the following expression is derived:

$$\varphi_D = \frac{1}{\eta_0}\frac{1}{\sqrt{1+\phi_t^2}} \tag{5.96}$$

Initial Thiele modulus for first-order reaction kinetics is given by

$$\phi_0 = \sqrt{k/D_0} \tag{5.97}$$

For long TOS, Thiele modulus can be defined as

$$\phi_t = \sqrt{k/D_t} \tag{5.98}$$

Relating Equations 5.97 and 5.98:

$$\frac{\phi_0}{\phi_t} = \frac{\sqrt{k/D_0}}{\sqrt{k/D_t}} = \sqrt{\frac{D_t}{D_0}} \tag{5.99}$$

By substituting Equation 5.99 into Equation 5.98, Equation 5.100 is derived:

$$\phi_t^2 = \frac{\phi_0^2}{D_t/D_0} \tag{5.100}$$

Idei et al. (1998) have proposed that restrictive diffusion due to deposits can be related to the initial facility of diffusion by

$$\frac{D_t}{D_0} = \beta\left(1 - \frac{t}{t_\infty}\right) \tag{5.101}$$

By setting β equal to the unity, Equation 5.101 is reduced to

$$\frac{D_t}{D_0} = 1 - \frac{t}{t_\infty} \tag{5.102}$$

Thus, Equation 5.96 can be rearranged as follows:

$$\varphi_D^2 = \frac{1}{\eta_0^2 + \frac{\eta_0^2 \phi_0^2}{D_t/D_0}} \tag{5.103}$$

By using Equations 5.95 and 5.102, Equation 5.103 is transformed to

$$\varphi_D^2 = \frac{1}{\eta_0^2 + \frac{1-\eta_0^2}{1-\frac{t}{t_\infty}}} \tag{5.104}$$

By taking logarithms of both sides of Equation 5.104 and solving for $\ln(\varphi_D)$:

$$\ln(\varphi_D) = \frac{1}{2}\ln\left(\frac{1}{\eta_0^2 + (1-\eta_0^2)/\left(1-\frac{t}{t_\infty}\right)}\right) \tag{5.105}$$

Equations 5.91 and 5.105 are substituted into Equation 5.92 and the following final expression is obtained:

$$\ln(\varphi) = \ln\left(1 + \frac{k_2}{k_1}e^{(\alpha_1-\alpha_2)t}\right) - \left[\ln\left(\frac{k_0}{k_1}\right) + \alpha_1 t\right] + \frac{1}{2}\ln\left(\frac{1}{\eta_0^2 + (1-\eta_0^2)/\left(1-\frac{t}{t_\infty}\right)}\right) \tag{5.106}$$

Equation 5.106 has seven parameters ($k_0, k_1, k_2, \alpha_1, \alpha_2, \eta_0$, and t_∞) that must be determined by using experimental information. For estimating transport and thermodynamic properties, correlations recommended previously can be employed (Mederos et al., 2009).

5.8.1.4 ***Model Solution*** Equation 5.106 is a nonlinear function. For estimating the model parameters, it is desirable to account for their initial values. For first-order kinetics and plug-flow reactor performance, intrinsic rate constant can be calculated by

$$k_0^{\text{in}} = \text{LHSV}\ln(C_f/C_p) \tag{5.107}$$

If there is intraparticle mass resistance, only apparent rate constant can be obtained from Equation 5.108:

$$k_0^{\text{app}} = \text{LHSV}\ln(C_{\text{f}}/C_{\text{p}}^*) \tag{5.108}$$

Relating constants from Equations 5.107 and 5.108, Equation 5.109 is derived:

$$\eta_0 = k_0^{\text{app}}/k_0^{\text{in}} \tag{5.109}$$

With Equation 5.109 and experimental information, the initial effectiveness factor can be obtained for each reaction.

Further analysis of Equation 5.106 allows for estimating the parameter α_1 as follows: At short TOS, excluding the third term, Equation 5.106 can be expressed as

$$(\varphi_{\text{A}}) = \ln\left(1 + \frac{k_2}{k_1}\right) - \ln\left(\frac{k_0}{k_1}\right) - \alpha_1 t \tag{5.110}$$

which can be rearranged as

$$(\varphi_{\text{A}}) = \ln\left(\frac{1 + k_2/k_1}{k_0/k_1}\right) - \alpha_1 t \tag{5.111}$$

Equation 5.111 is a linear function with slope $-\alpha_1$.

On the other hand, for enough long TOS, the first term of the right-hand side of Equation 5.106 can be approximated as follows:

$$\ln\left(1 + \frac{k_2}{k_1}e^{(\alpha_1-\alpha_2)t}\right) \approx \ln\left(\frac{k_2}{k_1}e^{(\alpha_1-\alpha_2)t}\right) = \ln\left(\frac{k_2}{k_1}\right) + (\alpha_1 t - \alpha_2 t) \tag{5.112}$$

and

$$-\left[\ln\left(\frac{k_0}{k_1}\right) + \alpha_1 t\right] = -\ln\left(\frac{k_0}{k_1}\right) - \alpha_1 t \tag{5.113}$$

Relating Equations 5.112 and 5.113 one arrives at

$$\ln(\varphi_{\text{A}}) \approx \ln\left(\frac{k_2}{k_1}\right) + (\alpha_1 t - \alpha_2 t) - \ln\left(\frac{k_0}{k_1}\right) - \alpha_1 t = \ln\left(\frac{k_2}{k_1}\frac{k_1}{k_0}\right) - \alpha_2 t \tag{5.114}$$

From which it can be concluded that Equation 5.114 is a linear function with an intercept of $\ln(k_2/k_0)$ and a slope $-\alpha_2$.

By plotting experimental values with Equations 5.111 and 5.114, it is possible to estimate initial values of some model parameters. These initial values together with the other values are optimized by means of minimization of differences between the experimental and calculated values of $\ln(\varphi)$ by using nonlinear analysis.

TABLE 5.18 Properties of Catalyst

Shape	Tetralobular Extrudate
Diameter (mm)	1.43
Length (mm)	3.2
Surface specific area (m^2/g)	175
Pore volume (cm^3/g)	0.56
Average pore diameter (Å)	127
Metal content (wt%)	
Mo	10.66
Ni	2.88

5.8.2 Experimental Procedure

The experimental setup consists of three sections, namely, the feeding section, the reaction section, and the separation section. The core of the setup is a fixed-bed catalytic reactor with an internal diameter of 2.54 cm. Such a reactor is equipped with electrical resistances to ensure isothermal operation. Details of the bench-scale plant are given elsewhere (Ancheyta et al., 2001).

Commercial NiMo–alumina-supported catalyst was used. Its properties are reported in Table 5.18. Catalyst activation was done by following a protocol reported previously (Rana et al., 2004).

Heavy oil (HCO in Table 5.5) was hydrotreated under the following reaction conditions: temperature of 400 °C, pressure of 6.9 MPa, LHSV of $1\,h^{-1}$, and hydrogen-to-oil ratio of $890\,m^3\ H_2/m^3$ oil.

Feed and products were characterized by the ASTM methods: Specific gravity, ASTM D70; total sulfur, ASTM D4294; and asphaltene concentration measured as insoluble in C_7, ASTM D3279. The concentration of nickel and vanadium were determined by atomic absorption. To study the effect of TOS on catalyst deactivation, product samples were collected for 500 h at regular intervals.

In separate experiments, the variation of catalyst size was carried out at the same operating conditions in order to define the suitable particle size of catalyst that minimizes intraparticle mass resistances and therefore to be able to determine the rate reaction constants (Perego and Paratello, 1999).

5.8.3 Results and Discussion

The size of catalytic particle under which kinetic control prevails was found to be 0.25 mm ($1/dp = 4$) for asphaltene conversion as shown in Figure 5.50. By using data of conversion of asphaltenes, sulfur, and metals, the rate constants were derived. Then employing catalyst of commercial size, the initial effectiveness factors were determined. Different values of effectiveness factors were found with this approach for metal and sulfur removal. These values together with graphical estimations of α_1,

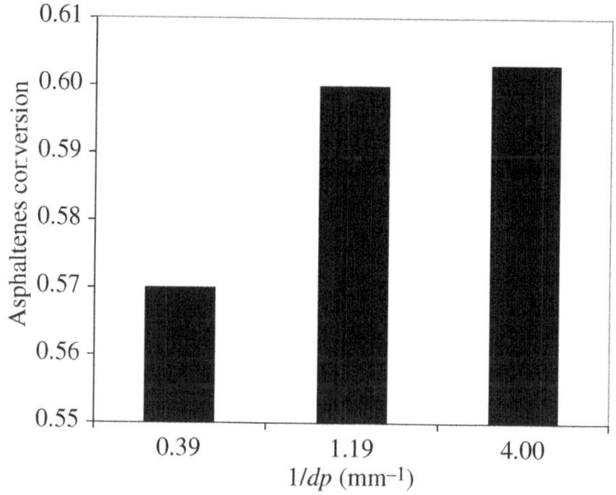

Figure 5.50 Effect of catalyst particle size on asphaltenes conversion at 400 °C, LHSV = 1 h^{-1}, 890 m^3/m^3 H_2-to-oil ratio and 6.9 MPa.

α_2, and $\ln(k_2/k)$ were used as initial values for Equation 5.106. An example of estimation of initial values for deactivation due to HDNi is shown in Figure 5.51. The approach for modeling the MOR by a single can be justified from the fact that metals are deposited at almost constant rate, so that they deactivate the catalyst in the same manner. Similar approaches can be used for HDS and HDM deactivation functions. From other HDT studies with similar feed, catalyst, and reaction conditions, it has been established a catalyst life (t_∞) of about 9–10 months (Ancheyta et al., 2009), hence 7000 h have been used as reference. Using all these information, the minimization of objective function was conducted.

Two approaches were followed to model the catalyst deactivation: In the first approach, it has been considered the deactivation of sites type I and II only (first and second terms of the right-hand side of Equation 5.106). In the second approach, Equation 5.106 with all terms was used. These approaches were solved for each reaction, namely, nickel and vanadium, and sulfur removal because of the assumption of independence on sites deactivation, while for asphaltenes only the first approach was employed due to important thermal effects on conversion could contribute to maintain the rate of reaction without significant changes besides the loss of catalytic sites at MOR.

The optimized model parameters of Equation 5.106 for HDNi conversion as a function of TOS are shown for the two approaches in Table 5.19. The rate constants of deactivation for sites type I and II for HDNi were very similar with both approaches and in general for the other reactions.

Experimental data of deactivation for HDNi together with simulated profiles are depicted in Figure 5.52. At short TOS, a rapid loss of activity is observed, whereas for time longer than 200 h almost constant catalyst activity is reached. This last behavior has been observed during HDT of heavy feedstocks with high asphaltene content

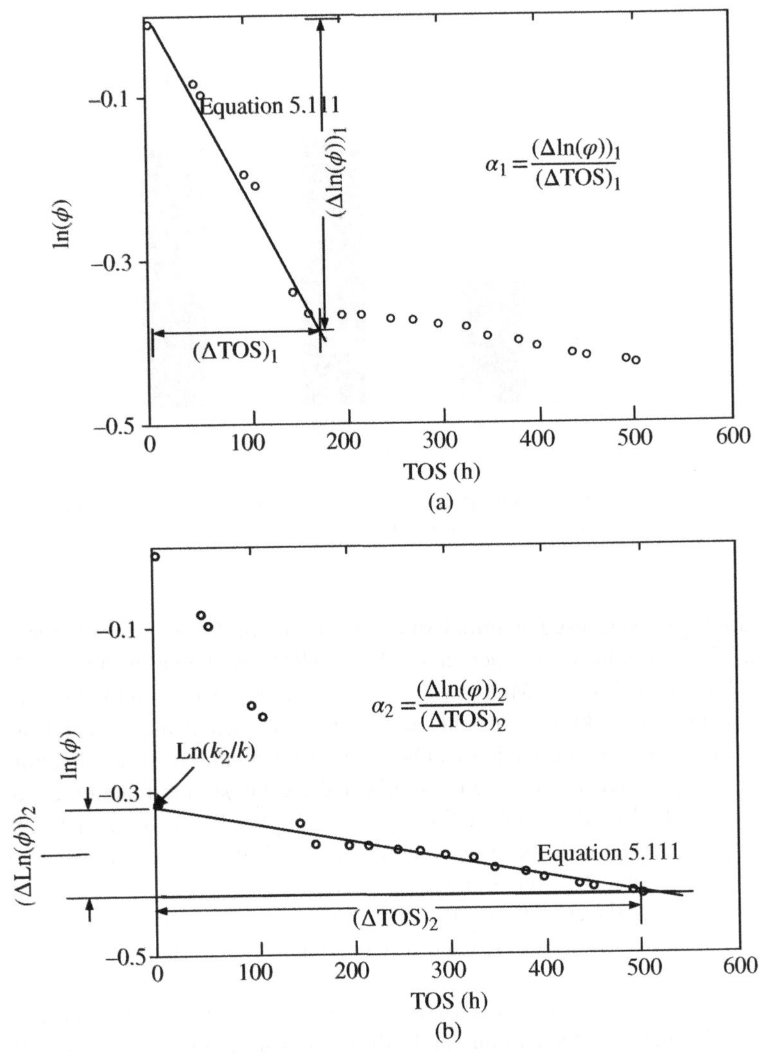

Figure 5.51 Graphical estimation of some model parameters for HDNi deactivation function.

(Higashi et al., 2002). The model captures in an acceptable way the trend of deactivation of HDNi reaction with TOS in this range.

The deactivation profiles for vanadium and sulfur removal in the range of experimental conditions are shown in Figure 5.53. Similar trends of deactivation at SOR for both reactions are observed, although the rate of deactivation is higher for HDS. This confirms that each reaction has its own effect due to catalyst deactivation at SOR, which justifies their separate analysis. Simulation closely follows these trends.

For asphaltenes, the profile of deactivation is very pronounced at SOR but after 200 h almost steady state is reached as observed in Figure 5.53.

TABLE 5.19 Initial and Optimal Model Parameters of Deactivation Function for HDNi by Using Two Approaches

Parameter	Initial Value	Optimal Values	
		Approach 1	Approach 2
α_1	2.3×10^{-3}	9.0×10^{-3}	9×10^{-3}
α_2	2.5×10^{-4}	4.0×10^{-5}	4.0×10^{-5}
k_2/k_1	–	0.53	0.50
$\ln(k_2/k_0)$	−0.32	−0.414	−0.40
η_0	0.73	–	0.73
t_∞ (h)	7000	–	7000

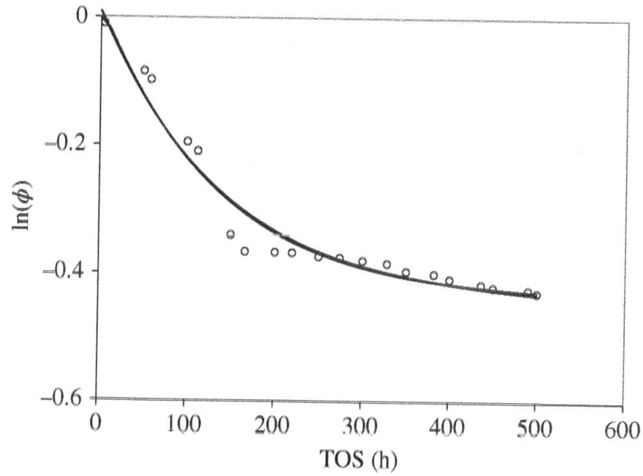

Figure 5.52 Simulation profiles of deactivation for HDNi for the range of experimental TOS: (symbols) experimental, (line) simulation.

To predict the loss of activity at long TOS for HDNi, the second approach was used. Also the catalyst life parameter was varied from 6000 to 8000 h. The simulations did not report appreciable changes in profiles of deactivation at short TOS. This, of course, is due to no dependency of the initial deactivation period on the middle and final deactivation stages. On the contrary, for longer TOS, expected differences were observed (Figure 5.54).

Simulations for predicting the performance of catalyst at SOR, MOR, and EOR for HDM and HDS were conducted with the optimal model parameters and using the second approach and a catalyst life of 7000 h. For HDAsph, the first approach was used for such a purpose. Simulated results are shown in Figure 5.55. According to the model approach, the loss of sites type I at EOR stage occurs quickly for all reactions. The activity for all reactions continues decreasing but at lower rate at MOR. For metal removal, almost constant rate of deactivation is observed in this stage. Such profiles

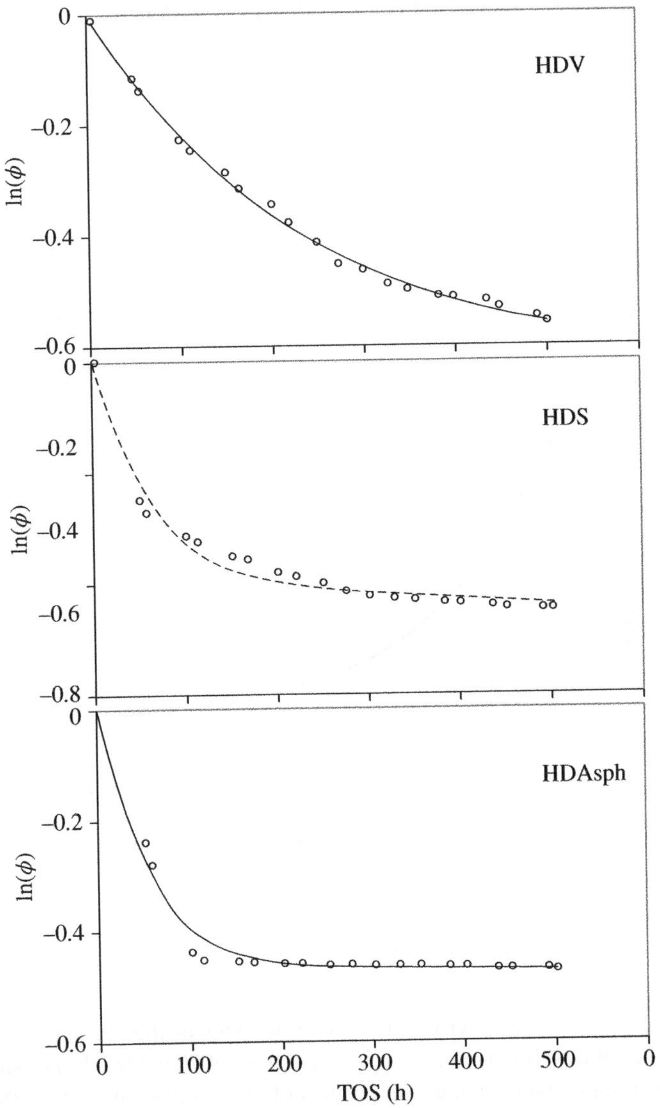

Figure 5.53 Simulation profiles of deactivation for HDV, HDS, and HDAsph for the range of experimental TOS: (symbols) experimental, (line) simulation.

are in agreement with the observation of Kam et al. (2005). The worst predicted performance of catalyst is to carry out the HDS reaction at MOR because the deactivation profile for such a reaction is more pronounced than for the others. That could be due to deposited metals at MOR that deactivate the sites (type II, more resistant to deactivation) for HDS. Metals cover the catalyst surface faster than pore mouth blocking

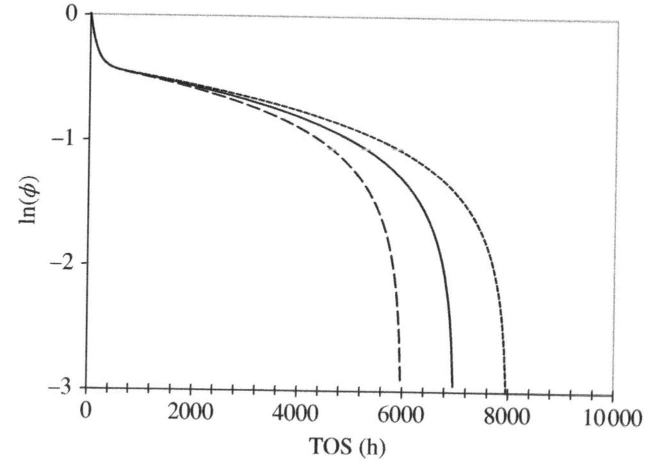

Figure 5.54 Profiles of deactivation for HDNi as function of TOS. Simulation for different catalyst life: (— —) 6000 h, (—) 7000 h, and (– – –) 8000 h, the three overlapping at short TOS.

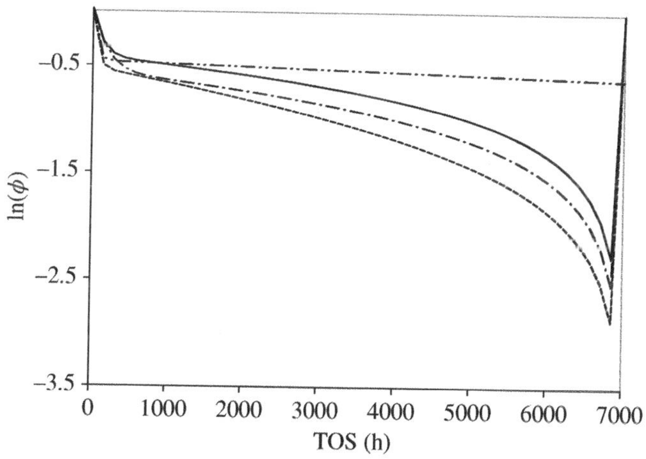

Figure 5.55 Predicted deactivation profiles for the complete deactivation period. (_____) HDNi, (_..._..._) HDAsph, (_.___._) HDV, (____) HDS. (_ _ _).

so that rapid diminution of HDS efficiency is expected compared with metal deactivation sites. Changes in activity of asphaltenes conversion are negligible for MOR stage.

The facility of diffusion depends on the diffusion length along with catalyst particle, catalyst pore size, and the size of molecule. Reports indicate that the radius of asphaltenes ranges between 2 and 15 nm while vanadium complexes possess a molecular size of 2–8 nm. Molecules of sulfur compounds are smaller than those of asphaltenes and hydrocarbon-containing metals. Very strong effect of pores diameter in the range of 14–100 nm on the diffusion coefficient of asphaltenes has been reported (Baltus and Anderson, 1983), while evidence at short TOS shows that the larger the pore size, the higher the activity for HDM.

It can be inferred from the results of simulations and experimental information that independence of deactivation of the different sites can be used as exploratory approach for modeling catalyst deactivation during HDT. As expected, this is artifice of modeling approach. It is known that activity can be selective for different reactions and also complex parallel–series reactions can provoke changes in the way how the catalyst is affected by separating each reaction and together (Marafi et al., 2010). Due to the difficulty of performing a fundamental study from the technical and economical point of view for exploratory studies, approximate approaches can be employed. Because simulations at longer TOS agree with literature reports, the model could be employed for further study including experimental information at longer TOS.

5.8.4 Conclusions

A three-stage deactivation model was used for simulation of deactivation of the main HDT reactions as a function of TOS. Simulated hydrodemetallization, hydrodesulfurization, and hydrodeasphaltenization profiles determined with the deactivation model as a function of TOS were compared with experimental data and, within the range of experimental conditions, acceptable agreement was found. The model properly predicts the profiles of deactivation for all HDT reactions at MOR and EOR conditions, which resulted in good agreement with literature reports.

REFERENCES

Alvarez, A., and Ancheyta, J. 2008. Modeling residue hydroprocessing in a multi-fixed-bed reactor system. Appl. Catal., A 351:148–158.

Alvarez, A., Ancheyta, J., and Muñoz, J.A.D. 2009. Modeling, simulation and analysis of heavy oil hydroprocessing in fixed-bed reactors employing liquid quench streams. Appl. Catal., A 361:1–12.

Almutairi, A., Bahzad, D., and Absi, H.M. 2008. A comparative study on the performance of a catalyst system for the desulfurization of two kinds of atmospheric residues. Catal. Today 125:203–210.

Ammus, J.M., and Androutsopoulos, G.P. 1987. HDS kinetic studies on Greek oil residue in a spinning basket reactor. Ind. Eng. Chem. Res. 26:494–501.

Ancheyta J., and Sotelo R. 2000. Estimation of kinetic constants of a five-lump model for fluid catalytic cracking process using simpler sub-models. Energy Fuels 14:1226–1231.

Ancheyta, J., Betancourt, G., Marroquin, G., Perez, A.M., Maity, S.K., Cortez, M.T., and del Río, R. 2001. An exploratory study for obtaining synthetic crudes from heavy crude oils via hydrotreating. Energy Fuels 15:120–127.

Ancheyta, J., Betancourt, G., Centeno, G., Marroquín, G., Alonso, F., and Garciafigueroa, E. 2002. Catalyst deactivation during hydroprocessing of heavy crude oil. (I). Evaluation at constant operating conditions. Energy Fuels 16:1438–1443.

Ancheyta, J., Betancourt, G., Centeno, G., and Marroquín, G. 2003. Catalyst deactivation during hydroprocessing of heavy crude oil. (II). Effect of temperature during time-on-stream. Energy Fuels 17:462–467.

Ancheyta, J., Muñoz, J.A.D., and Macías, M.J. 2005a. Experimental and theoretical determination of the particle size on hydrotreating catalysts of different shapes. Catal. Today 109:120–127.

Ancheyta, J., Centeno, G., Trejo, F., and Speight, J.G. 2005b. Asphaltene characterization as function of time on-stream during hydroprocessing of Maya crude. Catal. Today 109:162–166.

Ancheyta, J., Trejo, F., and Rana, M.S. 2009. Asphaltenes: chemical transformation during hydroprocessing of heavy oils, CRC Press, New York.

Baltus, R.E., and Anderson, J.L. 1983. Hindered diffusion asphaltenes through microporous membranes. Chem. Eng. Sci. 38:1959–1969.

Beckman, J.W., and Froment, G.F. 1980. Catalyst deactivation by site coverage and pore blockage-finite rate of growth of the carbonaceous deposit. Chem. Eng. Sci. 35:805–815.

Beggs, H.D., and Robinson, J.R. 1975. Estimating the viscosity of crude oil systems. J. Pet. Technol. 1:1140–1141.

Bej, S.K. 2002. Performance evaluation of hydroprocessing catalysts – a review of experimental techniques. Energy Fuels 16:774–784.

Bej, S.K., Dalai, A.K., and Adjaye, J. 2001. Comparison of hydrodenitrogenation of basic and nonbasic nitrogen compounds present in oil sands derived heavy gas oil. Energy Fuels 15(2):377–383.

Bollas, G.M., Lappas, A.A., Iatridis, D.K., and Vasalos, I.A. 2007. Five-lump kinetic model with selective catalyst deactivation for the prediction of the product selectivity in the fluid catalytic cracking process. Catal. Today 127:31–43.

Bondi, A. 1971. Handling kinetics from trickle-phase reactors. Chem. Technol. 1:185–188.

Callejas, M.A., and Martínez, M.T. 1999. Hydroprocessing of a Maya residue. Intrinsic kinetics of sulfur-, nitrogen-, nickel-, and vanadium-removal reactions. Energy Fuels 13:629–636.

Callejas, M.A., and Martínez, M.T. 2000a. Hydroprocessing of a Maya residue. I. Intrinsic kinetics asphaltene removal reactions. Energy Fuels 14:1304–1308.

Callejas, M.A., and Martínez, M.T. 2000b. Hydroprocessing of a Maya residue. II. Intrinsic kinetics of asphaltenic heteroatom and metal removal reactions. Energy Fuels 14:1309–1313.

Callejas, M.A., Martínez, M.T., Fierro, J.L.G., Rial, C., Jiménez-Mateos, J.M., and Gómez-García, F.J. 2001. Structural and morphological study of metal deposition on an aged hydrotreating catalyst. Appl. Catal., A 220:93–104.

Chang, J., Liu, J., and Li, D. 1988. Kinetics of resid hydrotreating reactions. Catal. Today 43:233–239.

Chen, J., Yang, H., and Ring, Z. 2005. Study of intra-particle diffusion effect on hydrodesulphurization of dibenzothiophenic compounds. Catal. Today 109:93–98.

Corella, J. 2004. On the modeling of the kinetics of the selective deactivation of catalysts: application to the fluidized catalytic cracking process. Ind. Eng. Chem. Res. 43:4080–4086.

Corella, J., and Monzón, A. 1988. Modeling the deactivation kinetics of solid catalyst by two or more simultaneous causes. Ind. Eng. Chem. Res. 27:369–374.

Corella, J., Bilbao, R., Molina, J.A., and Artigas, A. 1985. Variation with time of the mechanisms, observable order and activation energy of the catalyst deactivation by coke in FCC process. Ind. Eng. Chem. Process Des. Dev. 24:625–636.

Elizalde, I., and Ancheyta, J. 2011. On the detailed solution and application of the continuous kinetic lumping modeling to hydrocracking of heavy oils. Fuel 90:3542–3550.

Elizalde, I., and Ancheyta, J. 2014. Modeling catalyst deactivation during hydrocracking of atmospheric residue by using the continuous kinetic lumping model. Fuel Process. Technol. 123:114–121.

Elizalde, I., Rodríguez, M.A., and Ancheyta, J. 2009. Application of continuous kinetic lumping modeling to moderate hydrocracking of heavy oil. Appl. Catal., A 365:237–242

Froment, G.F., de Wilde, J., and Bischoff, K.B. 2010. Chemical reactor analysis and design, 3rd edition, John Wiley and Sons Inc., New York, USA.

Fuentes, G.A. 1985. Catalyst deactivation and steady-state activity: a general power-law equation model. Appl. Catal. 15:33–40.

Goto, S., and Smith, J.M. 1975. Trickle-bed reactor performance. Part I. Holdup and mass transfer effects. AIChE J. 21:706–713.

Haughey, D.P., and Beveridge, S.G. 1969. Structural properties of packed-beds: a review. Can. J. Chem. Eng. 47:130–140.

Hauser, A., Stanislaus, A, Marafi, A., and Al-Adwani, A. 2005. Initial coke deposition on hydrotreating catalysts. Part II. Structure elucidation of initial coke on hydrodemetallation catalysts. Fuel 84:259–269.

Higashi, H., Takashi, T., and Kai, T. 2002. The effect of start-up conditions on deactivation of hydrotreating catalyst for heavy residue with high asphaltene content. Catal. Surv. Jpn. 5:111–119.

Ho, T.C. 1984. Some aspects of the constant-conversion policy dealing with catalyst deactivation. J. Catal. 86:48–54.

Idei, K., Yamamoto, Y., and Yamazaki, H. 1995. Kinetic model and analysis method of catalyst deactivation in hydrodesulfurization of light and heavy oils. Kagaku Kogaku Ronbunshu 21:972–983.

Idei, K., Yamamoto, Y., and Takehara, S. 1998. Generalized deactivation equation of catalyst in hydrodesulfurization of light and heavy oils. Kagaku Kogaku Ronbunshu 24:653–659.

Idei, K., Takahashi, T., and Kai, T. 2003. Estimation of coke and metal deposition distribution within hydrodesulfurization catalyst pore at the last stage operation. J. Jpn. Pet. Inst. 46:45–52.

Juraidan, M., Al-Shamali, M., Qabazard, H., and Kam, E.K.T. 2006. A refined multicatalyst deactivation and reactor performance model-pilot-plant accelerated test applications. Energy Fuels 20:1354–1364.

Kabe, T., Aoyama, Y., Wang, D., Ishihara, A., Qian, W.H., Hosoya, M., and Zhang, Q. 2001. Effects of H_2S on hydrodesulfurization of dibenzothiophene and 4,6-dimethyldibenzothiophene on alumina-supported NiMo and NiW catalysts. Appl. Catal., A 209(1–2):237–247.

Kam, E.K., Al-Shamali, M., Juradain, M., and Qabazard, H. 2005. A hydroprocessing multicatalyst deactivation and reactor performance model-pilot plant life test applications. Energy Fuels 19:753–764.

Khorasheh, F., Rangwala, H.A., Gray, M.R., and Dalla Lana, I.G. 1989. Interactions between thermal and catalytic reactions in mild hydrocracking of gas oil. Energy Fuels 3:716–722.

Khorasheh, E., Zainali, H., Chan, E.C., and Gray, M.R. 2001. Kinetic modeling of bitumen hydrocracking reactions. Pet. Coal 43:208–218.

Kodama, S., Nitta, H., Takatsuka, T., and Yokoyama, T. 1980. Simulation of residue hydrodesulfurization reaction based on catalyst deactivation model. J. Jpn. Pet. Inst. 23:310–320.

Korsten, H., and Hoffmann, U. 1996. Three-phase reactor model pilot trickle-bed for hydrotreating in reactors. AIChE J. 42:1350–1360.

Kressman, S., Morel, F., Harlé, V., and Kasztelan, S. 1998. Recent developments in fixed-bed catalytic residue upgrading. Catal. Today 43:203–215.

Krishnaswamy, S., and Kittrell, J.R. 1981. Diffusional influences on deactivation rates. AIChE J. 27:120–130.

Laxminarasimhan, C.S., Verma, R.P., and Ramachandran, P.A. 1996. Continuous lumping model for simulation of hydrocracking. AIChE J. 42:2645–2653.

Macías, M.J., and Ancheyta, J. 2004. Simulation of an isothermal hydrodesulfurization small reactor with different catalyst particle shapes. Catal. Today 98:243–252

Magnico, P., and Fongarland, P. 2006. CFD simulations of two stirred tank reactors with stationary catalytic basket. Chem. Eng. Sci. 61:1217–1236.

Mahinsa, P., Sadeghi, M.T., and Ganji, H. 2012. Modeling and sensitivity analyses of hydrodesulfurization catalyst pellet. Pet. Coal 54:104–109.

Marafi, A., Fukase, S., Al-Marri, M., and Stanislaus, A. 2003a. A comparative study of the effect of catalyst type on hydrotreating kinetics of Kuwaiti atmospheric residue. Energy Fuels 17:661–668.

Marafi, A., Al-Bazzaz, H., Al-Marri, M., Maruyama, F., Absi, H.M., and Stanislaus, A. 2003b. Residual-oil hydrotreating kinetics for graded catalysts systems: effect of original and treated feedstocks. Energy Fuels 17:1191–1197.

Marafi, A., Hauser, A., and Stanislaus, A. 2007. Deactivation patterns of Mo/Al_2O_3, $Ni–Mo/Al_2O_3$ and $Ni–MoP/Al_2O_3$ catalysts in atmospheric residue hydrodesulphurization. Catal. Today 125:192–202.

Marafi, A., Kam, E., and Stanislaus, A. 2008a. A kinetic study on non-catalytic reactions in hydroprocessing Boscan crude oil. Fuel 87:2131–2140.

Marafi, A., Maruyama, F., Stanislaus, A., and Kam, E. 2008b. Multicatalyst system testing methodology for upgrading residual oils. Ind. Eng. Chem. Res. 47:724–741.

Marafi, A., Stanislaus, A., and Furimsky, E. 2010. Kinetics and modeling of petroleum residues hydroprocessing. Catal. Rev. Sci. Eng. 52:204–324.

Martínez, M.T., Benito, A.M., Callejas, M.A., and Trasobares, S. 1998a. Kinetics of sulfur removal from a liquid coal residue in thermal, hydrothermal, and hydrocatalytic cracking. Energy Fuels 12:365–370.

Martínez, M.T., Benito, A.M., Callejas, M.A., and Trasobares, S. 1998b. Kinetics of sulfur removal from a liquid coal residue in thermal, hydrothermal, and hydrocatalytic cracking. Energy Fuels 12:365–370.

McCoy, B., and Wang, M. 1994. Continuous-mixture fragmentation kinetics: particle size reduction and molecular cracking. Chem. Eng. Sci. 49:3773–3785.

Mederos, F.S., Ancheyta, J., and Chen, J. 2009. Review on criteria to ensure ideal behaviors in trickle-bed reactors. Appl. Catal., A 355:1–19.

Mitra-Kirtley, S., Mullins, O.C., van Elp, J., George, S.J., Chen, J., and Cramer, S.P. 1993. Determination of the chemical structures in petroleum asphaltenes using XANES spectroscopy. J. Am. Chem. Soc. 115:252–258.

Nakamura, H., Amemiya, M., Koide, R., Hensen, E.J.M., van Santen, R.A., and Ishida, K. 2008. Development of hydrodesulfurization catalyst using comparative model feed reactions and quantum chemical studies, 18th Saudi Arabia–Japan Joint Symposium Dhahran, Saudi Arabia, November 16–17.

Nuñez, I.M., Pachon, Z., Kafarov, V., and Resasco, D.E. 2000. Deactivation of $Ni–Mo/Al_2O_3$ catalysts aged in a commercial reactor during the hydrotreating of deasphalted vacuum residuum. Appl. Catal., A 199:263–273.

Oleck, S.M., and Sherry, H.S. 1977. Fresh water manganese nodules as a catalyst for demetallizing and desulfurizing petroleum residua. Ind. Eng. Chem. Proc. Des. Dev. 16:525–528.

Oyekunle, L.O., and Hughes, R. 1987. Catalyst deactivation during hydrodemetallization. Ind. Eng. Chem. Res. 26:1945–1950.

Oyekunle, L.O., Ikpekri, O.B., and Jaiyeola, A. 2005. Modelling of hydrodesulfurization catalysts: II. Effect of catalyst pore structures on deactivation by metal deposits. Catal. Today 109:128–134.

Papayannakos, N., and Marangozis, J. 1984. Kinetics of catalytic hydrodesulfurisation of a petroleum residue in a batch recycle trickle bed reactor. Chem. Eng. Sci. 39:1051–1061.

Perego, C., and Paratello, S. 1999. Experimental methods in catalytic kinetics. Catal. Today 52:135–145.

Philippopoulos, C., and Papayannakos, N. 1988. Intraparticle diffusional effects and kinetics of desulfurization reactions and asphaltenes cracking during catalytic hydrotreatment of a residue. Ind. Eng. Chem. Res. 27:415–420.

Pitault, I., Fongarland, P., Koepke, D., Mitrovic, M., Ronze, D., and Forissier, M. 2005. Gas–liquid and liquid–solid mass transfers in two types of stationary catalytic basket laboratory reactor. Chem. Eng. Sci. 60:6240–6253.

Rajagopalan, K., and Luss, D. 1979. Influence of catalyst pore size on demetallation rate. Ind. Eng. Chem. Process Des. Dev. 18:459–465.

Ramírez, S., Ancheyta, J., Centeno, G., and Marroquín, G. 2011. Non-catalytic hydrodesulfurization and hydrodemetallization of residua. Fuel 90:3571–3576.

Rana, M.S., Ancheyta, J., Rayo, P., and Maity, S.K. 2004. Effect of alumina preparation on hydrodemetallization and hydrodesulfurization of Maya crude. Catal. Today 98:151–160.

Riazi, M.R. 1997. A continuous model for C_{7+} fraction characterization of petroleum fluids. Ind. Eng. Chem. Res. 36:4299–4307.

Rodríguez, M.A., Elizalde, I., and Ancheyta, J. 2012. Modeling the performance of bench-scale reactor sustaining HDS and HDM of heavy crude oil at moderate conditions. Fuel 100:152–162.

Sánchez, S., and Ancheyta, J. 2007. Effect of pressure on the kinetics of moderate hydrocracking of Maya crude oil. Energy Fuels 21:653–661.

Sánchez, S., Rodríguez, M.A., and Ancheyta J. 2005. Kinetic model for moderate hydrocracking of heavy oils. Ind. Eng. Chem. Res. 44:9409–9413.

Scamangas, A., Papayannakos, N., and Marangozis, J. 1982. Catalytic hydrodesulfurization of a petroleum residue. Chem. Eng. Sci. 37:1810–1812.

Shah, Y.T., and Paraskos, J.A. 1975 Criteria for axial dispersion effects in adiabatic trickle bed hydroprocessing reactors. Chem. Eng. Sci. 30:1169–1176.

Toulhoat, H., Hudebine, D., Raybaud, P., Guillaume, D., and Kressmann, S. 2005. A new model for combined simulation of operation and optimization of catalysts in residue hydroprocessing units. Catal. Today 109:135–153.

Trasobares, S., Callejas, M.A., Benito, A.M., Martínez, M.T., Severin, D., and Brouwer, L. 1999. Upgrading of a petroleum residue. Kinetics of Conradson carbon residue conversion. Ind. Eng. Chem. Res. 38:938–943.

Trejo, F., and Ancheyta, J. 2005. Kinetics of asphaltenes conversion during hydrotreating of Maya crude. Catal. Today 109:99–103.

Tyn, M.T., and Calus, W.F. 1975. Diffusion coefficients in dilute binary liquid mixtures. J. Chem. Eng. Data 20(1):106–109.

INDEX

Deactivation of Heavy Oil Hydroprocessing Catalysts: Fundamentals and Modeling, First Edition. Jorge Ancheyta.

I

K

L

M

T

Printed and bound by CPI Group (UK) Ltd, Croydon, CR0 4YY

16/09/2026

14963424-0001